Healing Botanicals
From Tradition to Therapeutics

Healing Botanicals

From Tradition to Therapeutics

Edited by

Awanish Kumar

Department of Biotechnology, National Institute of Technology, Raipur, Chhattisgarh, India

Ch.K.V.L.S.N. Anjana Male

School of Pharmacy, ITM University, Gwalior, Madhya Pradesh, India

Kanaka Durga Devi Nelluri

Department of Pharmaceutics and Biotechnology, KVSR Siddhartha College of Pharmaceutical Sciences, Vijayawada, Andhra Pradesh, India

Grandhi Surendra

School of Pharmacy, ITM University, Gwalior, Madhya Pradesh, India

G.L. Viswanatha

Clinchoice Pvt Ltd & Department of Pharmacology, Manipal College of Pharmaceutical Sciences, Manipal, Karnataka, India

ELSEVIER

ACADEMIC PRESS

An imprint of Elsevier

Academic Press is an imprint of Elsevier
125 London Wall, London EC2Y 5AS, United Kingdom
50 Hampshire Street, 5th Floor, Cambridge, MA 02139, United States

ISBN: 978-0-443-44295-7

For Information on all Academic Press publications
visit our website at https://www.elsevier.com/books-and-journals

Publisher: Megan Ball
Acquisitions Editor: Elizabeth Brown
Editorial Project Manager: Nayanika Saikia
Production Project Manager: Jenela Priscy J.
Cover Designer: Raman Kumar

Typeset by MPS Limited, Chennai, India

Contents

PART 2 Botanicals for health and wellness

Part 5 Future directions and conclusion

List of contributors

Suneetha Achanti
Department of Pharmaceutical Analysis, KVSR Siddhartha College of Pharmaceutical Sciences, Vijayawada, Andhra Pradesh, India

Bhavana Alaparthi
Department of Pharmaceutics and Biotechnology, KVSR Siddhartha College of Pharmaceutical Sciences, Vijayawada, Andhra Pradesh, India

Manne Anupama Ammulu
Department of Mechanical Engineering, NIT, Warangal, Telangana, India

CH.K.V.L.S.N. Anjana Male
School of Pharmacy, ITM University, Gwalior, Madhya Pradesh, India

Ashok Kumar Balaraman
Research and Enterprise, University of Cyberjaya, Cyberjaya, Selangor, Malaysia

Radha Madhavi Baluguri
Department of Pharmaceutics and Biotechnology, KVSR Siddhartha College of Pharmaceutical Sciences, Vijayawada, Andhra Pradesh, India

Bhaskar Rao Bollineni
Great Eastern Medical School and Hospital, Srikakulam, Andhra Pradesh, India

Alekhya Dandamudi
College of Pharmacy, Krishna University, Machilipatnam, Andhra Pradesh, India

Lakshmi Lalitha Danturty
Great Eastern Medical School and Hospital, Srikakulam, Andhra Pradesh, India

Mohana Naga Rajeswari Dasaraju
Department of Chemistry, Siddhartha Academy of Higher Education, Deemed To Be University, Vijayawada, Andhra Pradesh, India

Sasi Priya Dodda
Department of Biotechnology, Florida Institute of Technology, Melbourne, FL, United States

Sai Prakash Donka
University of Wisconsin, Milwaukee, WI, United States

Ramana Gangireddy
Department of Pharmaceutics and Biotechnology, KVSR Siddhartha College of Pharmaceutical Sciences, Vijayawada, Andhra Pradesh, India

Indu Sree Golla
Department of Chemistry, Siddhartha Academy of Higher Education, Deemed to be University, Vijayawada, Andhra Pradesh, India

Grandhi Surendra
School of Pharmacy, ITM University, Gwalior, Madhya Pradesh, India

Vaishnavi Grandhi
Department of Pharmaceutics and Biotechnology, KVSR Siddhartha College of Pharmaceutical Sciences, Vijayawada, Andhra Pradesh, India

Bhagavan Jaladi
Johns Hopkins University, Department of Biotechnology, Baltimore, MD, United States

Anil Kumar Kakani
Department of Pharmaceutics and Biotechnology, KVSR Siddhartha College of Pharmaceutical Sciences, Vijayawada, Andhra Pradesh, India

Sahithi Kamepalli
Dublin City University, Dublin, Ireland

Pradeepthi Katragadda
Department of Pharmaceutics and Biotechnology, KVSR Siddhartha College of Pharmaceutical Sciences, Vijayawada, Andhra Pradesh, India

Arindam Kolay
NIMS Institute of Pharmacy, NIMS University, Jaipur, Rajasthan, India

Minakshi Kommoju
Department of Pharmaceutics and Biotechnology, KVSR Siddhartha College of Pharmaceutical Sciences, Vijayawada, Andhra Pradesh, India

Durga Prasad Kondeti
Department of Pharmaceutics and Biotechnology, KVSR Siddhartha College of Pharmaceutical Sciences, Vijayawada, Andhra Pradesh, India

Awanish Kumar
Department of Biotechnology, National Institute of Technology, Raipur, Chhattisgarh, India

Ravi Shankar Kunderu
Department of Pharmaceutics and Biotechnology, KVSR Siddhartha College of Pharmaceutical Sciences, Vijayawada, Andhra Pradesh, India

Surendra Kundurthi
Medical School, University of Birmingham, Edgbaston, Birmingham, United Kingdom

Mahima Swaroopa Mandava
Department of Pharmacy Practice, KL College of Pharmacy, KL Deemed to be University, Vijayawada, Andhra Pradesh, India

Vijaya Lakshmi Marella
Department of Pharmaceutical Analysis, KVSR Siddhartha College of Pharmaceutical Sciences, Vijayawada, Andhra Pradesh, India

Sai Vishnu Mopidevi
College of Pharmacy, Krishna University, Machilipatnam, Andhra Pradesh, India

Kanaka Durga Devi Nelluri
Department of Pharmaceutics and Biotechnology, KVSR Siddhartha College of Pharmaceutical Sciences, Vijayawada, Andhra Pradesh, India

Sunjana Nelluri
Great Eastern Medical School and Hospital, Srikakulam, Andhra Pradesh, India

Vineela Nirmala Nalam
Department of Pharmaceutics and Biotechnology, KVSR Siddhartha College of Pharmaceutical Sciences, Vijayawada, Andhra Pradesh, India

Bharghava Bhushan Rao Pathange
A.M.Reddy Memorial College of Pharmacy, Narasaraopeta, Andhra Pradesh, India

Vidya Perli
Department of Pharmaceutics and Biotechnology, KVSR Siddhartha College of Pharmaceutical Sciences, Vijayawada, Andhra Pradesh, India

S.T.V. Raghavamma
Department of Pharmaceutics, Chalapathi Institute of Pharmaceutical Sciences, Guntur, Andhra Pradesh, India

Padamata Sai Rohith
Department of Pharmaceutics and Biotechnology, KVSR Siddhartha College of Pharmaceutical Sciences, Vijayawada, Andhra Pradesh, India

Abdul Rahaman Shaik
Department of Pharmaceutical Sciences, Vignan's Foundation for Science, Technology and Research, Guntur, Andhra Pradesh, India

Arifa Begum Shaik
Department of Pharmaceutics and Biotechnology, KVSR Siddhartha College of Pharmaceutical Sciences, Vijayawada, Andhra Pradesh, India

Ashish Kumar Sharma
NIMS Institute of Pharmacy, NIMS University, Jaipur, Rajasthan, India

Indupalli Siddhartha
Department of Pharmaceutics and Biotechnology, KVSR Siddhartha College of Pharmaceutical Sciences, Vijayawada, Andhra Pradesh, India

Shivam Singh
NIMS Institute of Pharmacy, NIMS University, Jaipur, Rajasthan, India

Monica Singuluri
Department of Pharmaceutics and Biotechnology, KVSR Siddhartha College of Pharmaceutical Sciences, Vijayawada, Andhra Pradesh, India

Bantupalli Suranjan
University of Michigan, Ann Arbour, MI, United States

Anasuya Tanneeru
University of Glasgow, Glasgow, G12 8QQ, United Kingdom

P. Rao Tatineni
Department of Pharmaceutics and Biotechnology, KVSR Siddhartha College of Pharmaceutical Sciences, Vijayawada, Andhra Pradesh, India

Pallavi Vadlamudi
Department of pharmaceutics, Chalapathi Institute of Pharmaceutical Sciences, Guntur, Andhra Pradesh, India

Tanvitha Vadlamudi
Department of Health and Clinical Outcomes Research, Saint Louis University, Saint Louis, MO, United States

Sai Harsha Veganti
University of Wisconsin, Milwaukee, WI, United States

Praveen Kumar Vemuri
Department of Biotechnology, Koneru Lakshmaiah Education Foundation, Guntur, Andhra Pradesh, India

G.L. Viswanatha
Clinchoice Pvt Ltd & Department of Pharmacology, Manipal College of Pharmaceutical Sciences, Manipal, Karnataka, India

PART

1

Introduction to healing botanicals

CHAPTER

1 The evolution of herbaceuticals: from traditional medicine to modern therapies

Suneetha Achanti[1], Vijaya Lakshmi Marella[1], Kanaka Durga Devi Nelluri[2], Vaishnavi Grandhi[2] and Sahithi Kamepalli[3]

[1]*Department of Pharmaceutical Analysis, KVSR Siddhartha College of Pharmaceutical Sciences, Vijayawada, Andhra Pradesh, India,*
[2]*Department of Pharmaceutics and Biotechnology, KVSR Siddhartha College of Pharmaceutical Sciences, Vijayawada, Andhra Pradesh, India,*
[3]*Dublin City University, Dublin, Ireland*

1.1 Introduction

1.1.1 Definition and concept of herbaceuticals

"Herbaceuticals" is an emerging term that combines the indigenous wisdom of the ancient herbal medicine system with the scientific basis of the pharmaceutical disciplines. The term "herbaceutical" literally means "herb" + "pharmaceutical." The term "herbaceuticals" represents the products that are derived from the combination of "herb" and "pharmaceutical." Generally, "herbaceuticals" are based on bioactive botanicals. Bioactive botanicals denote bioactive compounds that derive their characteristic quality or feature from plants.

In contrast to other herbal products, which frequently rely on crude plant extracts such as leaves, roots, and barks, herbaceuticals remain subject to a rational scientific process of investigation and validation. These investigations include, but are not limited to, analysis, pharmacology, toxicology, and clinical validation. Accordingly, herbaceuticals emerge as neither herbal nor pharmaceutical; they constitute a middle class and a rising drug category recognized scientifically as natural drugs with therapeutic potential.

Moreover, the worldwide move towards "personalized/holistic medicine" has also contributed substantially to the acceptance of herbaceuticals. Essentially, "these types of products can promote health, enhance immune functions, control chronic degenerative diseases, and improve quality of life." Accordingly, it can be understood that "herbaceuticals constitute not an alternative, but one therapeutic option that has strong correspondence to current healthcare requirements" (Bubela et al., 2008).

Healing Botanicals. DOI: https://doi.org/10.1016/B978-0-443-44295-7.00010-0

1.1.2 Cultural heritage and historical roots

In fact, using medicinal herbs is an ancient tradition, rooted in the collective heritage of ancient cultural empires. It is evident that, long before the dawn of modern drug science, all human societies relied heavily on plant materials to prevent or cure various diseases. It is also evident from the documentary record that ancient human societies, including those in India, China, Egypt, and Mesopotamia, developed sophisticated traditions of herbal medicine.

Ayurveda is the medicinal system practiced in ancient India, now more than 5000 years old. Arguably, this represents the most advanced form of traditional medicine that utilizes herbs. Ayurveda is a medicinal system that emphasizes the need for harmony among the mind, body, and soul. Medicinal herbs such as ashwagandha, turmeric, and tulsi have been widely used to boost the immune system and treat stress-related ailments. In the same tradition, Siddha medicinal practice, primarily practiced in South India, utilizes herbs and minerals to promote harmony in the human body and longevity.

Another ancient form of medicine that emphasizes herbal preparations is traditional Chinese medicine (TCM). TCM considers health a state of balance between two opposing forces, Yin and Yang. The herbal preparations are designed and incorporated to restore or maintain balance and energy flow. The herbal preparations may contain more than one herb, rather than the "single entity therapy" concept.

These traditional systems form the foundations of modern herbaceuticals. Modern science is now validating many of the traditional claims made about the herbalism of another era. Advanced analytical methodologies, as well as scientific studies of modern clinical trials of these traditional herbs, have identified the phytochemicals they contain. Hence, modern herbaceuticals can be characterized as an advanced form of traditional herbal medicine.

The preservation of traditional knowledge has been instrumental in promoting the development of herbaceuticals. Ethnopharmacological research is continually exploring indigenous medicinal practices. This has opened the door to the discovery of new compounds with medicinal potency. It is essential to understand that herbaceuticals are a legacy of traditional knowledge that has been instrumental in promoting healthcare.

1.1.3 Comparison with traditional drugs

Herbal medicines differ in several ways, for example, source, mechanism of action, safety profile, therapeutic philosophy, etc., compared with conventional pharmaceutical products or medications.

1.1.3.1 Source

The main difference between the two classes of drugs is the source from which they are derived. Herbaceuticals are derived from natural plants, with parts of the plants, including leaves, roots, seeds, flowers, and bark, containing a variety of biological compounds, such as alkaloids, flavonoids, terpenes, and phenolic compounds, among others. Conventional drugs, on the other hand, are produced in human laboratories, where some are derived from natural sources but are modified with chemicals. Synthetic drugs are designed to target very specific targets within biological systems.

1.1.3.2 Mechanism of action

The reason herbaceuticals may have multiple mechanisms of action lies in the presence of various phytoactive substances within a single herb or herb combination. Phytoactive substances may have a synergistic effect in performing their functions, which may be useful in the management of diseases or other pathological manifestations, such as inflammation, oxidative stress, infections, or immune dysfunction.

In contrast, classic therapeutic drugs are typically isolated active ingredients that pharmacologically act on a specific target, such as a receptor or enzyme. Although this approach may be extremely effective, there is a risk of side effects, as other physiological processes may be affected (see Fig. 1.1).

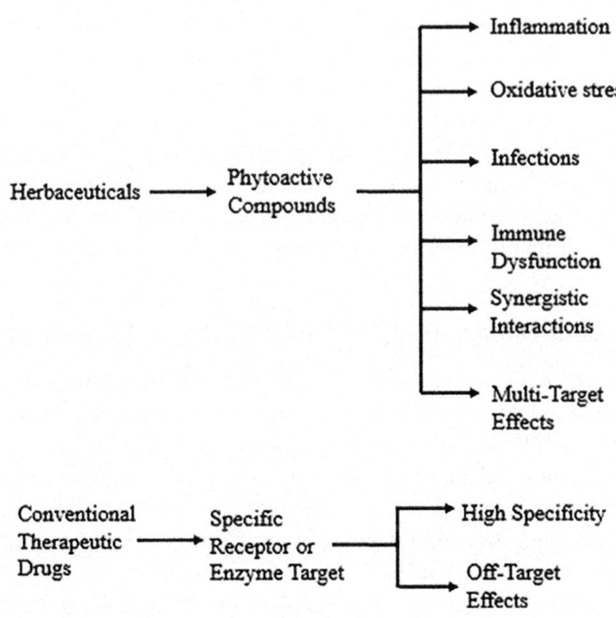

FIGURE 1.1

Mechanistic differences between herbaceuticals and conventional therapeutic drugs.

1.1.3.3 Side effects and safety

In general, it has been accepted that herbaceuticals are safe and well-tolerated, unlike conventional medicine, primarily because of their natural origin. However, it has been realized that the assumption that no side effects are associated with herbaceuticals is not necessarily true.

It has been observed that conventional medicine, despite being tested for side effects, has a propensity for a relatively greater number of side effects, especially during a prolonged phase of drug administration. Such a propensity has generated significant interest in herbaceuticals, especially for preventive medicine.

1.1.3.4 Subjective experience

Conventional medicine, on the other hand, is primarily intended to alleviate symptoms of a preexisting health condition. In contrast, herbaceuticals are intended for a holistic approach to health rather than disease treatment.

1.1.4 Standardization, quality control, and regulatory issues

Quality control, safety, and efficacy are some of the major challenges in the development and acceptance of herbaceuticals. This is because herbaceuticals are derived from natural sources. The chemical composition of natural sources varies from place to place depending on the location and processing of the herbaceuticals. The standardization of herbaceuticals is necessary to ensure consistent levels of bioactive compounds. The quality control of herbal products involves proper identification of plant materials, purity, absence of impurities, and stability.

The regulatory issues for herbal products vary from one country to another. In most countries, herbaceuticals are regarded as a combination of food supplements and pharmaceuticals. This has been a problem in public acceptance of herbaceuticals. There is a need to address regulatory issues with herbaceuticals to ensure strict adherence to quality control standards in their use in modern medicine.

1.1.5 Role of advanced analytical techniques in ensuring consistency

For the safe, effective, and reproducible use of herbaceuticals, it is essential to incorporate advanced analytical techniques. For example, high-performance liquid chromatography (HPLC), gas chromatography-mass spectrometry, nuclear magnetic resonance, and infrared spectroscopy are used to characterize bioactive compounds in herbal formulations. These analytical tools are also being applied to precisely quantify the active ingredients present, identify impurities/adulterants, and authenticate the chemical profiles in the formulations. In addition, the use of metabolomics and chemometrics has enabled the complete characterization of complex herbal formulations, thereby ensuring uniformity in therapeutic potential. The use of these analytical techniques, in combination with good manufacturing practices, ensures the uniformity in the nature and quality of the herbal

Table 1.1 Comparison between herbaceuticals and traditional drugs.

Aspect	Herbaceuticals	Traditional drugs
Source	Derived from plants and natural sources	Primarily synthesized or chemically manufactured in laboratories
Basis of development	Combines traditional herbal knowledge with modern scientific research	Developed through clinical trials, chemical synthesis, and pharmacological studies
Mechanism of action	Often have multiple mechanisms due to the presence of several bioactive compounds	Generally act on a single specific target or pathway
Therapeutic approach	Holistic, often aiming at balance and overall well-being	Targeted, disease-specific treatment
Side effects	Generally considered to have fewer or milder side effects, but not always risk-free	May have more pronounced or well-documented side effects
Regulation	Less strictly regulated in many regions	Tightly regulated by medical and drug authorities
Cultural roots	Deep roots in traditional medicine systems, such as Ayurveda, Siddha, and TCM	Rooted in modern pharmacology and evidence-based medicine
Formulation	Often in natural or semiprocessed form (powders, extracts, oils)	Typically, in refined forms, such as tablets, capsules, and injections

formulations, which, in turn, instils confidence in consumers and promotes their acceptance in the mainstream medical arena. The use of traditional herbal knowledge and modern analytical techniques in the preparation and characterization of herbaceutical formulations ensures their reproducibility and credibility, which are on par with conventional medicines (see Table 1.1).

1.2 Traditional systems and herbal origins

1.2.1 Ayurveda and Siddha

Ayurveda and Siddha are two traditional medical systems founded on the soil of our native land. They are among the most ancient and comprehensive medical systems to have emerged from antiquity. For as long as we can remember, these medical systems have employed herbs and vegetables extensively and effectively for myriad diseases and ailments, laying the foundation for the science of herbaceuticals itself.

The harmony between the human body, mind, and spirit under Ayurveda philosophy is a tradition that dates back more than 5000 years. Ayurvedic philosophy holds that human health is determined by a balance among three fundamental energies, or doshas: vata, pitta, and kapha. Any disorder of these doshas is believed to be associated with disease. Herbal medicine is used as a focal approach to restore health, combined with other remedial therapies.

Herbs such as ashwagandha, *Curcuma longa*, neem, and triphala, used in Ayurvedic medicine, possess significant immunomodulatory, antiinflammatory, antioxidant, and rejuvenating activities, among others, due to the diversity of bioactive compounds present within them, including alkaloids, glycosides, phenolic acids, flavonoids, sterols, essential oils, saponins, carbohydrates, and tannins, each of which has a distinct chemical structure, a characteristic feature of herbaceuticals.

The Siddha system of medicine, practiced in South India, shares roots with Ayurveda but also has distinct features. The emphasis in Siddha medicine lies in the effective use of minerals, metals, and herbs in its formulations, with an emphasis on complex alchemical techniques so that the bioactive components in the herbs may reach their full potential while minimizing their toxicity. The philosophy behind Siddha medicine states that it seeks balance in the physical, mental, and spiritual aspects.

Adhatoda vasica, *Acalypha indica*, and *Tinospora cordifolia* are among the herbs commonly used in Siddha medicine to treat respiratory system-related disorders, inflammatory conditions, and other debilitating diseases. The scientific documentation of these herbs and their medicinal uses has added considerable strength to the data pool for herbaceutical development. Ayurveda and Siddha serve as reservoirs of knowledge for modern researchers seeking potential drug targets (Mintii, 2023).

1.2.2 Traditional Chinese medicine

TCM, being an alternative medicine that is "holistic" in nature, is actually an ancient system of medicine that has been used for over 3000 years. "TCM is an ancient system of medicine that sees the body as an integrated system, where health is maintained by balancing energy within the body, 'Qi', and by the combination of contrasting forces found in nature, 'Yin' & 'Yang'." Herbal medicine is an inherent part of TCM.

It is evident that different categories of herbal medicine exist based on their attributes, such as whether they are hot, warm, cold, or cool. Their tastes are also classified into various categories: sweet, bitter, sour, pungent, or salty. Moreover, there is one meridian associated with a particular organ. This is what renders it very personal in application.

Unlike conventional pharmaceuticals, which are meant to treat one disease or another, herbal medicine compounds used in TCM are meant to treat all forms of physiological imbalances at once. Additionally, the compounds used differ by function, including the main drug, the assistant drug, and the harmonizer. This all shows that this drug has multitargeting abilities.

Famous herbs used in TCM, such as ginseng or Asian ginseng (*Panax ginseng*), *G. biloba*, root or leguminosae (*Glycyrrhiza glabra*), and *Astragalus membranaceus*, have actually gained popularity across the world as they have the potential to act as an adaptogenic agent, brain protector, or in boosting and/or increasing immune systems, etc.

Developments in Eastern herbal medicine, including herbal extracts and soft gel tablets, clearly describe the advancement of traditional medicine into modern herbaceutical medicine. Empirical traditional knowledge, when scientifically validated, is a critical factor in the development of herbal medicine worldwide, particularly in TCM (Evans et al., 2021).

1.2.3 Indigenous and folk medicinal practices

Indigenous and folk medicine represent one of the oldest branches of medicine, where traditional knowledge handed down from generation to generation is applied. The basis of this type of medicine is the application of local medicinal plants, herbs, and other natural products in order to prevent and cure illnesses. Extensive ethnobotanical knowledge about medicinal plants and their usage has been accumulated by indigenous people. Traditional medicines have made a significant contribution to sustaining good health of the community and have also been helpful in discovering bioactive substances. Many modern herbaceuticals and pharmaceutical substances owe their existence to traditional indigenous medical knowledge.

1.3 Traditional indigenous and folk medicinal practices

Indigenous and folk medicine practices are the backbone of traditional health systems. Most indigenous and folk medicinal practices are heavily dependent on local plant resources and are deeply anchored in a community's cultural and ecological knowledge. Local healers are found to have a better understanding of the natural environment of their area and are thus capable of selecting, collecting, and processing the medicinal plants.

The treatment modalities in folk medicine are usually simple and accessible, with leaves, roots, bark, and seeds of plants used to prepare remedies for common ailments. These remedies are usually passed down from generation to generation through oral tradition, part of the rich lore of traditional knowledge. Although most of them were never documented, many of these practices proved highly effective and stood the test of time.

There have been some modern medications and herbal products developed based on indigenous medical knowledge, too. For instance, painkillers, such as aspirin were developed from willow bark, while a commonly used herb in Chinese folk medicine, *Artemisia annua*, was initially used to produce the antimalarial medicine artemisinin.

Recently, research has been conducted to document and scientifically validate traditional uses of plants in ethnobotany and ethno-pharmacy, identifying the bioactive compounds that elicit therapeutic responses, thereby paving the way for a safe herbaceutical product; however, the ethical issues of intellectual property rights and the benefits to indigenous populations remain of concern.

Overall, indigenous and folk medicinal practices have greatly contributed to the development and evolution of herbaceuticals. Thus salvaging traditional knowledge and incorporating it into modern scientific methods means that herbaceutical development still benefits from generations of wisdom as it addresses contemporary healthcare challenges (Ray, 2017).

1.3.1 Influence of traditional herbal systems on modern herbaceutical development

The influence of the traditional systems of medicine in the development of the modern herbal industry has been substantial. The knowledge from traditional systems of medicine, such as Ayurveda, Siddha, TCM, and other healing traditions, has enabled the scientific community to identify plants with medicinal potential. Traditional systems of medicine have also enabled the scientific community to acquire knowledge of plant combinations, formulations, dosage forms, and preparations, which are now carried out using modern scientific techniques.

Advancements in analytical chemistry, pharmacognosy, and biotechnology have enabled the scientific community to identify the active phytoconstituents present in plants with medicinal potential. This has enabled the discovery of herbal medicines containing herbaceuticals, which are safer and more effective. Moreover, the traditional holistic approach to medicine and prevention has also been applied to the current approach of wellness and prevention.

1.3.2 Integration of traditional knowledge with modern drug discovery

The integration of traditional herbal knowledge and modern drug discovery techniques has led to remarkable improvements in the discovery and development of new herbaceuticals. Ethnopharmacology and traditional historical knowledge of herbal practices are valuable sources of information for selecting herbal medicines with medicinal properties. The application of modern drug discovery techniques has enabled the modern drug discovery scientist to identify the exact molecules in herbal medicines that possess medicinal properties. Moreover, the integration of traditional herbal knowledge on the combination of two herbs with modern pharmacology has led to remarkable advances in the discovery and development of synergistic medicines, with the potential to increase the efficacy of herbaceuticals without side effects. Thus it has not only validated the traditional knowledge of ancient peoples' practices but also led to the discovery of effective herbal medicines (see Table 1.2).

Table 1.2 Traditional medical systems: origins, features, and contributions to herbaceuticals.

Traditional system	Region of origin	Key features	Contribution to herbaceuticals
Ayurveda	India	Focuses on balancing doshas (vata, pitta, kapha) through herbs, diet, and lifestyle	Provides an extensive classification and use of herbs for maintaining health and treating ailments
Siddha	South India	Uses herbs, minerals, and metals with alchemical preparations	Offers unique formulations and processes involving both herbal and inorganic ingredients
Traditional Chinese medicine	China	Uses herbs based on energy (Qi), properties, and flavors to restore balance	Demonstrates the potential of multiherb combinations to address multiple conditions simultaneously
Indigenous and folk medicine	Global (local cultures)	Based on local knowledge, the use of native plants, passed down through generations	Serves as a foundation for discovering bioactive compounds used in herbaceutical development

1.4 Scientific basis and mechanism of action

1.4.1 Phytochemistry and bioactive constituents

Phytochemistry is the science that concerns the chemicals occurring naturally in the plant kingdom, especially those associated with their therapeutic/magical properties. Bioactivity is defined by the herbal bioactive compounds that ultimately constitute the concept of herbaceuticals. Knowledge of the constituent parts of any particular plant is described by understanding the herbal composition, aside from quality control, and the herbal constituent parts.

The major bioactive metabolite classes found in plant cells are flavonoids, alkaloids, and terpenoids. The bioactive metabolite class flavonoids comprises polyphenolics that have been well recognized for their antioxidant, antiinflammatory, antiviral, and heart-protective activities. These bioactive metabolites neutralize free radicals while providing cells with protection against oxidative injury.

Other phytochemicals that warrant mention are the group known as alkaloids, which essentially contain the element nitrogen in their structures. They are known to be highly biologically active compounds, with many having pain-relieving, antimalarial, antimicrobial, and anticancer properties. Due to their potent activity, alkaloids have attracted significant research for pharmaceutical development, including drug development.

Terpenoids, also known as isoprenoids, are a broad range of plant metabolites with diverse therapeutic activities, including antiinflammatory, antimicrobial, and anticancer effects. Some terpenoids are also involved in the formation of the distinctive scents and flavors of many medicinal plants, which can help improve patient compliance with prescribed medication. Knowledge of the chemical nature, biosynthesis, and activity of these plant metabolites is the foundation for the development of efficient and safe herbal medicines and, hence, is essential for the vital exchange between traditional herbal practices and the pharmaceutical sciences (Pazienza et al., 2022).

1.4.2 Pharmacodynamics and pharmacokinetics

Pharmacodynamics is an important concept in the efficacy of pharmaceutical drugs, describing the mechanisms by which a substance exerts its actions on the human body, including receptors, enzymes, and cell membranes. As herbaceutical drugs interact multiple times, diverse bioactive compounds consequently contribute to a wide range of therapeutic effects.

Pharmacokinetics, on the other hand, is used to describe the way that an organism takes up, stores, metabolizes, and excretes a drug. Unlike ordinary medicines, which typically contain a single active ingredient, herbaceuticals contain multiple active ingredients, making pharmacokinetics more complex. Their metabolisms can vary, affecting bioavailability.

The combination of all these may affect gastrointestinal absorption, enhance tissue distribution, or even alter metabolic stability. Under certain circumstances, they may even have an inhibitory or stimulating action in vivo on the metabolizing enzymes, hence influencing the in vivo excretion process for other components in their combination, an aspect that may be beneficial but at the same time poses its associated problems with the aspects of standardization, dose optimization, and safety (Ala-Juusela et al., 2021) (see Fig. 1.2).

1.4.3 Synergistic effects and therapeutic potentials

One of the most distinguishing factors of herbaceuticals is the synergies, which are described as a combination of a single herb or a combination of herbs with various bioactivity ingredients, where there can be a synergy between the various bioactivity ingredients, which can be described as a combination of their ability to produce effects beyond what they would individually produce. Synergies are present in all forms of medicine, especially in traditional medicine, where herbal formulations are used to treat certain diseases.

Using this approach, it is possible to enhance the drug's absorption and negate any side effects that may result from other compounds. Thus this approach to treating diseases, which combines synergistic effects, makes it a viable tool for managing disease.

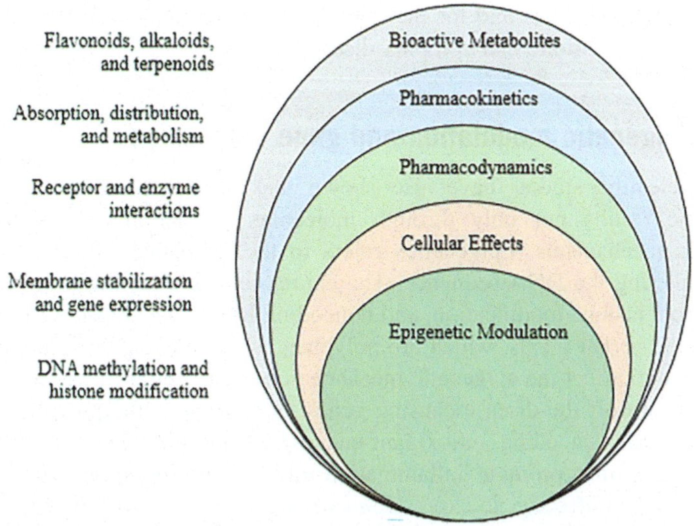

FIGURE 1.2

Scientific basis and mechanisms of action of herbaceuticals.

The evaluation of synergistic effects can largely aid in determining the therapeutic potential of herbaceutical compounds. Additionally, this approach can be used to validate herbal preparations, providing a means to produce both modern and conventional medicinal practices using herbaceutical compounds (Yu et al., 2020).

1.4.4 Molecular targets and cellular mechanisms

The therapeutic value of herbaceuticals can be understood in terms of their interactions with multiple cellular targets. Bioactive phytochemicals have been reported to interact with major signaling pathways involved in inflammation, oxidative stress, apoptosis, and immune modulation. The phytochemicals have been reported to interact with gene expression, enzymes, and receptors, thereby restoring cellular homeostasis. For example, some flavonoids have been reported to interact with the transcription factors that control the inflammatory response. In contrast, some alkaloids have been reported to interact with cell replication in abnormal cells.

At the cellular level, the herbaceuticals have been reported to interact with cell membrane stabilization, mitochondrial protection against oxidative stress, and the regulation of the cellular signaling pathways. The multitargeted action mechanism of the herbaceuticals distinguishes it from conventional drugs, which are generally considered single-targeted molecules. Knowledge of the cellular and molecular mechanisms of the herbaceuticals has provided a scientific rationale for their traditional medicinal use and for the rational formulation of the herbaceuticals for the treatment of chronic and complex diseases.

1.4.5 Epigenetic modulation and gene expression

Current scientific studies have also shown that these herbaceuticals can yield therapeutic results not only through molecular interactions but also through epigenetic mechanisms. Epigenetics refers to the regulation of gene expression without altering the DNA sequence. The expression includes the control of DNA methylation, histone modification, and noncoding RNAs. The bioactive compounds found in the herbal plants, which are polyphenols, flavonoids, and alkaloids, have been shown to affect the epigenetic mechanism, which could turn on the protective genes and turn off the disease-causing genes. For example, the polyphenols could upregulate genes encoding antioxidant enzymes, while the alkaloids could downregulate genes that promote inflammation and cell proliferation. The ability of herbaceuticals to affect gene expression explains how they work in the long term, which can even prevent the onset of diseases in the first place and maintain homeostasis in cells. The epigenetic effects of the herbaceuticals also open the door for the potential use of the compounds in the treatment and prevention of diseases, especially in the aspect of personalized medicine, where the individual's genes could be used as an indicator for the effect of the compound, which could lead to effective treatment and prevention (see Table 1.3).

1.5 Technological advancements in herbaceuticals

1.5.1 Standardization and quality control

Thus standardization and quality control are essential for the preparation/formulation of herbaceuticals, as they ensure the consistency of the drug's safety and efficacy. Herbaceuticals differ from synthetic drugs, as they are prepared from naturally occurring materials. Naturally occurring materials are variable due to factors such as geographical location, climatic conditions, soil conditions, and the time interval between harvests. In standardization, it is necessary to specify a particular parameter for the identification, purity, and concentration of the active ingredients present in botanicals.

In addition, it is necessary to use various modern analytical tools to confirm the authenticity and chemical nature of the raw materials and herbal preparations used to formulate the drug—for example, chromatographic fingerprinting, estimation of the bioactive ingredients present in the drug, etc. In addition, standardizing the methods used to extract the drug can help ensure uniformity in its quality.

Under these conditions, various government authorities worldwide are emphasizing standardized manufacturing processes and quality control testing. This move

Table 1.3 Fundamental concepts and their relevance in herbaceuticals.

Aspect	Description	Relevance to herbaceuticals
Phytochemistry	Study of plant-derived chemical compounds with medicinal properties	Identifies bioactive components, such as flavonoids, alkaloids, and terpenoids essential for therapeutic effects
Bioactive constituents	Naturally occurring compounds in herbs are responsible for their pharmacological activity	Understanding structure and function helps in the design of effective herbaceutical drugs
Pharmacodynamics	Mechanism of action – how a drug affects the body	Explains how herbaceuticals produce therapeutic effects, often involving multiple targets
Pharmacokinetics	How the body absorbs, distributes, metabolizes, and excretes a drug	Complex in herbaceuticals due to multiple constituents; essential for optimizing dosage and delivery
Synergistic effects	Combined effect of multiple compounds resulting in enhanced efficacy	Common in traditional formulations; explains why whole-herb preparations may outperform isolated compounds
Therapeutic potentials	Clinical applications and benefits of herbal compounds	Basis for integrating herbaceuticals into modern therapeutic systems and validating their use scientifically

towards quality must be adhered to by all the herbaceuticals to avoid any impurities, adulteration, or changes in the product. This ensures consumer confidence, proper assessment of the drug's practical application, and safe, practical consumption of the herbaceuticals.

1.5.2 Biotechnology and genetic engineering

In that regard, biotechnology and genetic engineering greatly improve the efficiency and quantity of herbaceutical production. These technologies can be used to improve herbal medicine, increasing yields, therapeutic value, and resistance to pests and diseases. Genetic modification and plant selection can also be used to increase the amounts of valuable active compounds, thereby enriching or enhancing pharmacologically active extracts used in herbal products.

The use of applications developed for genetic engineering helps identify specific genes used to produce bioactive compounds. The upregulation of some of these bioactive compounds may improve the efficacy of secondary metabolites used in their herbaceutical product. Moreover, this will ensure that raw materials for production are plentiful.

Apart from these plant-based changes, biotechnology offers other alternatives, including the production of microbial metabolites through microbial fermentation. These microbial metabolites can be produced by genetic engineering. Such a process is an environmentally friendly practice, economically viable, and breaks away from the limitations of bulk plant production, which is seasonal in most cases. Such biotechnological innovations contribute immensely to the scaling up of this herbaceutical practice (Yi et al., 2025).

1.5.3 Nanotechnology and novel drug delivery systems

Nanotechnology is an innovative technology that has entered the pharmaceutical industry for the development of value-added, advanced drug delivery systems. There are several disadvantages of herbal medicine, viz., low bioavailability due to poor solubility, instability, or rapid metabolism. All these disadvantages can be overcome through the application of nanotechnology, improving their absorption and increasing the targeting effect of active plant compounds.

These can be engineered to surround herbal constituents, protecting them from degradation and releasing them for controlled release at specified sites within the body. Therefore an increase in therapeutic potency, combined with minimum side effects, would be possible. This is because nanoparticle-based delivery systems enable deeper penetration than would normally be possible.

In addition, nanotechnology can formulate herbaceutical drugs into tailored formulations to meet specific medical needs. Therefore the use of nanotechnology enhances the therapeutic potential of herbaceutical drugs by increasing their solubility, stability, and bioavailability, opening new vistas for their clinical utility through incorporation into modern medicine. Nanotechnology can be very instrumental in the efficacy of herbaceutical drugs in the near future (Zhou et al., 2024).

1.5.4 Advanced extraction and processing technologies

Advanced technologies for extraction and processing have greatly contributed to the efficiency, safety, and quality of herbaceuticals. Conventional extraction technologies have been associated with low recoveries and the deactivation of heat-sensitive bioactive compounds. Advanced extraction technologies, including supercritical fluid extraction, microwave-assisted extraction, ultrasound-assisted extraction, and pressurized liquid extraction, may be used to extract bioactive compounds with minimal loss of activity selectively. These technologies may be used to optimize extraction efficiency with minimal solvent consumption and time.

Besides, drying, encapsulation, and stabilization technologies may be used to maintain the chemical stability of herbal compounds during the processing and storage of herbaceuticals. Advanced technologies in these areas may be used to optimize the

Table 1.4 Technological advancements and their impact on herbaceuticals.

Aspect	Description	Impact on herbaceuticals
Standardization and quality control	Establishing uniform standards for identification, purity, and active content of plant materials	Ensures consistency, safety, and efficacy of herbaceuticals; builds consumer trust and enables regulatory compliance
Biotechnology and genetic engineering	Genetic modification and biotechnological methods to improve plant yield, phytochemical content, and resistance	Enables production of high-potency herbs; allows eco-friendly biosynthesis of active compounds via microbial fermentation
Nanotechnology and novel drug delivery	Use of nanoparticles to improve bioavailability, targeted delivery, and controlled release of herbal drugs	Enhances therapeutic efficacy, reduces side effects, and expands clinical applications of herbaceuticals

reproducibility, shelf life, and bioavailability of herbaceuticals. It is therefore possible to produce high-quality herbaceuticals meeting the required standards by applying modern processing technologies. In conclusion, it is important to note that advanced extraction and processing technologies are required to develop modern herbaceuticals from conventional herbal compounds (see Table 1.4).

1.6 Therapeutic applications

1.6.1 Chronic diseases and lifestyle disorders

Some of the chronic health conditions that present a lifestyle component are diabetes mellitus, hypertension, cardiovascular diseases, and obesity, which have been the major contributors to health concerns in the general public in recent times. Over the years, these health conditions have been linked to poor nutritional habits, a sedentary lifestyle, oxidative stress, and chronic inflammation. However, in recent decades, considerable interest has emerged in the potential role of plant-based chemicals in the prevention and treatment of these health conditions.

The herbal compounds present in the herbal product play a role in the prevention and treatment of health conditions by virtue of their antioxidant, antiinflammatory, and other metabolic properties, which enable them to influence critical pathophysiological processes associated with these conditions. The phytochemicals present in the herbaceuticals may influence health conditions by improving insulin response, normalizing lipid profiles, and scavenging free radicals, which, in turn, reverse the pathologies associated with these conditions. Similarly, research done on one of the important herbal compounds, that is, curcumin, a compound found in turmeric, or "Curcuma longa," confirmed its antiinflammatory and insulin-sensitizing potential. Likewise, resveratrol, a phytochemical found in grapes and berries, was found to enhance carbohydrate and lipid metabolism and, hence, can prevent inflammation as well. The multifactorial beneficial effects of resveratrol emphasize the role of herbal medicines in managing lifestyle diseases.

1.6.2 Immunomodulation and antiinflammatory uses

Herbaceuticals have well-recognized attributes in the management of immune responses and inflammatory conditions. Although the inflammatory response of the human body is critical, a hyperinflammatory response may manifest as various pathological syndromes, including autoimmunity, allergy, and metabolic syndromes. There are different lists of herbal compounds, and it has been proven that these compounds are effective in controlling inflammatory conditions without major side effects, unlike chemical compounds.

There are different medicinal herbs, and these have shown promising immune response modulatory potential, meaning they can stimulate or modulate specific immune cells or substances. Herbs, such as echinacea and elderberry are highly recommended, as they can enhance immune response, especially in the management of respiratory tract infections. These herbs not only enhance immune response but also suppress inflammation. Besides, several scientific discoveries indicate that certain herbal extracts can regulate immune signaling pathways, and, as such, they are very important in the management and control of various autoimmune and inflammatory diseases. Herbaceuticals are used in the preparation of various formulas to improve an individual's immune system, reduce inflammatory conditions, and enhance health (Girotti et al., 2011).

1.6.3 Neurological and cognitive disorders

Neurological and cognitive disorders, like Alzheimer's, dementia, and age-related cognitive decline, are a serious therapeutic concern for clinicians worldwide. Accompanying the aforementioned neurological and cognitive disorders, recent findings affirm the role of oxidative stress, defective circulation, and neuroinflammatory reactions in the etiology of these disorders. Herbaceuticals are promising drug candidates for the management of neurological and cognitive disorders, because of the known neuroprotective, cognitive-enhancing effects of herbaceuticals themselves.

The herbs *Ginkgo biloba* and ginseng are among the most widely studied for their impact on the brain. It is well known that *G. biloba* stimulates blood flow to the brain, thus improving concentration and memory, and it also provides antioxidants that help prevent the degeneration of neuronal cells. The herb also shows adaptogenic properties, thus stimulating mental performance.

These herbaceuticals have the potential to help slow the progression of neurodegenerative disease by improving neurotransmission, reducing oxidative stress, or increasing neuronal survival. As such, plant-based preparations are becoming increasingly important in supporting therapies for cognitive or neurological pathologies (Luo et al., 2015).

1.6.4 Dermatological and cosmetic applications

Herbaceuticals are widely used in dermatological and cosmetic fields due to their natural composition and lack of side effects. Medicinal herbs offer health benefits that promote skin health. Herbal additives are becoming popular as skincare ingredients rather than chemical additives. The former can be employed for a longer period.

Normally, the herbal drug components comprise aloe vera, tea tree oil, chamomile, etc., which possess remarkable properties for soothing, wound healing, and antimicrobial activities. These drug components are widely used for various skincare applications, including acne and eczema. These pharmacological practices are employed to provide remarkable alternatives to herbaceuticals in dermatological sciences.

1.6.5 Gastrointestinal and metabolic health

The application of herbaceuticals is also important for promoting gastrointestinal and metabolic health. Gastrointestinal diseases such as indigestion, constipation, irritable bowel syndrome, and gastritis are often associated with dietary factors, stress, and the balance of the gastrointestinal microbiota. Some plant compounds have been found to possess carminative, laxative, and antiinflammatory properties, which could be advantageous for regulating the gastrointestinal system.

Some herbal medicines that can be utilized to enhance digestion, relieve intestinal spasms, and protect the gastric mucosa include ginger, peppermint, fennel, and liquorice. Moreover, some herbaceuticals were found to promote the growth of beneficial gut microbiota, which is important for proper nutrient absorption and immune regulation. Some plant polyphenols and fibers were found to be beneficial for regulating glucose absorption and lipid metabolism, thereby promoting metabolic activity (see Fig. 1.3).

1.7 Safety, toxicology, and regulatory aspects

1.7.1 Toxicological profiling and safety assessment

In the development and marketing of herbaceuticals, toxicological profiling and safety assessments play a very significant role. Although humans regard

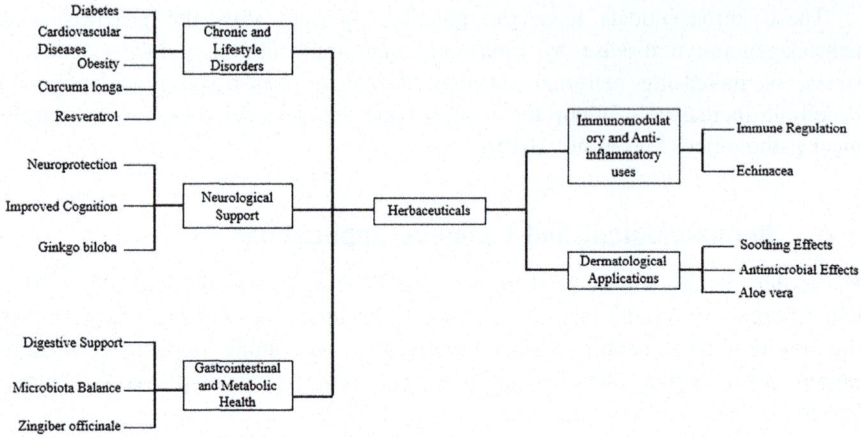

FIGURE 1.3

Therapeutic applications of herbaceuticals.

herbaceuticals as safe, as they are derived from a natural source, it is highlighted that without performing a proper evaluation of the safe use of herbaceuticals by humans, toxicological evaluation of these compounds is carried out with a view to assessing potential adverse effects, toxic dose, and long-term health hazards of herbal compounds.

The safety evaluation of a drug includes preclinical and clinical studies, which include acute, subacute, and chronic toxicity studies. The studies also included evaluations of organ toxicity, genotoxicity, carcinogenicity, reproductive toxicity, and human interest in herbal drug interactions, which have occupied a prominent place, especially in relation to the concurrent use of herbaceuticals. Several herbs have metabolic interactions with drug molecules, which reduce their efficacy or toxicity.

Various regulatory authorities in different countries around the world require stringent safety tests to establish the safe dosage of herbal products, which is crucial for consumer safety. Toxic fingerprinting facilities can be used to identify toxic compounds in the product, including metals, pesticides, and other toxins. Safety tests are crucial for enhancing consumers' confidence in the use of herbaceuticals in healthcare (Ma, 2007).

1.7.2 National and international regulatory frameworks

The regulatory framework for the use of herbaceuticals reveals varying degrees of divergence across nations. These divergent forms are a product of the divergent cultural, healthcare, and legal systems. In nations where the use of herbaceuticals is gaining prominence, a new trend is emerging toward the establishment of national and international regulatory systems that evaluate the quality, safety, and efficacy of products.

They have played a significant role in an international organization's capacity to establish guidelines for the evaluation and standardization of herbal medicines worldwide. WHO also emphasizes the establishment of guidelines for Good Agricultural and Collection Practices (GACP), quality assurance, and assessment-based evaluation of efficacy and safety. Other examples include the establishment of a regulatory system in which the Food and Drug Administration (FDA) is involved, and the development of guidelines for the manufacturing of herbaceutical products.

Across the globe, governments are adopting various regulatory policies to manage raw material procurement, manufacturing strategies, and quality control approaches. Proper regulations can prevent adulteration, misbranding, etc., involving herbs, making the product credible and facilitating the inclusion of herbaceuticals into mainstream healthcare systems.

1.7.3 Intellectual property and patent issues

One of the main challenges in the herbaceutical industry is intellectual property rights, which include patent rights. Most products in the herbal industry have originated from traditional practices of indigenous communities. These traditional practices, when patented, raise concerns. Patenting traditional practices, such as indigenous traditional medicinal practices, is a major challenge.

"The patenting of herbal formulations and bioactive compounds would involve showing the novelty, innovation, and industrial applicability of the compound/formulation in question, which is always difficult with herbal remedies." There is the problem of biopiracy and the illegal exploitation of resources by indigenous cultures, which brought about an international sense of awareness about the need for fair practices to be followed by all involved parties.

The ethical nature of business activities within the herbaceuticals sector highlights the importance of "benefit-sharing" and the need for profits from commercialized products to be shared with indigenous populations who have maintained and utilized their traditional knowledge over time. The Convention on Biological Diversity has been at the heart of promoting equitable access and "benefit-sharing" within the herbaceuticals sector (Williams, 2017).

1.7.4 Labeling, consumer awareness, and ethical marketing

Labeling, consumer awareness, and ethical marketing are important aspects associated with the safe use of herbaceuticals. The major difference between herbaceuticals and other conventional medicines is that they are a combination of complex bioactive compounds. Variations in the mixture's level can significantly affect the drug's safety and efficacy. Labeling the ingredients and interactions

Table 1.5 Regulatory and ethical aspects of herbaceutical development.

Aspect	Description	Relevance to herbaceuticals
Toxicological profiling and safety assessment	Evaluation of adverse effects, toxicity levels, and drug interactions of herbal products	Ensures public health safety; required for regulatory approval before commercialization
National and international regulations	Frameworks provided by authorities, such as WHO, FDA, and national bodies to govern herbaceutical standards	Promote quality control, efficacy validation, and safety compliance across countries
Intellectual property and patent issues	Concerns regarding patenting traditional knowledge, legal ownership of plant-based inventions, and benefit-sharing	Highlights the need for ethical commercialization and protection of indigenous knowledge rights

between drugs and herbs is an important step to ensure that the drug is not misused or results in adverse effects.

Consumer awareness is another important factor that bridges the gap between the conventional knowledge and the use of the drug. Consumer awareness can help consumers gain awareness of the benefits and the risks associated with the use of the drug. Ethical marketing can ensure that claims are based on scientific evidence for the drug and that it is not hyping the drug's benefits (see Table 1.5).

1.8 Market trends and global perspectives

1.8.1 World market growth and dynamics

Over the years, remarkable growth in the global herbaceuticals industry has been observed. This has been possible due to the consistent increase in the demand for more herb-based products, which play a crucial role in preventive health. This wave of preventive health has been playing a crucial role in encouraging consumers to adopt herbaceuticals as a more viable option than conventional pharmaceutical products.

Some of the key factors responsible for the growth of the herbaceuticals industry are: a rise in the level of income, particularly in developing countries, which has enabled consumers to spend more on health and wellness-related products, coupled with changes in lifestyles, particularly due to urbanization, which has increased lifestyle disorders, thereby fueling the growth of the herbaceuticals industry.

The markets in North America and Europe can be classified as mature for herbaceuticals, given the strong presence of well-defined regulatory systems, robust research conditions, and a high degree of consumer awareness of herbaceutical products. Ideally, in North America and Europe, herbaceutical products are used largely as dietary supplements, functional foods, and complementary medicine, thereby integrating well into the existing markets in these regions.

Asia-Pacific is expected to rise as a prominent growth market location for the herbaceuticals industry. Countries such as India, China, Japan, and Korea boast a long history of traditional herbal medicine, in addition to considerable biological diversity, a strong foundation upon which the herbaceutical industry can thrive. However, another reason for the increasing demand in the Asia-Pacific region is the rise in economic growth rates, the share of the market segment, and improved health consciousness. Yet another factor, which is considered a challenge, is the presence of divergent regulations in this market. However, there are many opportunities in this market segment (Williams, 2017).

1.8.2 Consumer awareness and preferences

Consumer awareness and trends are very strong influencers in shaping both the development and growth of the herbaceous market. Nowadays, consumers are more informed and take health-related issues very seriously. This has created strong demand for developing healthy ways to prevent or treat disease through more natural approaches. Such shifts appear to be based on the long-term side effects of artificial drugs.

This general consumer perception, among some, that herbaceuticals not only have a gentler effect on their systems but can also be used long term, is a contributing factor. The interest in organic and natural ingredients has increased manifold, supported by other macro drivers, such as clean labeling, natural, and green, which cover sustainability and going green, respectively. More products with no artificial ingredients, preservatives, or chemicals that could harm their systems can be expected in the future.

The low cost and prevalence of health information through media, social media, and educational campaigns have been synergistic with consumer awareness. The ability to access information on scientific studies, experts, and consumer reviews has empowered consumers to make informed choices about herbal products. Thus consumers will be more choosy, with quality, safety, and claims taking substantive importance in consumer purchase decisions.

The fact that people's needs keep evolving means that transparency is being given greater importance than ever before. In fact, when information about the specific sources of the substances used and the production methods for those products is clearly explained to customers, they will begin to trust these enterprises. Secondly, they provide education to increase awareness of the specific benefits of accessing herbaceuticals. The specific steps taken to ensure that customer needs are well addressed can therefore be regarded as a differentiating factor for these enterprises operating in a highly competitive market (Lestari and Quddoos, 2025).

1.8.3 Challenges in commercialisation

However, despite all these prospects for growth in the herbaceuticals industry, some key challenges have been identified as crucial to its commercialization success or failure. The first key challenge in the herbaceuticals industry is the complex regulatory environment governing herbaceuticals across various countries worldwide.

Moreover, the second key challenge in the herbaceuticals industry is the need for robust scientific data to validate the safety and efficacy of the herbaceuticals in question. Although the traditional knowledge in this industry is strong, there is still a need for further validation and standardized research data, which requires significant investments and technical expertize and may be daunting for small and medium-sized businesses. The issue of quality variability is a worrying factor in the pharmaceutical sector. Issues such as the quality of raw materials, varying growing conditions for herbs, and a lack of standardized processing may affect the efficacy of the developed product in various ways. This inconsistency in the efficacy of different herbal products may lead consumers to doubt their authenticity.

The competition from the pharmaceutical sector may be regarded as another challenge the herbaceutical sector may face in the near future. The pharmaceutical sector, over the years, has built its brands, conducted in-depth research, and made significant contributions to marketing. It is therefore essential that these challenges be addressed, not only to ensure the sector's authenticity but also to enable it to reach its potential on the world stage (Liu et al., 2022).

1.8.4 Emerging technologies and e-commerce in herbaceuticals

The advent of digital technology and e-commerce has affected the herbaceutical industry, providing an unprecedented opportunity to reach consumers worldwide. Digital technology has enabled small- and medium-scale industries to reach consumers worldwide through e-commerce. Digital marketing campaigns have also enabled consumers to make informed decisions about using herbaceuticals. The advent of technology has also enabled the elimination of pharmaceutical adulteration through blockchain technology, ensuring transparency in the supply chain and the authentication of raw materials. Blockchain technology has been considered a way to eliminate the possibility of raw material adulteration, as consumers are becoming increasingly health-conscious about their food habits. Apart from this, AI-generated recommendations are also being considered to help consumers make informed decisions about using herbaceuticals based on their needs. The advent of technology, as well as e-commerce, is thus affecting the marketing of the herbaceuticals (see Table 1.6).

Table 1.6 Market trends and commercialization challenges in the herbaceutical industry.

Aspect	Description	Impact on herbaceutical industry
World market growth and dynamics	Rising demand for natural, plant-based products, driven by health trends, higher disposable incomes, and wellness awareness	Strong growth in North America, Europe, and Asia-Pacific; expanding opportunities for global herbaceutical firms
Consumer awareness and preferences	Shift toward organic and alternative medicine due to concerns over side effects of synthetic drugs; fueled by media and educational outreach	Encourages clean-labeling, transparent sourcing, and increased consumer trust and loyalty
Challenges in commercialization	Barriers include regulatory complexity, lack of standardized efficacy data, market competition, and product quality inconsistencies	Requires robust R&D, strict quality control, and effective communication to gain a competitive edge

1.9 Future directions and integrative approaches

1.9.1 Personalized herbaceutical therapy

The future of herbaceutical medicine is more in line with other developments in the medical field. This is especially true in the development of personalized medicine, which tailors therapeutic interventions to the individual. The "one-size-fits-all" approach is the tendency in conventional medicine. Individuals react differently to herbal medicine formulations based on their genotype, metabolism, exposure to specific environmental factors, diet, and proneness to specific diseases. The developments in genomics, proteomics, metabolomics, and biotechnology provide unprecedented opportunities to tailor herbaceutical formulations to fit the individual's physiological needs.

Genetic profiling can provide information on differences in drug-metabolizing enzyme activity, inflammatory responses, or antioxidant potential. These can then be used in selecting the herbal constituents. For example, if a person has a predisposition to oxidative stress, they may require more herbal supplements containing polyphenolic compounds. If they have inflammatory gene expressions, they may require herbaceuticals with immunomodulatory properties. This is more effective in therapeutic responses, with a minimal risk of adverse effects, interactions, or the use of ineffective agents.

In addition, factors such as diet, exercise, stress levels, and sleep habits contribute to the efficacy of the herbal intervention. This includes the design of individualized herbal therapies that incorporate multiple interventional factors. These factors include the design of individualized drug therapies that support preventive care, as well as the treatment of diseases such as diabetes, cardiovascular diseases, and neurodegenerative diseases.

In addition, digital health devices such as wearables and artificial intelligence-based analytical tools have the potential to fine-tune the efficacy of a herbal approach. Indeed, it is possible to monitor health markers in real time continuously. As such, it is possible to continuously modify an herbal approach. As more is discovered about biomarkers of efficacy for herbaceutical interventions, a personalized herbaceutical approach could become a mainstay of integrative medicine (Peredy et al., 2024).

1.9.2 Traditional-modern systems interaction

To ensure the universal acceptance and incorporation of herbaceuticals in the mainstream healthcare system, there should exist a synergistic relationship between traditional medicine systems and modern scientific research. Traditional medicine systems such as Ayurveda, TCM, and Unani, as well as traditional practices, have accumulated centuries of empirical data on the therapeutic potential of herbs. However, most of this information is well left unutilized owing to the lack of standardized scientific validation with a mechanistic basis. Modern biomedical research offers robust tools, such as randomized clinical trials, molecular pharmacology, and bioinformatics, to evaluate traditional herbal formulations. The traditional knowledge of herbs, as viewed in light of modern scientific research, offers tremendous potential for new bioactive agents, therapeutic targets, and delivery systems.

The collaboration among medical practitioners from the west, pharmacologists, and traditional healers is vital to bridging this epistemological gap. This would certainly facilitate mutual respect's point of view so that herbaceuticals do not become either discredited as being uninformed or idealized without solid grounding in reality. Such collaborative educational efforts, research initiatives, or even a hybrid approach to medicine would certainly facilitate this collaboration with tremendous benefits to patient-oriented medicine.

Furthermore, this shift in regulation would facilitate the evolution of existing regulations to incorporate an all-encompassing approach to herbaceuticals by developing quality, safety, and efficacy parameters that are herbaceutical-friendly. The marriage of traditional and contemporary approaches has tremendous potential to elevate herbaceuticals from an alternative medicine to a legitimate medical discipline that espouses a holistic, socially conscious, and scientifically evidence-based approach to healthcare (Farooqui et al., 2020).

1.9.3 Environmental and sustainable sourcing impact

One issue that has come to characterize environmental sustainability for the herbaceuticals industry is environmental concern, driven by increasing awareness among herbaceuticals consumers. The increasing demand for herbaceutical raw materials presents a challenge to plant resources. Some plant resources used as herbal medicines are typically collected from the wild. The sustainability of such behavior in the environment can lead to environmental destruction, thereby affecting the sustainability of herbal medication. The practice of sustainable sourcing involves balancing economic demands with environmental friendliness. This includes cultivation, collection, rotation, and renewal. In addition, biodiversity conservation is often a priority for medicinal plants, especially when they are important to the environment at certain times. This is where cultivation, rather than collection, comes into play to reduce the industry's footprint. Other important factors, such as the application of fair trade practices, are useful in developing a sustainable environment that provides fair remuneration to the indigenous inhabitants of the regions for the knowledge they have accumulated in developing the concerned plant species. The majority of the herbal plant species used in the development of medicine originate from regions with low standards of prosperity.

In addition, more environmentally friendly processing, packaging, and transportation strategies can reduce carbon footprints in the long run. Finally, the adoption of symbiotic technologies, biodegradable products, and environmentally conservative processing technologies can also help the herbaceutical industry move closer to achieving sustainability goals and forces. Therefore sustainability in the herbaceutical industry can be regarded more as a necessity for the long-term viability and credibility of the sector (Guan, 2025).

1.9.4 Digital health integration and smart herbaceuticals

The integration of digital health is poised to revolutionize the herbaceutical industry by facilitating real-time tracking, personalized recommendations, and compliance. Wearable technology, mobile health applications, and biosensors would facilitate tracking of physiological processes such as blood glucose, blood pressure, heart rate, and sleep patterns, which would be extremely beneficial for improving the herbal treatment process. Moreover, integrating artificial intelligence and machine learning would provide the necessary inputs to inform adjustments to drug dosages, potential herb–drug interactions, and poly-herb treatment options for patients.

Furthermore, implementing smart herbaceutical packaging technologies, such as QR code/Nutraceuticals and Functional Foods Centre (NFC) chip technology, would provide consumers with the necessary product information, thereby promoting transparency and trust.

Table 1.7 Innovative approaches and sustainability in herbaceutical development.

Aspect	Description	Significance for herbaceuticals
Personalized herbaceutical therapy	Use of genomic data, lifestyle factors, and biotechnology to tailor herbaceuticals to individual needs	Enables precision medicine, maximizes therapeutic efficacy, and minimizes side effects
Traditional-modern systems interaction	Integration of traditional knowledge with modern scientific validation	Fosters the development of evidence-based therapies, bridges the gap between traditional healers and modern practitioners
Environmental and sustainable sourcing	Focus on eco-friendly cultivation, biodiversity conservation, and fair trade practices	Aligns with green consumerism, ensures long-term availability of medicinal plants, and builds brand trust

The integration of digital health innovation with conventional and modern herbal knowledge would provide a fascinating new paradigm in integrative medicine, where critical insights would enhance the therapeutic efficacy, safety, and sustainability of herbaceutical treatment processes (see Table 1.7).

1.10 Conclusion

The evolution of herbaceuticals, from the ancient days of herbals to the modern validation of their efficacy, can be considered a blend of ancient wisdom and modern scientific validation. The role that medicinal plants have played in the healing practices of various civilizations throughout the course of history can be considered immense, and this has provided the basis for the evolution of various traditional systems of medicine, including Ayurveda, TCM, Unani, and other indigenous systems of medicine that are practiced throughout the world. The modern validation of this ancient wisdom, which has dominated human healing practices for centuries, can be considered the basis for the concept of herbaceuticals. The modern-day interest in herbaceuticals can be attributed to advancements in pharmacology, phytochemistry, and biotechnology. Modern research methods can aid in the identification and characterization of bioactive compounds that produce therapeutic effects. The scientific validation and proof of concept obtained through HPLC, mass spectrometry, and molecular docking studies have helped elucidate the mechanisms and transform ancient herbal formulations into standardized formulations with predictable results, thereby increasing their acceptability and credibility in modern-day healthcare systems.

Clinical research has played an important role in legitimizing the concept of herbaceuticals in modern medical practice. Clinical trials to assess the efficacy, safety, dose-response relationship, and therapeutic uses of herbal formulations are now integral to the assessment and validation process. These clinical trials not only validate the traditional uses of medicinal plants but also discover novel uses for their prevention and treatment of chronic diseases, metabolic disorders, immunological disorders, and psychiatric illnesses. The acceptability and integration of the concept of herbaceuticals into mainstream medical practices are exemplified by the formal inclusion of traditional medicine in the national healthcare systems of several countries. For example, in China, TCM is routinely combined with modern medical treatments, offering integrative therapeutic approaches that blend herbal formulations with pharmaceuticals and advanced medical technologies. Despite these advancements, the mainstream acceptance of herbaceuticals faces several challenges. One significant limitation is the insufficient understanding of the pharmacokinetics and pharmacodynamics of many herbal compounds. Unlike synthetic drugs, herbal products often contain complex mixtures of multiple active constituents, making it difficult to predict absorption, distribution, metabolism, and excretion. This complexity also raises concerns regarding herb–drug interactions, particularly for patients who use herbaceuticals alongside conventional medications. Inadequate awareness among healthcare professionals and patients about these interactions can compromize treatment safety. Regulatory and quality control issues further complicate the integration of herbaceuticals into conventional medicine. Variability in plant species, cultivation conditions, harvesting methods, and processing techniques can lead to inconsistencies in product quality and potency. While regulatory frameworks are increasingly being developed to address these concerns, global harmonization of standards remains limited. Ensuring quality assurance, proper labeling, and postmarketing surveillance is essential for building trust among healthcare providers and consumers. Cultural and perceptual barriers also influence the acceptance of herbaceuticals.

In some regions, traditional medicine is viewed as unscientific or inferior to conventional medicine, while in others it is uncritically accepted without adequate scientific scrutiny. Bridging this divide requires educational initiatives that promote mutual respect between traditional healers and modern healthcare practitioners. Participatory research models that involve both scientific experts and custodians of traditional knowledge can foster collaboration, ethical knowledge sharing, and culturally sensitive innovation. Sustainability represents another critical challenge confronting the herbaceutical industry. The growing demand for medicinal plants has intensified pressure on natural resources, leading to overharvesting, habitat destruction, and the depletion of rare species. Unsustainable practices threaten biodiversity and undermine the long-term availability of herbal raw materials. Addressing these issues necessitates conservation strategies, sustainable cultivation practices, and community-based resource management. Ethical sourcing and fair trade practices can also support the livelihoods of indigenous communities while promoting environmental stewardship.

In conclusion, the future of herbaceuticals in conventional medicine depends on achieving a harmonious balance between honoring traditional knowledge and applying rigorous scientific standards. By addressing research gaps, strengthening regulatory frameworks, fostering interdisciplinary collaboration, and promoting sustainability, herbaceuticals can realize their full therapeutic potential. Their successful integration into modern healthcare systems offers an opportunity to enrich medical practice with holistic, culturally grounded, and scientifically validated approaches to health and disease management.

References

Ala-Juusela M, Rehman Hur, Hukkalainen M, Reda F. Positive energy building definition with the framework, elements and challenges of the concept. Energies 2021;14(19):6260. Available from https://doi.org/10.3390/en14196260.

Bubela T, Boon H, Caulfield T. Herbal remedy clinical trials in the media: a comparison with the coverage of conventional pharmaceuticals. BMC Medicine 2008;6(1). Available from https://doi.org/10.1186/1741-7015-6-35.

Evans AL, Bulla AJ, Kieta AR. The precision teaching system: a synthesized definition, concept analysis, and process. Behavior Analysis in Practice 2021;14(3):559–76. Available from https://doi.org/10.1007/s40617-020-00502-2.

Farooqui SJ, Qureshi J, Zeeshan M, Soomro KA. Bays international private limited: evolving through market dynamics. International Journal of Experiential Learning & Case Studies 2020;5(2). Available from https://doi.org/10.22555/ijelcs.v5i2.314.

Girotti A, Fernández-Colino A, López IM, Rodríguez-Cabello JC, Arias FJ. Elastin-like recombinamers: biosynthetic strategies and biotechnological applications. Biotechnology Journal 2011;6(10):1174–86. Available from https://doi.org/10.1002/biot.201100116.

Guan B. Research on the factors contributing to the popularity of luxury goods in the chinese market. Advances in Economics, Management and Political Sciences 2025;187(1):7–12. Available from https://doi.org/10.54254/2754-1169/2025.bl23697.

Lestari DST, Quddoos MU. The influence of market research, consumer behavior and word of mouth communication for sustainable marketing of electric vehicle in Indonesia. Multidisciplinary Indonesian Center Journal (MICJO) 2025;2(2):2247–57. Available from https://doi.org/10.62567/micjo.v2i2.809.

Liu W, Hao Z, Florkowski WJ, Wu L, Yang Z. A review of the challenges facing global commercialization of the artificial meat industry. Foods 2022;11(22):3609. Available from https://doi.org/10.3390/foods11223609.

Luo C-H, Shanmugam V, Yeh C-S. Nanoparticle biosynthesis using unicellular and subcellular supports. NPG Asia Materials 2015;7(8):e209. Available from https://doi.org/10.1038/am.2015.90.

Ma X. Special issue on microtechnology and nanotechnology in medicine. Proceedings of the Institution of Mechanical Engineers, Part H: Journal of Engineering in Medicine 2007;221(2):i. Available from https://doi.org/10.1177/095441190722100201.

Mintii IS. Blended learning: definition, concept, and relevance. Educational Dimension 2023;8. Available from https://doi.org/10.31812/ed.539.

Pazienza M, De Jong M, Schoenmaker D. Clarifying the concept of corporate sustainability and providing a unified definition. Sustainability 2022;14(13):7838. Available from https://doi.org/10.3390/su14137838.

Peredy Z, Vigh L, Lorenzoni P, Ghisaura R. Impacts of global trends on the European Union competitiveness. Acta Periodica 2024(31):4–28. Available from https://doi.org/10.47273/ap.2024.31.4-28.

Ray PP. An introduction to dew computing: definition, concept and implications. IEEE Access 2017;6:723–37. Available from https://doi.org/10.1109/ACCESS.2017.2775042, http://ieeexplore.ieee.org/xpl/RecentIssue.jsp?punumber=6287639.

Williams DB. Application of pharmaceutical sciences to modern pharmacy practice. Journal of Pharmacy Practice and Research 2017;47(6):416–7. Available from https://doi.org/10.1002/jppr.1404, http://onlinelibrary.wiley.com/journal/10.1002/%28ISSN%292055-2335.

Yi W, Hu S, Qian X, Yan W, Li Y. Synthetic biology-based engineering cells for drug delivery. Exploration 2025;5(2). Available from https://doi.org/10.1002/exp.20240095.

Yu C, Xu X, Yu S, Sang Z, Yang C, Jiang X. Shared manufacturing in the sharing economy: concept, definition and service operations. Computers & Industrial Engineering 2020;146:106602. Available from https://doi.org/10.1016/j.cie.2020.106602.

Zhou J, Wang Y, Wang J, Sheng M. An in-depth analysis of intravenous oncology drug compounding in pharmacies: ensuring safety, enhancing efficiency, driving innovations, and maintaining stability. Medicine Insights 2024;1(7):45–57. Available from https://doi.org/10.70088/nwjtvz03.

CHAPTER

2 The science behind healing botanicals: phytochemistry and pharmacology

Ashok Kumar Balaraman[1], Kanaka Durga Devi Nelluri[2], Durga Prasad Kondeti[2], Padamata Sai Rohith[2] and Indupalli Siddhartha[2]

[1]*Research and Enterprise, University of Cyberjaya, Cyberjaya, Selangor, Malaysia,* [2]*Department of Pharmaceutics and Biotechnology, KVSR Siddhartha College of Pharmaceutical Sciences, Vijayawada, Andhra Pradesh, India*

2.1 Introduction

2.1.1 Importance of medicinal plants

Medicinal plants hold a pivotal position in shaping and managing healthcare throughout human history, from the advent of civilization, and continue to play a major role in modern medicine. They constitute an inestimable treasure trove of therapeutic agents that harbor a bewildering array of potent, diverse bioactive compounds to guard against, manage, and cure a range of diseases. They are credited with developing complex bioactive compounds, including alkaloids, glycosides, terpenes, phenolics, and essential oils, some of which are extremely useful for treating diseases due to their potent biological activities and pharmacological properties (Fabricant and Farnsworth, 2001).

Nearly 80% of people worldwide use traditional medicine, primarily herbal medicine, as their main source of healthcare, according to the World Health Organization, especially in underdeveloped nations where facilities of modern medicine are scarce and beyond economic reach. Even in civilized nations, there is a growing use of medicinal plants as an adjunct or an alternative replacement along with allopathic medicines. The increasing demand for natural-source-derived medicines may be due to their safety, economic advantages, cultural acceptability, and lower frequency of adverse effects compared with synthetic medicines. Moreover, medicinal plants lay the base upon which modern drug development takes place or will be carried out in the near future because it has been clearly observed that many of the available medicinal or drug compounds available up to this point in time get their origins from medicinal plants or are at least based upon some compounds or elements present in plants (Ekor, 2014).

Healing Botanicals. DOI: https://doi.org/10.1016/B978-0-443-44295-7.00013-6

2.1.2 Traditional perspective of herbal medicine

The therapeutic use of plant products has an ancient history, steeped in the traditional medicine of the oldest civilizations of Egypt, China, India, Greece, and Rome. The medical literature of the earliest civilizations contains voluminous writings on plant products for the management of various diseases and ailments. In fact, the Ebers Papyrus (c.1550 BCE) predates the earliest known works on the subject from Egypt itself, while the works of Hippocrates and Dioscorides formed the basis for rational therapy by the scientific study of herbal drugs (Petrovska, 2012).

Over time, such knowledge was passed on by word of mouth or in written texts, which, with time, came to be recognized as traditional medicines such as Siddha, Ayurveda, Unani, traditional Chinese medicine, etc. These are holistic treatments diseases, emphasizing the concept of balance in the body concerning the mind and environment rather than disease symptoms. The use of a complex mix of plant extracts used in Ayurvedic medicine, etc., demonstrates their synergistic effect. As modern medicine evolved, the practice of traditional herbal medicine became the most popular and widely followed. Its overall use and recognition established it not only as crucial but also as potentially curative. Thus scientific evaluation became essential to establish the efficacy and safety of such drugs (Patwardhan and Mashelkar, 2009). Fig. 2.1 highlights that medicinal plants are at the core of many scientific fields, including the identification of therapeutic molecules, traditional medicine, phytochemistry, and pharmacology. The wide range of applications underscores the key contribution of medicinal plants to modern healthcare practices worldwide.

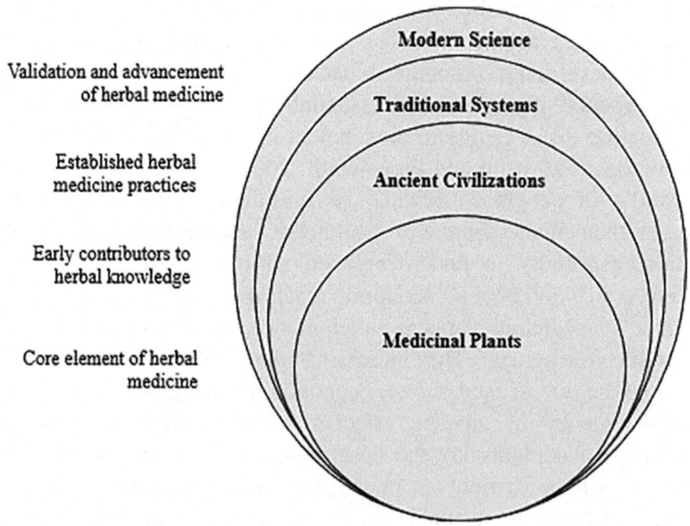

FIGURE 2.1 Traditional perspective and scientific evolution of herbal medicine.

Herbals scientific perspective.

2.1.3 Phytochemistry and pharmacology relevance today

Phytochemistry has gained recognition in recent decades as a significant field of study that helps understand the therapeutic effects of plants and their potential applications in medicine. This branch of science has been based on the chemical components of plants, which have been implicated in their defense mechanisms and play essential ecological roles without affecting their growth. Pharmacology, as a complementary field to phytochemistry, also seeks to determine how these compounds influence biological processes and to define their medicinal importance. Modern approaches, combined with traditional ethnobotany, have enabled the discovery of drug candidates through instrumental techniques, such as spectroscopy and chromatography, as well as biological methods, including the isolation of pure compounds from crude extracts. Owing to this interdisciplinary approach, not only has it helped prove several traditional medicinal claims, but it has also helped identify several pharmacologically significant molecules. Thus phytochemistry and pharmacology collectively provide the scientific basis for the development of herbal drugs and traditional medicine (Atanasov et al., 2015).

2.2 Introduction to phytochemistry

2.2.1 Scope and definition

Phytochemistry is a branch of chemistry devoted to the study of phytochemicals, naturally occurring bioactive compounds produced by plants. Phytochemicals are generally divided into two classes: primary metabolites, which include the nutrients plants require to survive, such as proteins, fats, carbohydrates, and nucleic acids, and secondary metabolites, which have been shown to play significant roles in plant biology. The scope of phytochemical studies includes identification, isolation, purification, and elucidation, as well as the investigation of biological activity and therapeutic use. Phytochemical studies integrate traditional knowledge with the pharmaceutical sciences, thereby significantly impacting drug, nutraceutical, and functional food research (Gurib-Fakim, 2006). Fig. 2.2 symbolically represents the various stages in the evolution of phytochemistry, from the early stages of plant development to the mature phase of bioactive compound development for therapeutic use.

2.2.2 Phytochemicals classification

Phytochemicals can be categorized into two types: primary metabolites and secondary metabolites. Primary metabolites are substances, such as carbohydrates, amino acids, proteins, lipids, and nucleic acids, that are directly involved in plant growth, energy, and reproduction. In the case of secondary metabolites, they do not represent the category of actual needs but are essential to the plant and its overall environment. These can be categorized into various groups or subgroups, including

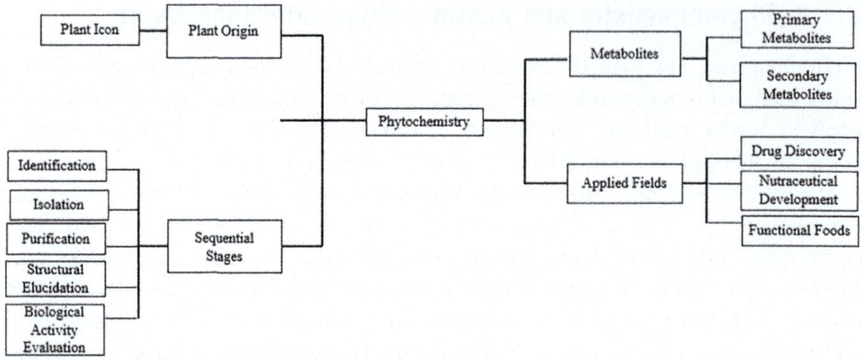

FIGURE 2.2 Evolution and scope of phytochemistry.

Phytochemistry evolution.

the pharmacological properties and medicinal uses of compounds such as terpenoids, alkaloids, and others. Depending on their chemical composition, they fall under one of the following categories, namely, flavonoids, phenolic, terpenoids, saponins, tannins, and glycosides. This will therefore allow the researcher to grasp the concept of chemical diversity in plant constituents and which compounds will be used for medicinal purposes (Koschier, 2008).

2.2.3 Plant secondary metabolite biosynthesis

The biosynthetic pathways of several secondary compounds in higher plants are complex, influenced by many variables, and controlled at multiple levels: genetic, developmental, and environmental. Various extrinsic stimuli, such as photosynthetically active radiation, temperature stress, nutrient limitation, infections, and herbivory, act as important drivers of higher plant biosynthetic processes. Flavonoid biosynthesis is stimulated under high-intensity UV radiation and herbivory-induced alkaloid biosynthesis. These biosynthetic pathways have been found to play an important role in optimizing the production of valuable phytochemicals. Advances in modern biotechnology, including genetic engineering, tissue culture, and related techniques, have helped optimize levels of secondary phytochemicals. These techniques are now being used to meet the ever-growing demand for phytochemicals as medicine (Verpoorte and Memelink, 2002).

2.2.4 Isolation and extraction techniques

Isolation and extraction are significant steps in phytochemical investigation and in the preparation of herbal drugs. Because of their simplicity, maceration, solvent extraction, percolation, and steam distillation are among the traditional methods of extraction commonly employed in phytochemical investigations. Some major drawbacks of these methods include their efficiency, selectivity, and thermal degradation during extraction. Some modern methods have been introduced to overcome these drawbacks, even as some extraction methods are still commonly practised, thereby increasing efficiency, yield, and purity. These modern methods include supercritical fluid extraction, microwave-assisted extraction, and ultrasound-assisted extraction. Various complex studies on isolated phytochemicals from plants, including their separation, identification, and structural elucidation, are carried out using techniques such as thin-layer chromatography (TLC), high-performance liquid chromatography (HPLC), gas chromatography (GC), UV, IR, NMR, and MS. The extraction method has played a significant role in phytochemical investigation. Various methods can be used to obtain the required information for phytochemical investigation (Azmir et al., 2013).

2.3 Main phytochemical classes of bioactivity

Medicinal plants produce various bioactive phytochemical compounds, which enhance their medicinal value. Therefore various medicinal plants contain various biologically active compounds. The major classes of chemicals with established biological activity are discussed in detail below.

2.3.1 Alkaloids

Alkaloids constitute a general group of chemical compounds consisting of a great number of different chemical compounds, most of which fall into a group of biogenic products of various amino acids, including tryptophan, tyrosine, phenylalanine, ornithine, and lysine. Concerning their general chemical structure, it should be noted that a key feature of many alkaloid compounds is that their general formula contains at least one nitrogen atom. Therefore the presence of a nitrogen atom is a main factor in their alkalinity and biological activity. Alkaloids can be generally subdivided into true alkaloids, protoalkaloids, and pseudoalkaloids, depending upon their biogenesis and chemical structures (Wink, 2003).

Alkaloids constitute one large and pharmacologically significant class of secondary plant metabolites that, together, account for particularly wide occurrence among higher plant species. Alkaloids tend to accumulate in particular plant tissues, including roots, bark, seeds, or leaves, where these substances may serve ecological functions, especially in protecting those same species from various herbivores, insects, or microbial pathogens. Their structure, especially, is sufficiently complex and diverse to allow interactions with various other molecules, including enzymes, ions, receptors, and nucleic acids. From a biological and therapeutic point of view, alkaloids have been found to display a wide spectrum of activities. Some of the alkaloids display analgesic, antimalarial, antiparasitic, antihypertensive, antiarrhythmic, antitumor, and central nervous system effects, including stimulating, sedative, and cytoprotective activities. Given their potent activities, alkaloids have been a valuable source for drug development, and several important drugs have resulted from them (Cordell et al., 2001). Besides their bioactivity, which underlies their potential therapeutic use, this largely accounts for the narrow "therapeutic index" most of these compounds are known to have. It is therefore necessary to determine the dose levels, the toxicological potential demonstrated, and the regulatory requirements. Ultimately, while they clearly have immense value in pharmacology and medicinal chemistry, there must be a balance between their therapeutic utility and potential risks (Asadi et al., 2010).

2.3.2 Flavonoids

These compounds include flavonoids, a wide range of phytochemicals. They occur widely in fruits, vegetables, and floral products of the plant kingdom, as well as in beverages such as tea, cocoa, and wine. These compounds are very important in plant physiology, protecting plants from ultraviolet light, pathogens, and other environmental factors, and, in addition, they contribute to pigmentation that attracts pollinating insects and seed-dispersing animals. These pigments include a wide range of colors, from yellow and red to blue and purple. Flavonoids are characterized by a common molecular core, a 15-carbon backbone (C6–C3–C6), formed by two phenolic structures attached to heterocyclic rings, denoted as ring C and ring A, respectively. The various modifications of flavonoids, in terms of their oxidation state, hydroxylation, methoxylation, and glycosylation, give rise to various flavonoid subclasses, such as isoflavones, flavones, flavonols, flavanones, flavonals or catechins, and anthocyanidins (Panche et al., 2016).

Flavonoids are especially known for their high antioxidative properties, which give them a defensive role: they can donate hydrogen atoms or electrons and chelate metal ions. Apart from their antioxidative properties, flavonoids are known to be involved in various biochemical processes, including antiinflammatory properties through modulation of biochemical pathways and cytokine cascades, antiviral properties, antiallergic properties, and cardioprotective properties through modulation of endothelial function and suppression of platelet aggregability. Epidemiological research has clearly shown that flavonoid intake can lead to greater reductions in the incidence of chronic diseases, especially cardiovascular disease and cancer. This can directly be attributed to the antioxidant, antiinflammatory, and cell-regulatory properties of flavonoids. Thus their importance is quite evident in human nutrition (Pietta, 2000).

2.3.3 Terpenoids and essential oils

Terpenoids are also called isoprenoids. This particular group not only contains the highest number of constituents but also has a significant degree of quality diversity because all the members come from pentagonal isoprene units (C_5H_8) bonded together in a great variety of forms from light molecules to very heavy ones.Depending upon the total number of isoprene units in their backbones, monoterpenes, sesquiterpenes, diterpenes, triterpenes, tetraterpenes, and polyterpenes are the names given to terpenoids (Gershenzon and Dudareva, 2007).

In plant physiology, terpenoids have important functional roles. Terpenoids account for the prominent fragrance, taste, or even aroma of aromatic or essential oil-containing herbs or plant species, fruits, or spices. Such qualities define the purpose of most aromatic or essential oil plants or spices. In plant ecology, terpenoids play important roles as plant protection chemicals against herbivores and pathogens, pollination attractants, and in plant interactions with other flowering plants and microorganisms. In summary, based on pharmacological evidence, terpenoids are generally reported to exhibit a wide range of biological activities. Numerous scientific studies have demonstrated antimicrobial, antiinflammatory, antimalarial, anticancer, and antioxidant activities, among others, for terpenoids. These biological activities are attributed to their ability to interact with various biological targets. For this reason, they have gained popularity as potential leads for pharmaceutical development. A particularly important characteristic of terpenoids is their role as important members of essential oils. Essential oils, which consist of volatile terpenoids, hold a critical place in traditional and modern pharmacology due to their bioactivity, lipophilicity, and membrane-permeability. The bioactivity of terpenoids in essential oils, however, has justified paying critical attention to dosage, safety, and toxicity, which reflect their importance in pharmacology and natural product research, given their potency (Bakkali et al., 2008).

2.3.4 Glycosides

Glycosides represent a large class of organic compounds in which a carbohydrate unit, usually called the glycone, is covalently bonded to some noncarbohydrate substance, the aglycone or genin. This glycosidic O-, N-, S-, or C-bond then becomes the decisive factor in the compound's stability, solubility, and biological activity. They are distributed among medicinal plants, and within their ecological niches, they may play key physiological or ecological roles, such as defending the plants against herbivores or pathogens (Koschier, 2008).

The bioactivity of glycosides, in most cases, stems pharmacologically from the aglycone part of the glycoside, as it directly interacts with specific targets in man. On the other hand, it should be recognized that, although pharmacoinactive, the glycone portion of glycosides plays a significant role in the absorption, biotransformation, and bioavailability of the active aglycones. In addition, glycosides are prodrugs, in that they are hydrolyzed in vivo to yield bioactive aglycones at their target sites. One well-known and perhaps the most important category of these glycosides to consider is the cardiac glycosides, particularly those in the *Digitalis* species, commonly known as foxgloves. These drugs are powerful in their interactions with cardiac muscle, particularly the Na+/K+-ATPase pump, which increases calcium levels to strengthen heart contraction and increase cardiac contractility. Outside heart-related glycosides, other important subclasses of glycosides with various pharmacological attainabilities have been found within nature. Laxative glycosides, such as anthraquinone glycosides, have been found to enhance intestinal activity; diuretic glycosides have been found to increase urine volume, while their antimicrobial potential contributes considerably to plant defense systems and to general therapeutics (Eagle et al., 1992).

2.3.5 Phenolics and polyphenols

The presence of one or more hydroxyl (-OH) groups as a directly attached moiety to an unsaturated six-membered carbon-heteroatom cycle characterizes phenolics, a diverse class of secondary metabolite compounds. This specific structure confers phenolics' unique reactivity and enables them to serve as the active moiety. Phenolics are abundant in the plant kingdom, and their metabolism is a natural characteristic of the plant life cycle. Moreover, phenolics are produced when plants experience both abiotic and biotic stress (Dai and Mumper, 2010).

Phenolic compounds are central in their functions. First, they have been involved in plant defense mechanisms against different plant-herbivorous animal pathogens. They have been involved in enhancing plant cell walls by producing lignin. Additionally, phenolic compounds have been shown to protect plants against various unfavorable environmental factors, such as ultraviolet radiation, drought, and oxidative stress. The roles of phytochemicals have enabled one to understand and gain insight into the use of phenolic compounds across different plant species. The most important phenolic compounds are polyphenols, including phenolic acids and tannins: the phenolic acids, hydroxybenzoic and hydroxycinnamic acids, control plant defense mechanisms and growth. Different types of hydroxybenzoic acid and a hydroxycinnamic acid compound are classified under the phenolic acid category.

Regarding the human health perspective, the antioxidant properties of phenolic compounds are widely accepted. Their significant role in regulating low levels of oxidative stress by scavenging free radicals, chelating metals, and modulating antioxidant enzymes has been recognized as one of the most important factors in the prevention of chronic diseases. Various epidemiological and experimental studies have already established an important association, whether direct or indirect, between foods containing phenolic compounds and the prevention of cardiovascular, neurodegenerative, and cancer disorders. Apart from their antioxidant properties, phenolic compounds have significant roles in antiinflammatory, antimicrobial, and cell-signaling modulation, which are of great concern (Manach et al., 2004).

2.3.6 Saponins

Saponins represent a special group of carbohydrate-containing secondary metabolites characterized by their soap-like properties, which reduce the surface tension of fluids and produce stable froths upon agitation. These activities depend on the unusual chemical structures of saponins. Compounds have a hydrophobic aglycone portion, called sapogenin, and, as with saponosides, one or more hydrophilic carbohydrate side chains. Conservative data mining and review revealed their wide occurrence among the plant kingdom, including legumes and several medicinal herb species. Regarding its structural properties, the aglycone portion of a saponin is usually a steroid/triterpenoid derivative. In contrast, the glycosidic chain can consist of various monosaccharides, such as glucose, galactose, rhamnose, and xylose. The variety of sapogenins, glycans, and attaching units yields a wide range of saponin compounds in biological systems (Francis et al., 2002).

In addition, the biological effects of saponins vary widely. Saponins are well known to possess immunomodulatory effects. Saponins can stimulate both innate immunity and adaptive immune response. The hypocholesterolemic activity of saponins can result from their capability to produce insoluble compounds with cholesterol during their stay in the gastrointestinal tract. Moreover, antiinflammatory, antimicrobial, and anticancer activities of saponins could result from their modulation of these biological activities. Besides their therapeutic properties, saponins have received much interest for pharmaceutical use. The ability of saponins to react with the cell membrane, which makes the cells permeable to the saponins. This makes saponins useful for the formulation of medicines, for example, by improving the activity of the drugs they are used with. This improving property makes saponins instrumental in contemporary medicines (Podolak et al., 2010).

2.3.7 Tannins

Tannins are defined as relatively large, high-molecular-weight polyphenolic substances that can precipitate or bind to proteins. Tannins have an important biochemical property: they precipitate or bind to proteins, which produces an unusual sensation known as astringency when consumed in certain amounts. Tannins have numerous phenolic hydroxyl groups that have a high affinity for, or bind well to, proteins, carbohydrates, and metals. Tannins have been divided into hydrolysable tannins, consisting of gallic or ellagic acid derivatives, and condensed or proanthocyanidins, depending on their structures (Scalbert, 1991).

Tannins occur widely in the plant kingdom, including the leaves of teas, fruits like grapes and berries, wines, unripe fruits, nuts, the bark, roots, and leaves of many valuable medicinal plants, among others. Tannins play many defensive roles in the plant kingdom, including protecting plant parts from herbivores, maintaining low microbial growth, protecting against stress, etc. Tannin buildup in plant bark provides much-needed protection to the plant. The compound has scored well in pharmacology. In this case, the compound has been shown to have beneficial biological activities important to human health. Therefore the compound expresses its antimicrobial activity as a stable microbial protein enzyme called the tannin compound. Consequently, the growth of microorganisms is inhibited.

Additionally, tannins have antiinflammatory properties by altering different proinflammatory mediators and thereby stabilizing injured tissues. On the same note, tannins exhibit astringent properties, which play a crucial role in wound healing by promoting contraction of injured tissues, reducing exudation, and forming a protective crust on wound surfaces. Similarly, tannins have antidiarrheal properties by decreasing intestinal motility and excretion, as a result of their precipitation with protein compounds. Tannins have also been shown to exhibit antioxidative activity, enabling them to scavenge free radicals and chelate prooxidant metal ions, thereby preventing oxidative stress-related disorders, such as cardiovascular disease, as well as various inflammatory and degenerative conditions (Chung et al., 1998).

2.3.8 Coumarins

The naturally aromatic family of compounds that are characterized by the chemical constituents of coumarins are defined as the derivatives of 1,2-benzopyrone. The most notable quality of coumarins is seen to be their characteristic stability, which is achieved through the fusion of a benzene ring with an α-pyrone ring. Perhaps one of the most characteristic qualities of coumarins is seen to be their sweet hay odor (Venugopala et al., 2013).

Naturally occurring coumarins occur naturally in the plant kingdom, as may be expected, and tend to occur in aromatic plants and plant products, such as cinnamon, sweet clover, and tonka beans; in plant perfumes, such as lavender; in citrus fruits; in grasses, as may also be expected; and in other places. Coumarins perform important ecological roles in the plant kingdom, protecting from herbivory, disease, and ultraviolet light. As might be expected, the production of naturally occurring coumarins in plants is linked to other plant products in the general category of phenolic compounds, which also provide protection. From the pharmacological point of view, coumarins show a wide range of biological activity. Anticoagulant, antiinflammatory, antimicrobial, antiviral, antioxidant, and anticancer are a few examples of activity. Divergence in the range of activity can be attributed to differences in hydroxylation, methoxylation, and substitution on the benzopyrone skeleton. In clinical usage, coumarins are more closely associated with their utilization as anticoagulant agents. Several synthetic coumarins that function as potent anticoagulants have been identified; however, their main mechanism of action involves inhibition of vitamin K-dependent clotting factors. They are generally used to prevent and treat thromboembolic syndromes. Nonetheless, owing to their strong biological activity, both coumarins warrant dosage control measures to ensure greater efficacy. The significance of coumarins can therefore not be overstated (Vieira Dos Santos and Rey, 2006).

2.3.9 Lignans

Lignans are classified under another class of phytochemicals known as phenolics. It is further classified under a group of phytochemicals known as phytoestrogens based on their structural characteristics that resemble or are similar to a class of hormones known as estrogens. Lignans are chemical compounds that are formed by two phenylpropanoids coupled together. This phenylpropanoid group comprises specific functional groups, known as phenols, which are characteristic of these compounds' ability to mimic or act like a class of hormones, the estrogens, when ingested by humans (Cloos et al., 1998).

One of the interesting features of lignans is their capacity to exhibit estrogenic and antiestrogenic activities under different physiological and hormonal conditions. In individuals with low levels of estrogen, as in postmenopausal women, lignans are known to display weak estrogenic activity, thereby providing relief from hormonal imbalances in the body. On the contrary, in individuals with high estrogen levels, these compounds compete with estrogen, thereby showing antiestrogenic activity. Using these compounds, hormonal imbalances can be efficiently contained in an individual. Once ingested as part of a diet, plant lignans are converted by intestinal microbes into enterolignans. The most representative ones are enterodiol and enterolactone. These are collectively known to perform most of the health-promoting functions associated with lignans. These include antioxidant, antiinflammatory, and anticancer activities. Epigenetic evidence suggests that regularly consuming lignans could play a key role in protecting against hormone-related cancers, heart-related problems, and metabolic issues. Sources of lignans are diverse; some include flax seeds, whole grains, legumes, beans, seeds, fruits, and a variety of vegetables. Among all food sources of lignans, one should pay particular attention to flax seeds because this source of lignans has been found to contain the greatest quantity of lignans of all food sources. Lignans are considered to be a very important class of bioactive phytochemicals with significant implications for nutrition and preventive medicine (De Clercq et al., 2007). Table 2.1 summarizes the major classes of bioactive phytochemicals present in medicinal plants, highlighting their chemical characteristics.

2.4 Pharmacological actions of phytoconstituents

Phytoconstituents exert diverse pharmacological effects by interacting with multiple molecular targets, thereby affecting various physiological and pathological processes. Phytoconstituents exhibit a multifaceted mode of action, unlike synthetic medicaments, which generally target only one molecule at a time. It is these multifaceted properties of phytochemicals that, according to some investigators, make these compounds particularly relevant for the protection and treatment of multifactorial illnesses. The pharmacological effects of phytoconstituents are believed to be mostly mediated through enzyme regulation, changes in gene expression, reduction of oxidative stress, and antiinflammatory effects. This highlights the essential role of phytochemicals in multifaceted therapy and preventive interventions (Wagner and Ulrich-Merzenich, 2009).

Table 2.1 Main phytochemical classes of bioactivity and their pharmacological significance.

Phytochemical class	Core chemical features	Plant distribution and ecological role	Major biological/ pharmacological activities	Therapeutic significance and remarks
Alkaloids	Nitrogen-containing compounds; mostly heterocyclic; derived from amino acids (tryptophan, tyrosine, phenylalanine, ornithine, lysine)	Accumulate in roots, bark, leaves, and seeds; protect plants from herbivores, insects, and microbes	Analgesic, antimalarial, antiparasitic, antihypertensive, antiarrhythmic, anticancer, CNS stimulant or depressant	Highly potent bioactivity with a narrow therapeutic index; important drug sources, but require strict dose and toxicity control
Flavonoids	Polyphenolic compounds with C6–C3–C6 skeleton; subclasses include flavones, flavonols, flavanones, anthocyanidins, isoflavones	Found in many fruits, vegetables, flowers, tea, cocoa, and wines; helps with UV protection and pigmentation	Strong antioxidant, antiinflammatory, antiviral, antiallergic, cardioprotective	Important dietary phytochemicals associated with a lower risk of cancer and cardiovascular disease
Terpenoids and essential oils	Classified as mono-, sesqui-, di-, tri-, tetra-, and polyterpenes, they are composed of isoprene (C–H–) units	Responsible for aroma, flavor, and fragrance; act as plant defense chemicals and pollinator attractants	Antimicrobial, antiinflammatory, antimalarial, anticancer, antioxidant	used extensively in both conventional and modern medicine; due to its volatile and lipophilic nature, safety and dosage considerations are necessary
Glycosides	Consists of glycone (sugar) linked to aglycone via O-, N-, S-, or C-bonds	Occur widely in medicinal plants; act as storage and defense compounds	Cardiotonic, laxative, diuretic, antimicrobial	Often act as prodrugs; aglycone provides activity while glycone improves solubility and bioavailability
Phenolics and polyphenols	Phenolic acids, flavonoids, and tannins are examples of compounds that contain one or more phenolic hydroxyl groups	Ubiquitous in plants; involved in defense, UV protection, oxidative stress response	Antioxidant, antiinflammatory, antimicrobial, anticancer	Strong correlation between dietary intake and reduced chronic diseases, including cardiovascular and neurodegenerative disorders

Saponins	Amphiphilic molecules with sapogenin (steroidal or triterpenoid) linked to sugar chains	Abundant in legumes and medicinal plants; protect plants from pathogens	Immunomodulatory, hypocholesterolemic, antiinflammatory, antimicrobial, anticancer	Useful as drug delivery enhancers due to membrane-permeabilizing properties
Tannins	High molecular weight polyphenols capable of precipitating proteins; hydrolysable and condensed types	Found in bark, leaves, fruits, seeds, tea, and wine; deters herbivory	Antimicrobial, antiinflammatory, antioxidant, antidiarrheal, and wound healing	Astringent properties beneficial in gastrointestinal disorders and tissue repair
Coumarins	Benzopyrone derivatives (1,2-benzopyrone nucleus)	Present in aromatic plants, grasses, and citrus fruits, they provide UV and herbivore protection	Anticoagulant, antiinflammatory, antimicrobial, antiviral, antioxidant	Synthetic analogs widely used as anticoagulants, require careful dose regulation
Lignans	Phenolic compounds formed by dimerization of phenylpropanoids; classified as phytoestrogens	Found in flaxseed, whole grains, legumes, fruits, and vegetables	Estrogenic/ antiestrogenic, antioxidant, antiinflammatory, anticancer	Converted to enterolignans in the gut; protective against hormone-related cancers and metabolic disorders

2.4.1 Antiinflammatory activity

Inflammation is a biological process, and in most cases, it is considered a protective mechanism for the host's health. However, if the inflammation process is not controlled, it has been discovered to be a major factor in the development of various diseases, including arthritis, heart disease, asthma, and inflammatory bowel diseases. Therefore the study of phytochemicals that modulate inflammation pathways is, in most cases, a major theme in the field, due to their ability to control inflammation (Medzhitov, 2010).

There are several phytochemicals, mainly plant-derived compounds, with major antiinflammatory properties. Curcumin is one such compound, mainly derived from plants. It has been shown to inhibit proinflammatory cytokines, including IL-1, IL-6, and TNF. These proinflammatory cytokines can have profound effects on the inflammatory process, as they are major components of the proinflammatory cascade. The inhibitory role of these proinflammatory cytokines is to reduce tissue damage. Resveratrol is a plant-derived compound mainly found in grapes. It has actually demonstrated the capacity to inhibit proinflammatory cytokines such as IL-1, IL-6, and TNF (Aggarwal and Harikumar, 2009).

2.4.2 Antioxidant effects

Oxidative stress is a condition arising from an imbalance between the amount of reactive oxygen species generated and the ability to defend against them, leading to cellular damage. It has been mentioned as a causative factor in aging and some diseases such as cancer, diabetes, and cardiovascular disease. Phytochemicals such as flavonoids, phenolics, and polyphenols have been reported to have antioxidant properties. This property is important in maintaining the integrity of the cell and preventing oxidative damage to macromolecules such as DNA, proteins, and lipids. The antioxidant properties of phytochemicals have been exerted through scavenging of free radicals and activation of antioxidant enzyme systems such as superoxide dismutases, catalases, and glutathione peroxidases (Lobo et al., 2010).

In addition to their capacity to scavenge free radicals, phytochemicals also exhibit redox properties, including the regulation of signaling pathways involved in inflammation, apoptosis, and cell survival. The majority of plant-based antioxidants have been shown to exert their beneficial effects by regulating transcription factors, such as nuclear factor erythroid 2-related factor 2 (Nrf2), thereby enhancing the expression of phase II detoxifying and antioxidant enzymes. By inhibiting the activation of proinflammatory mediators and lipid peroxidation induced by oxidative stress, phytochemicals have been demonstrated to play a pivotal role in regulating cellular homeostasis. In addition, their chelating capacity for metal ions and ability to stabilize reactive intermediates have been linked to the prevention of oxidative stress at the cellular and tissue levels. Owing to the diverse properties of phytochemicals, the significance of a phytochemical-based diet is now recognized as a preventive measure against chronic oxidative stress, diseases, and age-related disorders (Kensler et al., 2007).

2.4.3 Antimicrobial and antiviral properties

Antimicrobial resistance has led to increased efforts to develop potent alternatives to existing antimicrobials. Some plant-derived antimicrobials have been shown to exhibit broad-spectrum antimicrobial activity against various infectious organisms, including bacteria, viruses, and fungi. Bioactive constituents present in these oil compounds, derived from various plants including thyme and oregano, include bioactive terpenoids responsible for antitubercular activities through disruption of microbial cell membranes, inhibition of microbial enzymes, and disruption of microbial metabolism. Additionally, alkaloids such as berberine have been shown to have antiviral activity through replication inhibition and immunomodulation, including inhibition of HIV and influenza viruses (Burt, 2004).

There has been growing interest in plant-based antimicrobials due to their multiple modes of action and the lower risk of bacterial resistance compared to conventional synthetic antimicrobials. Phytochemicals like essential oils, phenolics, alkaloids, and flavonoids bring about antimicrobial effects by making the cell membrane more permeable, which ultimately causes the leakage of some important intracellular components and death of the microbial cell. One plant metabolite has been found to disrupt bacterial quorum-sensing, leading to the inability to form biofilms and lowering the microbes' virulence. In addition, the combination of phytochemicals and existing antimicrobial agents has yielded results that are even more therapeutically effective and have the potential to overcome the barrier of drug resistance. These characteristics make plant-based antimicrobials a good source for either alternative or supplementary medicines in the fight against drug-resistant infectious diseases (Hemaiswarya et al., 2008).

2.4.4 Anticancer potential

Cancer is a multifactorial disease encompassing genetic mutations, oxidative stress, inflammation, and dysregulated cellular signaling. Phytochemicals fight cancer in various ways; thus they may be the most promising candidates for cancer prevention and adjuvant therapy. These compounds, including flavonoids, alkaloids, and terpenoids, trigger apoptosis, arrest cell cycle progression, inhibit angiogenesis, and prevent tumor metastasis. The most common flavonoid, quercetin, has been widely studied for its anticancer properties, inhibiting cancer cell growth and modulating several signaling pathways involved in tumor growth and survival. These multitargeted actions reduce the possibility of drug resistance and generally give better therapeutic outcomes (Aggarwal and Shishodia, 2006).

In addition to their direct cytotoxic effects, phytochemicals also exhibit chemopreventive effects that help regulate oxidative stress and chronic inflammation, well-recognized major contributing factors in the development of tumors. Various plant compounds can regulate important intracellular signaling pathways, such as phosphatidylinositol 3-kinase/protein kinase B (PI3K/Akt), mitogen-activated protein kinase (MAPK), and nuclear factor κB (NF-κB), thereby inhibiting cell growth and sensitizing cells to apoptosis. Moreover, phytochemicals have been shown to distinguish between cancer and normal cells, thus having lower systemic toxicity than conventional chemotherapeutic agents (Surh, 2003).

2.4.5 Cardiovascular effects

Cardiovascular diseases are a major cause of morbidity and mortality in many countries; they are associated with oxidative stress, chronic inflammation, endothelial dysfunction, and disturbances in lipid metabolism. This has led to various pathophysiological complications such as hypertension, atherosclerosis, myocardial infarction, and coronary artery disease. In recent times, there has been a growing emphasis on exploring the cardioprotective actions of dietary and medicinal phytochemicals used in therapy, given their multifaceted activity and good safety profile (Upadhyay, 2015).

Various phytochemicals, including flavonoids, polyphenolic compounds, and omega-3 fatty acids, have been shown to have significant cardiovascular protective effects via multiple pathways. They improve endothelial function by increasing nitric oxide bioavailability, thereby promoting vasodilation and improved vascular tone. They have potent antioxidant activity, thereby reducing oxidative stress by scavenging free radicals, inhibiting lipid peroxidation, and protecting vascular tissues from oxidative damage and injury. Furthermore, phytochemicals have shown antiinflammatory activity by inhibiting proinflammatory cytokines and pathways involved in vascular tissue damage and injury. Furthermore, these bioactive agents regulate lipid metabolism by reducing LDL cholesterol, lowering triglycerides, and increasing HDL cholesterol. All of this helps curb the progression of atherosclerosis and the buildup of arterial plaques. Epidemiological studies have demonstrated the effectiveness of phytochemical intake in reducing hypertension and the decreased incidence of coronary heart disease. The regular consumption of phytochemical-rich foods such as fruits, vegetables, nuts, fatty fish, cereals, legumes, and grains helps maintain optimal cardiovascular health without increasing the risk of high blood pressure or hypertension (Noll et al., 2009). Table 2.2 shows the pharmacological actions of phytoconstituents, illustrating their molecular targets, mechanisms of action, and therapeutic relevance.

2.5 Mechanisms of action

2.5.1 Receptor binding and enzyme inhibition

The medicinal properties of phytochemicals are believed to result from their ability to bind specific biological targets, such as receptors and enzymes. Most bioactive

Table 2.2 Major pharmacological actions of phytoconstituents and their mechanisms.

Pharmacological action	Key pathophysiological targets	Major classes of phytoconstituents involved	Primary mechanisms of action	Therapeutic/preventive significance
Antiinflammatory activity	COX enzymes, NF-κB signaling, and proinflammatory cytokines (TNF-α, IL-1β, and IL-6).	Curcuminoids, flavonoids, polyphenols, stilbenes	Suppression of cytokine release, inhibition of inflammatory mediators, and modulation of transcription factors	Useful in chronic inflammatory disorders such as arthritis, asthma, cardiovascular diseases, and inflammatory bowel diseases
Antioxidant effects	Oxidative DNA damage, lipid peroxidation, and reactive oxygen species	Flavonoids, phenolics, polyphenols, tannins	Metal ion chelation, free-radical scavenging, antioxidant enzyme activation (SOD, catalase, GPx), and Nrf2 pathway activation	Protects against aging, cancer, diabetes, cardiovascular, and neurodegenerative diseases
Antimicrobial and antiviral properties	Microbial cell membranes, enzymes, biofilm formation, viral replication machinery	Terpenoids, essential oils, alkaloids, flavonoids, phenolics	Membrane disruption, enzyme inhibition, quorum sensing interference, and immunomodulation	Alternative or adjunct therapy against drug-resistant bacteria, fungi, and viruses
Anticancer potential	Cell cycle regulators, apoptotic pathways, angiogenesis, metastasis	Flavonoids, alkaloids, terpenoids, polyphenols	Apoptosis induction, cell-cycle arrest, angiogenesis and metastasis inhibition, and PI3K/Akt, MAPK, and NF-κB pathway modulation	Acts as chemopreventive and adjuvant agents with lower toxicity and reduced drug resistance
Cardiovascular effects	Endothelial dysfunction, oxidative stress, inflammation, dyslipidemia	Flavonoids, polyphenols, omega-3 fatty acids	Enhancement of nitric oxide bioavailability, antioxidant and antiinflammatory actions, lipid profile regulation	Reduces risk of hypertension, atherosclerosis, myocardial infarction, and coronary artery disease

phytochemicals have high binding affinity for various receptors, including G-protein-coupled receptors, nuclear hormone receptors, and ion channels. For example, flavonoids and alkaloids can modulate the function of these receptors, leading to physiological effects such as antiinflammatory responses, analgesia, and modulation of neurotransmission. Another mechanism of action by which phytochemicals can exert their effects is through enzyme inhibition. Curcumin, for example, is a polyphenolic compound that is present in turmeric. It inhibits the action of the COX enzymes, which are involved in the inflammatory process. Curcumin inhibits COX enzymes, thereby reducing levels of proinflammatory prostaglandins (Santa-María et al., 2023). Likewise, many phytochemicals are enzyme system inhibitors of carcinogenesis, for instance, histone deacetylases, which are targeted by caffeic acid phenethyl ester (Omene et al., 2016).

2.5.2 Modulation of cell signaling pathway

Phytochemicals can modulate multiple cell signaling pathways, thereby altering cellular responses. For example, the activation of the proteic pathway by NF-κB is a common target for most phytochemicals since NF-κB is a master regulator of the inflammatory and immune responses. Resveratrol and quercetin are among the compounds reported to inhibit NF-κB activation, thereby reducing inflammation and promoting cell survival (Nair et al., 2007). Additionally, phytochemicals have also been demonstrated to modulate the MAPK pathways, which play a very important role in cell proliferation, differentiation, and survival. For example, the terpenoid ginsenoside Rg3 from ginseng was shown to inhibit the MAPK pathway, thereby inhibiting cancer cell proliferation (Hassan et al., 2019). This modulation of the signal pathways is one area where the potential of phytochemicals as drugs under disease conditions such as cancer, neurodegenerative disease and cardiovascular disease can be realized.

2.5.3 Gene expression and epigenetic modifications

Phytochemicals can influence gene expression through epigenetic mechanisms, in which gene function is modified without altering the DNA sequence. Curcumin, for instance, blocks DNA methyltransferases (DNMTs) and thus induces demethylation of tumor suppressor genes and reactivation of their expression in cancer cells. This epigenetic control will reverse abnormal gene expression patterns in cancer and other diseases. Additionally, phytochemicals are known to modulate histone modifications, such as acetylation and methylation, which are significant in chromatin re-modeling and gene transcription. For example, resveratrol was found to increase histone acetylation, thereby increasing the expression of antioxidant defense and apoptosis genes. These epigenetic mechanisms provide phytochemicals with a potential role in cancer prevention and treatment, as they can normalize gene expression and halt cancer progression (Muñoz et al., 2016). Table 2.3 describes the

Table 2.3 Mechanisms of action of bioactive phytochemicals.

Mechanism	Key phytochemicals	Biological outcome/therapeutic relevance
Receptor binding and enzyme inhibition	Flavonoids, alkaloids, curcumin, CAPE	Modulate GPCRs, ion channels, nuclear receptors; inhibit COX and HDAC enzymes
Cell signaling modulation	Resveratrol, quercetin, ginsenoside Rg3	Suppress NF-κB and MAPK pathways; reduce inflammation, cell proliferation
Gene expression and epigenetics	Curcumin, resveratrol	Inhibit DNMTs, modulate histone acetylation/methylation; reactivate tumor suppressor genes

major mechanisms of action through which phytochemicals execute their therapeutic potential, including receptor binding, modulation of signaling, and gene regulation.

2.6 Pharmacological properties and bioavailability of pharmaceuticals

2.6.1 Absorption, distribution, metabolism, and excretion

The pharmacokinetics of phytochemicals include their absorption, distribution, metabolism, and excretion, which are virtually synonymous with determining their bioavailability and effectiveness. When given orally, phytochemicals reach the gastrointestinal tract, where they get absorbed. Generally, the solubility and molecular weight of phytochemicals can affect their bioavailability, along with the consumption of foodstuffs. After they are assimilated, phytoelements are transported from the site of absorption through the bloodstream to their target site, where they can execute their function. An example is the binding of several flavonoids to plasma proteins, which determines their distribution and bioavailability in the target site (Manach et al., 2005). Phytochemical metabolism involves the liver, where phytochemicals undergo phase I and phase II reactions, giving rise to metabolites of either increased or decreased potency compared to their natural forms. Finally, excretion is the final part of the pharmacokinetic process (Li et al., 2016).

2.6.2 Bioavailability determining factors

The bioavailability of phytochemicals is influenced by factors such as their chemical composition, formulation, and interactions with other nutrients. For example, certain lipophilic phytochemicals may show increased uptake, while hydrophilic phytochemicals may require different mechanisms for transport. The formulation method used to administer phytochemicals may significantly affect their solubility, thereby enhancing their bioavailability. Nutrient components may also affect phytochemical bioavailability (McClements et al., 2015). For example, the simultaneous consumption of fats could enhance the bioavailability of fat-soluble compounds, and conversely, some fiber could interact with phytochemicals, thereby decreasing their bioavailability. All these aspects are important for fully realizing the therapeutic potential of phytochemicals (Bohn, 2014).

2.6.3 Methods to improve bioavailability

To increase the bioavailability of phytochemicals, several strategies can be employed. These include the use of nanotechnology in drug delivery systems, which can increase the bioavailability and stability/solubility of phytochemicals. For example, nanoparticles can carry phytochemicals, which would not be broken down and would increase their absorption in the digestive system (Sanna and Sechi, 2012). The second strategy involves the use of bioenhancers, compounds that increase the bioavailability of phytochemicals. For instance, piperine, obtained from black pepper, can substantially increase curcumin bioavailability by inhibiting its metabolism.

Additionally, coadministration of phytochemicals with bioavailability enhancers can increase bioavailability. For instance, fats and surfactants may be used (Shoba et al., 1998). Fig. 2.3 schematically illustrates the pharmacokinetic pathway described by LADME, outlining the release, absorption, distribution, metabolism, and elimination of phytochemicals. Various parameters affecting bioavailability, such as solubility, permeability, interactions with gut microbes, and first-pass metabolism, are emphasized as important contributors to bioactivity.

2.7 Analytical phytochemistry methods

2.7.1 Chromatography methods (thin-layer chromatography, high-performance liquid chromatography, gas chromatography)

Analytical methods are essential tools for phytochemical investigations, as they help identify, quantify, and characterize the active agents. TLC is a cost-effective, less complicated approach used for the preliminary investigation of phytochemicals, separating compounds based on polarity. HPLC is a more sophisticated approach that includes high-resolution separation and quantitation of phytochemicals. In addition, HPLC is commonly used for analyzing phytochemicals due to its sensitivity and ability to separate complex mixtures. GC is another extremely

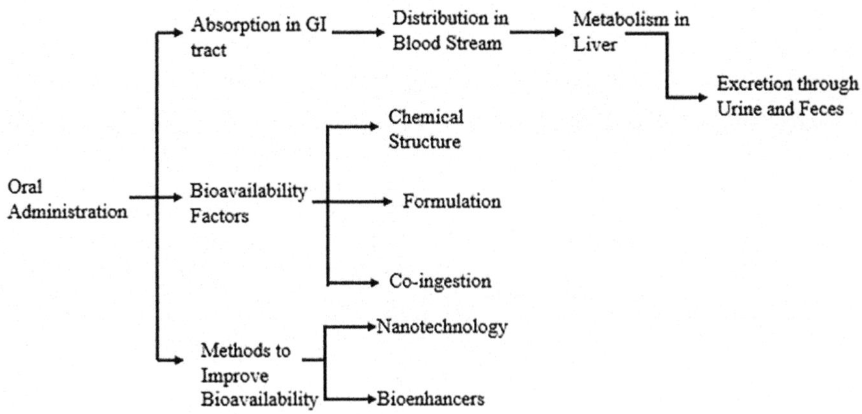

FIGURE 2.3

Pharmacokinetics and bioavailability of phytochemicals.

powerful approach, best used for analyzing volatile compounds and separating and analyzing essential oils and other volatile phytochemicals (Kuete et al., 2011).

2.7.2 Spectroscopic techniques (UV, IR, NMR, MS)

Spectroscopy is an important tool for identifying and determining the structures of phytochemicals. Moreover, ultraviolet-visible spectroscopy is used to determine the concentration of phytochemicals, which are identified by their absorbance. Infrared spectroscopy is employed to identify the functional groups of phytochemicals, thereby enabling identification. Moreover, nuclear magnetic resonance spectroscopy is an important technique for determining and identifying the structural chemistry of complex phytochemicals. It provides detailed information on the molecular conformation. Finally, mass spectrometry is commonly coupled with chromatography and has great capability for analyzing phytochemicals by identifying molecular weights (Khoddami et al., 2013).

2.7.3 Bioassay-informed fractionation

One method is bioassay-guided fractionation for isolating and characterizing plant bioactive constituents based on their bioactivity. The method consists of an initial step, in which plant extracts are screened for the target biological activity, followed by fractionation of active extracts using chromatographic methods (Álvarez-Maqueda et al., 2004). The fractions are tested for activity, allowing the constituents to be separated and identified as they contribute to the effect. The system of testing fractions for activity is most valuable when searching for new drugs from medicinal plants (Shuja, 2024). Table 2.4 summarizes the principal analytical techniques

Table 2.4 Key analytical methods used in phytochemistry.

Method category	Techniques	Purpose/application
Chromatography methods	TLC, HPLC, GC	Separation, identification, and quantification of phytochemicals; GC suited for volatiles
Spectroscopic techniques	UV-Vis, IR, NMR, MS	Structural elucidation, functional group analysis, and molecular weight determination
Bioassay-guided fractionation	Bioactivity-directed chromatography	Isolation and identification of bioactive compounds based on observed biological effects

employed in phytochemistry, including chromatography, spectroscopy, and bioassay-guided fractionation methods used for compound isolation and characterization.

2.8 Case studies of therapeutically relevant botanicals

2.8.1 *Curcuma longa* (turmeric)

C. longa, more generally speaking, also goes by the name of turmeric, and it is defined as a perennial, rhizomatous plant. The plant occurs in the tropical regions of Southern Asia and Eastern Asia. The dried root of the turmeric plant serves as a spice, coloring agent, and component of the Ayurvedic, Unani, and Chinese systems of pharmacotherapy. The antiquity of *C. longa's* ethnopharmacological action makes the plant highly interesting from a pharmacological perspective. While curcumin is recognized as the active compound responsible for the COSAS composition's biological activities, it's only the active offshoot related to turmeric's phenolic compounds. Indeed, curcumin's activities contribute to the beneficial outcomes seen in relation to the composition. The biological activities attributed to curcumin primarily involve antiinflammatory effects. These activities have been recognized in the majority of the related studies (Gupta et al., 2013).

It acts on a molecular level by modulating various pathways of inflammation and cellular interfaces. First, it suppresses NF-κB, one of the major targets in inflammation pathways. NF-κB regulates or controls the production of various inflammation-related cytokines, such as interleukins IL-1β and IL-6, COX-2, tumor necrosis factor-alpha (TNF-α), and inducible nitric oxide synthase. Curcumin does not target a single pathway but can act on multiple pathways by modulating NF-κB signaling in inflammatory pathways. Curcumin can target multiple pathways by acting on mitogenic pathways, such as MAP kinases; these pathways have distinct effects on cellular interfaces, including growth and immunity (Hewlings and Kalman, 2017).

Another important activity of curcumin is its potent antioxidant activity, which involves scavenging free radical species and enhancing the activity of endogenous antioxidants. These intrinsic antioxidants include SOD, catalase, and glutathione peroxidase. Then, the multifunctional antiinflammatory and antioxidant potential of curcumin contributes significantly to the prevention of cell damage caused by oxidative stress. Various clinical studies have documented the impact and potential of curcumin in alleviating or curing diseases such as rheumatoid arthritis, osteoarthritis, metabolic syndrome, cardiovascular diseases, and certain bowel disorders. Though curcumin's bioavailability is poor, and hence impairs its clinical activity in many cases, new approaches to improving bioavailability have shown improved clinical effectiveness. Overall, *C. longa* can be considered a scientifically established medicinal plant having great potential in modern clinical practice (Anand et al., 2008).

2.8.2 *Withania somnifera* (ashwagandha)

W. somnifera, also known as ashwagandha or Indian ginseng, is a popular medicinal plant in India and is valued as a key herb in adaptogenic medicine. It has traditionally been prescribed as a rasayana, an agent that promotes longevity, vitality, and resistance to stress. Ashwagandha is especially appreciated for its ability to enhance the body's capacity to counteract physical, chemical, and biological forms of stress (Singh et al., 2011).

The therapeutic activities of *W. somnifera* have been largely supported by the presence of certain active pharmacological agents, namely withanolides. Withanolides have been shown to bear antiinflammatory, antioxidant, immunomodulatory, and neuroprotective activities. Experimental studies have also shown that withanolides can modulate various inflammatory mediators, thereby inhibiting certain inflammatory cytokines and reducing oxidative stress by elevating endogenous antioxidants (Mishra et al., 2000).

The effects of ashwagandha supplementation were assessed in relation to the hypothalamic-pituitary-adrenal axis, which is considered the primary system regulating stress responses. Preclinical- and clinical studies showed that ashwagandha supplementation decreases serum cortisol levels, alleviating the effects of stress on the human body. Enhanced cognitive functions, including memory, attention, and executive functions, were observed in several studies, indicating the positive effects of ashwagandha on neurocognitive functions. Furthermore, it is also proven via clinical trials that the effectiveness of the herb is not only as a remedy, but it is also good for therapy, especially for anxiety disorders, depression, and even stress. It is also proven that the herb is highly effective and works as a tonic, especially to enhance physical performance and muscle strength. Overall, *W. somnifera* is an imperative therapeutic botanical agent with substantial scientific evidence supporting its role in the management of stress, mental wellness, and comprehensive wellness, making the herb priceless within the dynamics of integrative mental therapy (Chandrasekhar et al., 2012).

2.8.3 *Panax ginseng* (ginseng)

One of the most prescribed and studied plants among Chinese herbal practitioners, *P. ginseng*, has extraordinary potential for the practice of traditional Chinese medicine and alternative medicine. The plant, where the vital energy and spiritual essence of an individual could be found, has been used as a tonic to promote vitality and agelessness, and to augment an individual's physical and mental capabilities. Among its pharmacological activities, *P. ginseng* exhibits adaptogenic effects, enabling the body to manage and respond to stressors, both physical and chemical (Attele et al., 1999).

The pharmacological effects of the compounds of *P. ginseng* extract are primarily thought to be mediated by the active components of drugs containing these compounds, which are triterpenoids. These compounds in the ginseng extract have been shown to exhibit various biological effects. These compounds have the potential to modulate the immune system and help control diseases caused by oxidation, fatigue, etc. These experimental studies indicate that ginsenosides enhance the body's immune function, thereby protecting it against infectious diseases (Christensen, 2008).

Another instance is that ginseng is also highly instrumental in positively affecting mental processes, such as cognition. For instance, research has shown that ginsenosides can improve memory, concentration, and learning. Another advantage of ginseng is its ability to reduce fatigue. Therefore physical endurance and recovery from activities are highly enhanced. It is this advantage, linked to ginsenosides, that has made this plant so popular, especially among physically active individuals. Other studies on ginseng further established its ability to affect human metabolism linked to diabetes mellitus positively. For example, ginseng capsules have been shown to reduce blood glucose levels and improve insulin sensitivity in humans. Other studies focused on its use in the enhancement of sexual activities. For instance, this plant was shown to improve sexual libido in humans. Therefore it can be asserted that this plant, *P. ginseng*, is very highly valued as a medicinal herb, as reflected by its use in medicine (Vuksan et al., 2000).

2.8.4 *Ginkgo biloba*

A native species, the *G. biloba* plant is one of the most ancient living tree species, described as the "living fossil." Its place within the field of phytotherapy is special. The plant has found numerous uses in herbal medicine, specifically for treating cognitive decline, impairments, and neuropsychiatric diseases. Its leaf extract has developed into one of the more common herbal formulations currently prescribed. G. biloba extract possesses therapeutic value due to its high levels of bioactive compounds, including flavonoid glycosides and terpenoids such as ginkgolides and bilobalide. *G. biloba* flavonoids and terpenoids are potent antioxidants. They prevent brain cells from oxidative damage caused by free radical formation. *G.*

biloba's terpenoids are strong vasodilators, which increase oxygen delivery to brain tissues by reducing blood viscosity (Rantakokko et al., 2009).

The neuroprotective activity at the molecular level is of great significance for *G. biloba*. Its neuroprotective mechanism acts by regulating neurotransmitter function, platelet-activating factor, and mitochondrial function, thereby reducing apoptosis or cell death. All this is linked to neurodegenerative diseases involving the aging process. Research by specialists indicates that extracts derived from *G. biloba* have shown positive effects, particularly on cognitive function, attention, and memory, in Alzheimer's disease and mild-to-moderate dementia. It is noteworthy that relief concerning symptoms of depression, anxiety, and mental fatigue is achieved, implying its implications concerning enhanced mental wellness. Impacts concerning age-related cognitive impairment, as well as cerebrovascular insufficiencies, have equally positive implications, though variable. However, ongoing research is still aimed at determining the full potential of G. biloba in preventing or delaying brain deterioration associated with the aging process or other neurological disorders. Based on the analysis, it can be stated conclusively that *G. biloba* is a botanical agent of considerable clinical relevance (Ahlemeyer and Krieglstein, 2003).

2.8.5 *Cannabis sativa*

The plant *C. sativa* is also referred to as cannabis and marijuana and is accepted as one of the earliest and first recorded medicinal plants for humans across various continents of Asian, Middle Eastern, and African origin. Cannabis is accepted as being traditionally applied to cure pain relief, inflammation, gastrointestinal problems, and neurological problems; however, there is a revived scientific curiosity as a result of bioactive compounds being identified and as a result of its move towards legal acceptance, as is the case within its area of application. The plant contains more than 100 active biocompounds, including cannabinoids, terpenes, and flavonoids, which may affect pharmacological action. In all, cannabidiol (CBD) and delta-9-tetrahydrocannabinol (THC) are explored out of all cannabinoids. The principal psychoactive compound is usually derived from the action of delta-9-THC on cannabinoid receptors in the CNS. Nevertheless, CBD does not exhibit any psychological activity, but rather has garnered significant attention for possible therapeutic benefits in a nonintoxicating manner (Pertwee, 2006).

Cannabinoids exert their effects by activating the endocannabinoid system, which is crucial for modulating pain perception, inflammation, mood, appetite, and neuronal excitability. Indeed, both preclinical and clinical evidence indicate that cannabinoids exert pain-relieving and antiinflammatory effects and, hence, have proven efficacy in various chronic pain syndromes, for example, neuropathic pain, rheumatoid arthritis, and cancer pain. Recently, CBD has shown promise in exerting anxiolytic, anticonvulsant, and neuroprotective effects. Indeed, evidence of its efficacy in anxiety and resistant epilepsy in human subjects is available. The promising effects of the therapies concerning the symptomatic relief of multiple sclerosis, nausea related to chemotherapy treatments, and sleep disorders were shown to be present in the case of cannabis-based therapies. Factors such as safety, the need to standardize doses, the level of dependency, and the long-term effects of this medicine highlight the need to establish proper regulations governing its use. With a constantly altering legal position and a higher degree of societal acceptability concerning medical cannabis use, scientific investigations into cannabis pharmacology, therapeutic utility, and risk/benefit assessment have been bolstered (Whiting et al., 2015).

2.8.6 *Aloe vera*

It is the perennial form of the succulent class, highly popular for its immense medicinal and esthetic uses. The herb has been used over the last several centuries in systems of indigenous practices such as Ayurvedic and Chinese medicine, and in indigenous medicine, mostly to treat skin ailments and wounds. The need for the product stems from its soothing, healing, and moisturizing effects (Surjushe et al., 2008).

Many experimental studies have confirmed the utility of *A. vera* in accelerating wound and burn healing and relieving skin irritations such as psoriasis and dermatitis. *A. vera* is also effective in preventing further infections by its antimicrobial activity and in retaining moisture on the skin surface and increasing its suppleness. Apart from its dermatological uses, *A. Vera* is also considered useful for its beneficial effects on the gastrointestinal system. The latex component of the *A. Vera* plant contains anthraquinone glycosides, which act as laxatives, enhancing intestinal motility and increasing intestinal water secretion. In general, *A. vera* appears to be a multifaceted and practical herbal medicine that, without losing its uniqueness, has been scientifically confirmed for its therapeutic potential and continues to have numerous applications across different fields of healthcare (Hamman, 2008).

2.9 Limitations and challenges in herbal drug research

2.9.1 Standardization and quality assurance

One of the main hindrances to the scientific investigation of traditional herbal medicines is the lack of standardization and quality control. Variation in the content of active constituents in herbal medicines, driven by differences in plant source, production processes, cultivation, and processing, can result in variation in the intensity of therapeutic activity (Balekundri and Mannur, 2020). Standardized cultivation practices, harvesting, processing, and production of medicinal plants should be implemented to ensure the efficacy and safety of herbal drugs (Guo et al., 2017).

Table 2.5 Limitations and challenges in herbal drug research.

Challenge area	Key issues	Implications
Quality control and standardization	Variability in phytochemical content brought on by variations in harvesting, processing, and cultivation	Inconsistent efficacy and safety; requires standardized practices
Safety, toxicity, and herb-drug interactions	Risk of adverse effects when combined with conventional drugs; unknown toxicity of some formulations	Necessitates a thorough toxicological evaluation and interaction studies
Clinical trials and regulatory issues	Limited regulatory frameworks; most herbal products classified as dietary supplements	Hinders robust clinical validation; leads to weak evidence for therapeutic claims

2.9.2 Safety, toxicity, and herb-drug interactions

The safety and toxicity of herbal formulations are crucial when they are administered alongside conventional drug therapy. Herbal formulations can interact with drugs, and thus alteration of drug metabolism can lead to adverse effects (Wamil and Seckl, 2007). The safety profile of herbal formulations, including their potential toxicity and drug interactions, should be well demonstrated by studies to support clinical practice and ensure patient safety (Alghamdi et al., 2023).

2.9.3 Clinical trials and regulatory issues

Herbal medicine regulatory policies vary widely across countries, governing their approval and use in healthcare settings. Herbal extracts are considered dietary supplements rather than drugs in most countries, thereby limiting the clinical testing required for their marketing (Patwardhan and Mashelkar, 2009). This is an unfavorable circumstance for conducting extensive clinical studies, which are required to establish the safety profile and efficacy of herbal products (Segers et al., 2020). As highlighted in Table 2.5, major challenges in herbal drug research include the lack of standardization, safety concerns arising from herb-drug interactions, and regulatory challenges affecting clinical trial development.

2.10 Hopes in the future of herbal pharmacology

2.10.1 Ancient knowledge and contemporary science come together

The convergence of ancient wisdom and modern scientific methodology has vast potential for the growth of herbal pharmacology. Joint involvement of scientists and herbal healers shall pave the way for the discovery of novel drugs from nature and for elucidating the mechanisms of action of herbal drugs. Joint research has great potential for ushering in safe and effective herbal medicine (Campetelli et al., 2013).

2.10.2 Computation phytochemistry

Innovation in computational phytochemistry is revolutionizing the field of medicinal herbalism. Simulations using in silico techniques and bioinformatics software are employed to predict the interactions, properties, and pharmacology of phytochemicals. These tools can streamline drug discovery and efficiently automate herbal research (Gan et al., 2023).

2.10.3 Individualized herbal medicine and systems biology

The future of herbal pharmacology could perhaps lie in personalized medicine, with one that uses the patient's genetic constellation to determine the herbal remedy to be administered. Systems biology techniques elucidate the complex interactions between phytochemicals and biological systems, thereby enabling personalized herbal therapy with maximal therapeutic gain.

2.11 Conclusion

The examination of medicinal plants in light of their pharmacology and phytochemistry suggests that these plants have significant potential in contemporary medicine, as knowledge of their pharmacokinetics, mechanisms, and therapeutic applications enables their use in clinical settings. Phytochemistry is considered a very important division of drug discovery due to the availability of information on the biologically active molecules of medicinal plants. As herbal medicine practice becomes more advanced, an evidence-based system is always the priority—good scientific inquiry, coupled with centuries of experience, can serve as the impetus for creating useful and safe herbal therapies to meet the demands of modern-day patients amid the evolving healthcare setting.

Acknowledgments

The authors express their sincere gratitude to the principal and management of KVSR Siddhartha College of Pharmaceutical Sciences in Vijayawada for their unwavering encouragement and support.

References

Aggarwal BB, Shishodia S. Molecular targets of dietary agents for prevention and therapy of cancer. Biochemical Pharmacology 2006;71(10):1397–421. Available from https://doi.org/10.1016/j.bcp.2006.02.009, www.elsevier.com/locate/biochempharm.

Aggarwal BB, Harikumar KB. Potential therapeutic effects of curcumin, the anti-inflammatory agent, against neurodegenerative, cardiovascular, pulmonary, metabolic, autoimmune and neoplastic diseases. The International Journal of Biochemistry & Cell Biology 2009;41(1):40–59. Available from https://doi.org/10.1016/j.biocel.2008.06.010.

Ahlemeyer B, Krieglstein J. Neuroprotective effects of Ginkgo biloba extract. Cellular and Molecular Life Sciences (CMLS) 2003;60(9):1779–92. Available from https://doi.org/10.1007/s00018-003-3080-1.

Alghamdi W, Al-Fadel N, Alghamdi EA, Alghamdi M, Alharbi F. Signal detection and assessment of herb–drug interactions: Saudi food and drug authority experience. Drugs - Real World Outcomes 2023;10(4):577–85. Available from https://doi.org/10.1007/s40801-023-00388-w.

Álvarez-Maqueda M, Bekay RE, Alba G, Monteseirín J, Chacón P, Vega A, et al. 15-Deoxy-Δ12,14-prostaglandin J2 induces heme oxygenase-1 gene expression in a reactive oxygen species-dependent manner in human lymphocytes. Journal of Biological Chemistry 2004;279(21):21929–37. Available from https://doi.org/10.1074/jbc.m400492200.

Anand P, Sundaram C, Jhurani S, Kunnumakkara AB, Aggarwal BB. Curcumin and cancer: an "old-age" disease with an "age-old" solution. Cancer Letters 2008;267(1):133–64. Available from https://doi.org/10.1016/j.canlet.2008.03.025.

Asadi F, Shahriari A, Chahardah-Cheric M. Effect of long-term optional ingestion of canola oil, grape seed oil, corn oil and yogurt butter on serum, muscle and liver cholesterol status in rats. Food and Chemical Toxicology 2010;48(8-9):2454–7. Available from https://doi.org/10.1016/j.fct.2010.06.012.

Atanasov AG, Waltenberger B, Pferschy-Wenzig E-M, Linder T, Wawrosch C, Uhrin P, et al. Discovery and resupply of pharmacologically active plant-derived natural products: a review. Biotechnology Advances 2015;33(8):1582–614. Available from https://doi.org/10.1016/j.biotechadv.2015.08.001.

Attele AS, Wu JA, Yuan CS. Ginseng pharmacology. Biochemical Pharmacology [Internet] 1999;58(11):1685–93. Available from https://doi.org/10.1016/s0006-2952(99)00212-9.

Azmir J, Zaidul ISM, Rahman MM, Sharif KM, Mohamed A, Sahena F, et al. Techniques for extraction of bioactive compounds from plant materials: a review. Journal of Food Engineering 2013;117(4):426–36. Available from https://doi.org/10.1016/j.jfoodeng.2013.01.014, http://www.sciencedirect.com/science/journal/02608774.

Bakkali F, Averbeck S, Averbeck D, Idaomar M. Biological effects of essential oils – a review. Food and Chemical Toxicology 2008;46(2):446–75. Available from https://doi.org/10.1016/j.fct.2007.09.106.

Balekundri A, Mannur V. Quality control of the traditional herbs and herbal products: a review. Future Journal of Pharmaceutical Sciences 2020;6(1). Available from https://doi.org/10.1186/s43094-020-00091-5.

Bohn T. Dietary factors affecting polyphenol bioavailability. Nutrition Reviews 2014;72(7):429–52. Available from https://doi.org/10.1111/nure.12114.

Burt S. Essential oils: their antibacterial properties and potential applications in foods—a review. International Journal of Food Microbiology 2004;94(3):223–53. Available from https://doi.org/10.1016/j.ijfoodmicro.2004.03.022.

Campetelli AN, Monesterolo NE, Previtali G, Santander VS, Amaiden MR, Arce CA, et al. Activation of H+-ATPase by glucose in *Saccharomyces cerevisiae* involves a membrane serine protease. Biochimica et Biophysica Acta - General Subjects 2013;1830(6):3593–603. Available from https://doi.org/10.1016/j.bbagen.2013.03.012.

Chandrasekhar K, Kapoor J, Anishetty S. A prospective, randomized double-blind, placebo-controlled study of safety and efficacy of a high-concentration full-spectrum extract of Ashwagandha root in reducing stress and anxiety in adults. Indian Journal of Psychological Medicine 2012;34(3):255–62. Available from https://doi.org/10.4103/0253-7176.106022.

Christensen LP. Chapter 1: Ginsenosides. Advances in Food and Nutrition Research [Internet] 2008;55:1–99.. Available from https://doi.org/10.1016/s1043-4526(08)00401-4 Available from.

Chung KT, Wong TY, Wei CI, Huang YW, Lin Y. Tannins and human health: a review. Critical Reviews in Food Science and Nutrition 1998;38(6):421–64. Available from https://doi.org/10.1080/10408699891274273, www.tandf.co.uk/journals/titles/10408398.asp.

De Clercq PJ, De Buysser F, Minne G, Schepens W, Vrielynck F, Van Haver D, et al. The development of CD-ring modified analogs of 1α,25-dihydroxyvitamin D. The Journal of Steroid Biochemistry and Molecular Biology 2007;103(3-5):206–12. Available from https://doi.org/10.1016/j.jsbmb.2006.12.023.

Cloos, Wahl, Hasslacher, Traber, Kistner, Jurkuhn, et al. Urinary glycosylated, free and total pyridinoline and free and total deoxypyridinoline in diabetes mellitus. Clinical Endocrinology 1998;48(3):317–23. Available from https://doi.org/10.1046/j.1365-2265.1998.00374.x.

Cordell GA, Quinn-Beattie ML, Farnsworth NR. The potential of alkaloids in drug discovery. Phytotherapy Research 2001;15(3):183–205. Available from https://doi.org/10.1002/ptr.890.

Dai J, Mumper RJ. Plant phenolics: extraction, analysis and their antioxidant and anticancer properties. Molecules (Basel, Switzerland) 2010;15(10):7313–52. Available from https://doi.org/10.3390/molecules15107313, http://www.mdpi.com/1420-3049/15/10/7313/pdf.

Eagle K, Jay ME, Silver MT. Acute bacterial endocarditis. New England Journal of Medicine [Internet] 1992;327(24):1734. Available from https://doi.org/10.1056/nejm199212103272407.

Ekor M. The growing use of herbal medicines: issues relating to adverse reactions and challenges in monitoring safety. Frontiers in Pharmacology 2014;4. Available from https://doi.org/10.3389/fphar.2013.00177.

Fabricant DS, Farnsworth NR. The value of plants used in traditional medicine for drug discovery. Environmental Health Perspectives 2001;109(suppl 1):69–75. Available from https://doi.org/10.1289/ehp.01109s169.

Francis G, Kerem Z, Makkar HPS, Becker K. The biological action of saponins in animal systems: a review. British Journal of Nutrition 2002;88(6):587–605. Available from https://doi.org/10.1079/BJN2002725.

Gan X, Shu Z, Wang X, Yan D, Li J, Ofaim S, et al. Network medicine framework reveals generic herb-symptom effectiveness of traditional Chinese medicine. Science Advances 2023;9(43). Available from https://doi.org/10.1126/sciadv.adh0215, https://www.science.org/doi/full/10.1126/sciadv.adh0215.

Gershenzon J, Dudareva N. The function of terpene natural products in the natural world. Nature Chemical Biology 2007;3(7):408–14. Available from https://doi.org/10.1038/nchembio.2007.5.

Guo S, Guo T, Cheng N, Liu Q, Zhang Y, Bai L, et al. Hepatoprotective standardized EtOH–water extract from the seeds of *Fraxinus rhynchophylla* Hance. Journal of Traditional and Complementary Medicine 2017;7(2):158–64. Available from https://doi.org/10.1016/j.jtcme.2016.05.001.

Gupta SC, Patchva S, Aggarwal BB. Therapeutic roles of curcumin: lessons learned from clinical trials. AAPS Journal 2013;15(1):195–218. Available from https://doi.org/10.1208/s12248-012-9432-8.

Gurib-Fakim A. Medicinal plants: traditions of yesterday and drugs of tomorrow. Molecular Aspects of Medicine 2006;27(1):1–93. Available from https://doi.org/10.1016/j.mam.2005.07.008.

Hamman JH. Composition and applications of Aloe vera leaf gel. Molecules (Basel, Switzerland) 2008;13(8):1599–616. Available from https://doi.org/10.3390/molecules13081599, http://www.mdpi.org/molecules/papers/13081599.pdf.

Hassan FU, Rehman MSU, Khan MS, Ali MA, Javed A, Nawaz A, et al. Curcumin as an alternative epigenetic modulator: mechanism of action and potential effects. Frontiers in Genetics 2019;10. Available from https://doi.org/10.3389/fgene.2019.00514, https://www.frontiersin.org/journals/genetics#.

Hemaiswarya S, Kruthiventi AK, Doble M. Synergism between natural products and antibiotics against infectious diseases. Phytomedicine: International Journal of Phytotherapy and Phytopharmacology 2008;15(8):639–52. Available from https://doi.org/10.1016/j.phymed.2008.06.008.

Hewlings S, Kalman D. Curcumin: a review of its effects on human health. Foods [Internet] 2017;6(10):92. Available from https://doi.org/10.3390/foods6100092.

Kensler TW, Wakabayashi N, Biswal S. Cell survival responses to environmental stresses via the Keap1-Nrf2-ARE pathway. Annual Review of Pharmacology and Toxicology 2007;47(1):89–116. Available from https://doi.org/10.1146/annurev.pharmtox.46.120604.141046.

Khoddami A, Wilkes MA, Roberts TH. Techniques for analysis of plant phenolic compounds. Molecules (Basel, Switzerland) 2013;18(2):2328–75. Available from https://doi.org/10.3390/molecules18022328Australia, http://www.mdpi.com/1420-3049/18/2/2328/pdf.

Koschier EH. Essential oil compounds for thrips control - a review. Natural Product Communications 2008;3(7):1171–82. Available from https://doi.org/10.1177/1934578x0800300726, http://members.naturalproduct.us/Secure/ViewDoc.aspx?docId=1356.

Kuete V, Ngameni B, Wiench B, Krusche B, Horwedel C, Ngadjui B, et al. Cytotoxicity and mode of action of four naturally occuring flavonoids from the genus dorstenia: gancaonin Q, 4-hydroxylonchocarpin, 6-prenylapigenin, and 6,8-diprenyleriodictyol. Planta Medica 2011;77(18):1984–9. Available from https://doi.org/10.1055/s-0031-1280023.

Li Y, Zhang JJ, Xu DP, Zhou T, Zhou Y, Li S, et al. Bioactivities and health benefits of wild fruits. International Journal of Molecular Sciences 2016;17(8). Available from https://doi.org/10.3390/ijms17081258, http://www.mdpi.com/1422-0067/17/8/1258/pdf.

Lobo V, Patil A, Phatak A, Chandra N. Free radicals, antioxidants and functional foods: impact on human health. Pharmacognosy Reviews/Bioinformatics Trends/Pharmacognosy Review [Internet] 2010;4(8):118. Available from https://doi.org/10.4103/0973-7847.70902.

Manach C, Scalbert A, Morand C, Rémésy C, Jiménez L. Polyphenols: food sources and bioavailability. American Journal of Clinical Nutrition 2004;79(5):727–47. Available from https://doi.org/10.1093/ajcn/79.5.727, http://www.ajcn.org/contents-by-date.2005.shtml.

Manach C, Williamson G, Morand C, Scalbert A, Rémésy C. Bioavailability and bioefficacy of polyphenols in humans. I. Review of 97 bioavailability studies. The American Journal of Clinical Nutrition 2005;81(1):230S. Available from https://doi.org/10.1093/ajcn/81.1.230s.

McClements DJ, Li F, Xiao H. The nutraceutical bioavailability classification scheme: classifying nutraceuticals according to factors limiting their oral bioavailability. Annual Review of Food Science and Technology 2015;6(1):299–327. Available from https://doi.org/10.1146/annurev-food-032814-014043.

Medzhitov R. Inflammation 2010: new adventures of an old flame. Cell [Internet] 2010;140(6):771–6. Available from https://doi.org/10.1016/j.cell.2010.03.006.

Mishra LC, Singh BB, Dagenais S. Scientific basis for the therapeutic use of *Withania somnifera* (ashwagandha): a review. Alternative Medicine Review 2000;5(4):334–46.

Muñoz P, Aschrafi A, Moya PR. Connecting synaptic activity with plasticity-related gene expression: from molecular mechanisms to neurological disorders. Neural Plasticity 2016;2016. Available from https://doi.org/10.1155/2016/7149527, http://www.hindawi.com/journals/np/.

Nair S, Li W, Kong ANT. Natural dietary anti-cancer chemopreventive compounds: redox-mediated differential signaling mechanisms in cytoprotection of normal cells versus cytotoxicity in tumor cells. Acta Pharmacologica Sinica 2007;28(4):459–72. Available from https://doi.org/10.1111/j.1745-7254.2007.00549.x.

Noll C, Hamelet J, Matulewicz E, Paul JL, Delabar JM, Janel N. Effects of red wine polyphenolic compounds on paraoxonase-1 and lectin-like oxidized low-density lipoprotein receptor-1 in hyperhomocysteinemic mice. Journal of Nutritional Biochemistry 2009;20(8):586–96. Available from https://doi.org/10.1016/j.jnutbio.2008.06.002.

Omene C, Patel M, Kannan K, Heguy A, Barcellos-Hoff MH. Abstract P1-06-06: mammary stem cell modulation of wildtype and Trp53 null stem cells by CAPE (caffeic acid phenethyl ester), a potential therapeutic agent. Cancer Research 2016;76(4_Supplement). Available from https://doi.org/10.1158/1538-7445.sabcs15-p1-06-06.

Panche AN, Diwan AD, Chandra SR. Flavonoids: an overview. Journal of Nutritional Science 2016;5. Available from https://doi.org/10.1017/jns.2016.41, http://journals.cambridge.org/action/displayJournal?jid=JNS.

Patwardhan B, Mashelkar RA. Traditional medicine-inspired approaches to drug discovery: can Ayurveda show the way forward? Drug Discovery Today 2009;14(15-16):804–11. Available from https://doi.org/10.1016/j.drudis.2009.05.009.

Pertwee RG. Cannabinoid pharmacology: the first 66 years. British Journal of Pharmacology 2006;147(1):S163. Available from https://doi.org/10.1038/sj.bjp.0706406.

Petrovska B. Historical review of medicinal plants′ usage. Pharmacognosy Reviews/Bioinformatics Trends/Pharmacognosy Review [Internet] 2012;6(11):1. Available from https://doi.org/10.4103/0973-7847.95849.

Pietta PG. Flavonoids as antioxidants. Journal of Natural Products [Internet] 2000;63(7):1035–42. Available from https://doi.org/10.1021/np9904509.

Podolak I, Galanty A, Sobolewska D. Saponins as cytotoxic agents: a review. Phytochemistry Reviews 2010;9(3):425–74. Available from https://doi.org/10.1007/s11101-010-9183-z.

Rantakokko P, Kiviranta H, Rylander L, Rignell-Hydbom A, Vartiainen T. A simple and fast liquid–liquid extraction method for the determination of 2,2′,4,4′,5,5′-hexachlorobiphenyl (CB-153) and 1,1-dichloro-2,2-bis(p-chlorophenyl)-ethylene (p,p′-DDE) from human serum for epidemiological studies on type 2 diabetes. Journal of Chromatography. A 2009;1216(6):897–901. Available from https://doi.org/10.1016/j.chroma.2008.12.013.

Sanna V, Sechi M. Nanoparticle therapeutics for prostate cancer treatment. Maturitas 2012;73(1):27–32. Available from https://doi.org/10.1016/j.maturitas.2012.01.016.

Santa-María C, López-Enríquez S, Montserrat-de la Paz S, Geniz I, Reyes-Quiroz ME, Moreno M, et al. Update on anti-inflammatory molecular mechanisms induced by oleic acid. Nutrients 2023;15(1). Available from https://doi.org/10.3390/nu15010224, http://www.mdpi.com/journal/nutrients/.

Scalbert A. Antimicrobial properties of tannins. Phytochemistry 1991;30(12):3875–83. Available from https://doi.org/10.1016/0031-9422(91)83426-L.

Segers S, Pennings G, Mertes H. The ethics of ectogenesis-aided foetal treatment. Bioethics 2020;34(4):364–70. Available from https://doi.org/10.1111/bioe.12715.

Shoba G, Joy D, Joseph T, Majeed M, Rajendran R, Srinivas P. Influence of piperine on the pharmacokinetics of curcumin in animals and human volunteers. Planta Medica 1998;64(04):353–6. Available from https://doi.org/10.1055/s-2006-957450.

Shuja N. The emerging role of lncRNAs in cancer therapy. Developmental Medico-Life-Sciences 2024;1(9):1–3. Available from https://doi.org/10.69750/dmls.01.09.086.

Singh N, Bhalla M, de Jager P, Gilca M. An overview on Ashwagandha: a rasayana (rejuvenator) of Ayurveda. African Journal of Traditional, Complementary and Alternative Medicines 2011;8(5):208–13. Available from https://doi.org/10.4314/ajtcam.v8i5S.9, http://www.bioline.org.br/pdf?tc11087.

Surh YJ. Cancer chemoprevention with dietary phytochemicals. Nature Reviews. Cancer 2003;3(10):768–80. Available from https://doi.org/10.1038/nrc1189, http://www.nature.com/cancer/.

Surjushe A, Vasani R, Saple DG. Aloe vera: a short review. Indian Journal of Dermatology 2008;53(4):163. Available from https://doi.org/10.4103/0019-5154.44785.

Upadhyay RK. Emerging risk biomarkers in cardiovascular diseases and disorders. Journal of Lipids 2015;2015:1–50. Available from https://doi.org/10.1155/2015/971453.

Venugopala KN, Rashmi V, Odhav B. Review on natural coumarin lead compounds for their pharmacological activity. BioMed Research International 2013;2013. Available from https://doi.org/10.1155/2013/963248.

Verpoorte R, Memelink J. Engineering secondary metabolite production in plants. Current Opinion in Biotechnology 2002;13(2):181–7. Available from https://doi.org/10.1016/S0958-1669(02)00308-7, http://www.elsevier.com/locate/copbio.

Vieira Dos Santos C, Rey P. Plant thioredoxins are key actors in the oxidative stress response. Trends in Plant Science 2006;11(7):329–34. Available from https://doi.org/10.1016/j.tplants.2006.05.005.

Vuksan V, Stavro MP, Sievenpiper JL, Beljan-Zdravkovic U, Leiter LA, Josse RG, et al. Similar postprandial glycemic reductions with escalation of dose and administration time of American ginseng in type 2 diabetes. Diabetes Care 2000;23(9):1221–6. Available from https://doi.org/10.2337/diacare.23.9.1221.

Wagner H, Ulrich-Merzenich G. Synergy research: approaching a new generation of phytopharmaceuticals. Phytomedicine: International Journal of Phytotherapy and Phytopharmacology 2009;16(2-3):97–110. Available from https://doi.org/10.1016/j.phymed.2008.12.018.

Wamil M, Seckl JR. Inhibition of 11ß-hydroxysteroid dehydrogenase type 1 as a promising therapeutic target. Drug Discovery Today 2007;12(13-14):504–20. Available from https://doi.org/10.1016/j.drudis.2007.06.001.

Whiting PF, Wolff RF, Deshpande S, Di Nisio M, Duffy S, Hernandez AV, et al. Cannabinoids for medical use. JAMA: The Journal of the American Medical Association 2015;313(24):2456. Available from https://doi.org/10.1001/jama.2015.6358.

Wink M. Evolution of secondary metabolites from an ecological and molecular phylogenetic perspective. Phytochemistry 2003;64(1):3–19. Available from https://doi.org/10.1016/s0031-9422(03)00300-5.

PART

2

Botanicals for health and wellness

CHAPTER

3

Adaptogenic herbs: stress relief and resilience

Pallavi Vadlamudi[1], Bhavana Alaparthi[2], Anil Kumar Kakani[2], Kanaka Durga Devi Nelluri[2] and Bhagavan Jaladi[3]

[1]*Department of pharmaceutics, Chalapathi Institute of Pharmaceutical Sciences, Guntur, Andhra Pradesh, India,*

[2]*Department of Pharmaceutics and Biotechnology, KVSR Siddhartha College of Pharmaceutical Sciences, Vijayawada, Andhra Pradesh, India,*

[3]*Johns Hopkins University, Department of Biotechnology, Baltimore, MD, United States*

3.1 Introduction

3.1.1 Concept of adaptogens

Adaptogens are a distinctive group of natural compounds, mostly of plant origin, thought to enhance the body's capacity to withstand stress and support overall wellness. The concept of an "adaptogen" was coined by Russian scientist N.V. Lazarev in the 1940s when he described these compounds as agents that enable the body to withstand physical, emotional, or environmental stressors.

3.1.1.1 Characteristics of adaptogens

Adaptogens are defined by their capacity to:

Promote homeostasis: They help maintain the body's balance, especially in stress response.

Enhance resilience: Adaptogens are believed to support the body's resistance to stress, fatigue, and anxiety.

Nonspecific effects: They have a wide range of effects on the body rather than targeting a single condition or symptom.

3.1.1.2 Mechanisms of action

Research shows that adaptogens may work through a variety of physiological mechanisms, including:

Regulation of hormones: They can affect the hypothalamic-pituitary-adrenal (HPA) axis, a critical system in the body's response to stress.
Modulation of neurotransmitters: Adaptogens also affect levels of neurotransmitters such as serotonin and dopamine, which are crucial to mood regulation.

Healing Botanicals. DOI: https://doi.org/10.1016/B978-0-443-44295-7.00007-0

Antiinflammatory effects: Some adaptogens have been found to reduce inflammation, a condition typically induced by chronic stress (Panossian and Efferth, 2022; Wróbel-Biedrawa and Podolak, 2024).

3.1.2 Historical perspectives—Ayurveda, traditional Chinese medicine, Russian studies

The use of adaptogens dates back to some of the oldest health belief systems, such as Ayurveda and traditional Chinese medicine (TCM). Adaptogenic plants such as ashwagandha (*Withania somnifera*) and brahmi (*Bacopa monnieri*) have been used for millennia in Ayurveda to optimize energy levels, enhance mental acuity, and reduce stress (Santhanu and Senthil, 2021). TCM also utilizes herbal medicines such as ginseng (*Panax ginseng*) and *Schisandra* (*S. chinensis*), believed to enhance the body's resilience to stress and overall well-being (Bhardwaj and Kumar, 2024).

Later, in the 20th century, Russian scientists conducted further investigations into the adaptogenic principle, especially during the Cold War, when the Soviet Union sought to improve the performance and resistance of its soldiers. Investigations by researchers like Lazarev and, subsequently, by Panossian and Efferth pointed to the pharmacological effects of many adaptogenic herbs, and it became scientifically established that they should be used in contemporary health management (Panossian et al., 2021).

3.1.3 Modern relevance of adaptogens in health and wellness

In the early modern era, the role of adaptogens has become even more important as stress-related diseases are on the rise. With a hectic life, occupational stress, and environmental stress, the world has been awaiting natural products to support mental and physical health (Machín et al., 2023). Adaptogens have been found to counteract chronic stress, improve cognitive function, and establish resilience (Panossian, 2017).

Contemporary research has also established that adaptogens have promise for the treatment of disorders such as anxiety, depression, and fatigue and are sought-after elements in integral health and wellness programs (Panossian et al., 2003). The outreach to food supplements and healthcare products is part of a broader trend toward preventive and natural care, as consumers increasingly focus on safe, effective nondrug therapies (Agarwal, 2023).

3.1.4 Definition criteria of adaptogens

Brekhman and Dardymov initially established three major criteria for a compound to be considered an adaptogen: (1) it must increase nonspecific resistance to a variety of physical, chemical, or biological stressors rather than simply focusing on one mechanism; (2) it should have a stabilizing or homeostatic effect on physiological functions, thus it will help to restore changed parameters to their normal level, regardless of the direction of the change; and (3) it should be safe, and at therapeutic doses, it should cause very little physiological disturbance or toxic effects. These ideas highlight broad stress-modulating and regulatory activities rather than narrow pharmacological antagonism. Among a handful of traditionally used plants that can be considered as a good fit for this model are *P. ginseng*, *Rhodiola rosea*, and Eleutherococcus (Siberian ginseng), all of which have been reported to influence the stress response mechanisms such as the hypothalamic-pituitary-adrenal axis and cellular stress pathways, while exhibiting relatively low acute toxicity in their traditional use and several modern studies. Even though the terms adaptogens, tonics, and stimulants are frequently mixed up in everyday language, they actually relate to different therapeutic concepts. Adaptogens refer to herbs that have been demonstrated to increase nonspecific resistance to stress and to help regulate physiological functions under different stress conditions; the effects of adaptogens are generally regulatory, pleiotropic, and become more perceptible with continuous usage, influencing neuroendocrine immune networks and cellular stress responses to raise resilience and lower the occurrence of harmful responses to stressors.

On the other hand, tonics are a more general category in traditional medicine: a tonic is any substance that can strengthen, restore, or rejuvenate the overall vital energy or a specific organ system. Tonics might be nutritive (providing missing nutrients), restorative (supporting the healing process), or organ-specific, and thus do not necessarily have to exhibit the properties of an adaptogen. An iron supplement can be a tonic for anemia, but it is not an adaptogen.

Stimulants are pharmacologically distinct: they lead to almost immediate increases in physiological or mental activity (e.g., heightened alertness, heart rate, or metabolic rate) by activating specific pathways, such as the adrenergic or dopaminergic systems. Stimulants normally produce short-term performance enhancements after a single dose, have the potential to cause tolerance or rebound effects when used chronically, and do not inherently function as normalizers (Panossian and Wikman, 2009; Todorova et al., 2021).

3.1.5 Global rise of stress-related disorders

Stress, related conditions such as anxiety, depression, and occupational burnout have dramatically increased over the last few decades, and this rise has been repeatedly pointed out by public and health agencies that continue to push for better prevention and treatment strategies; conventional therapies are still not perfect for many patients, and thus there is an increasing interest in more comprehensive approaches to stress management (Wheaton et al., 2021). Rapid urbanization and change of living conditions, such as overcrowded housing, commuting stress, and social isolation, when combined with constant digital connection, give rise to new, chronic psychosocial stressors. For example, social media and the constant flow of information increase the contagion of emotions and the amount of anxiety, provoking content, which in turn may result in higher levels of anxiety and stress in the general population during epidemics. Environmental and lifestyle factors, including excessive sitting, unhealthy eating, and rising metabolic disorders, not only influence stress regulation but are also influenced by it. Moreover, epigenetic and molecular investigations have demonstrated a connection between chronic environmental stress and changes in cellular and neuroendocrine stress responses at the population level. These converging trends explain the growing popularity of adaptogens in integrative medicine: their combination of features makes them highly attractive agents to complement lifestyle and psychological interventions. Firstly, they are marketed as broad-spectrum, low-toxicity substances that regulate neuroendocrine-immune stress pathways and facilitate homeostasis instead of causing sharp stimulation or single-target pharmacology (Fan et al., 2013; Inaba et al., 2022; Knezevic et al., 2023).

3.2 Understanding stress and adaptation

3.2.1 Physiological and psychological stress: acute vs chronic

Stress can be classified into two major types: acute and chronic.

Acute stress: This occurs in response to an immediate perceived threat or danger. Acute stress triggers a quick response in the body, often called the "fight or flight" response, which releases stress hormones such as adrenaline and cortisol. Acute stress is desirable because it prepares the body to respond to imminent threats, sharpening concentration and energy.

Chronic stress: Chronic stress, on the contrary, is ongoing and caused by long-term pressures such as job stress, financial pressures, or relationship pressures. Chronic stress may result in serious health problems, including anxiety, depression, cardiovascular disease, and weakened immunity. Chronic stress is characterized by persistently elevated cortisol levels that may disrupt many of the body's systems and lead to long-term health consequences.

3.2.2 Hypothalamic-pituitary-adrenal axis and stress response

The HPA axis is part of the body's stress response system. It is a complex interaction between the hypothalamus, pituitary gland, and adrenal glands:

Hypothalamus: Upon stress, the hypothalamus releases corticotropin-releasing hormone (CRH).
Pituitary Gland: CRH triggers the pituitary gland to secrete adrenocorticotropic hormone (ACTH).
Adrenal Glands: ACTH stimulates the adrenal glands to secrete cortisol, the major stress hormone.

Cortisol is also important for maintaining metabolism, the immune response, and blood pressure. While acute cortisol increases are needed for stress response, chronic elevation can cause several health disorders such as metabolic syndrome, anxiety disorders, and neurodegenerative diseases.

3.2.3 Concept of "allostasis" and resilience

Allostasis is the mechanism by which the body maintains stability through change, coming to terms with stressors by altering physiological parameters. It highlights the dynamic nature of the body's response to stress, in contrast to the traditional view of homeostasis as stabilizing invariant internal states.

Resilience is the ability to bounce back from adversity and to adjust to stress. Psychological and physiological components contribute to resilience, enabling an individual to withstand and overcome stressors. Resilience can be developed through techniques such as social support, coping skills, and lifestyle factors (Misiak, 2020; Polick et al., 2024).

Allostasis refers to the active mechanisms that maintain bodily stability by adjusting cardiovascular, metabolic, neuroendocrine, immune, and behavioral parameters in response to various challenges. If these adaptive responses are constantly used or are poorly regulated, the cumulative physiological burden becomes known as allostatic load: a general wear and tear that spans systems and predicts disease and death. Persistent activation of key agents, such as cortisol, catecholamines, and inflammatory cytokines, and metabolic risk factors, contributes to the development of allostatic load. Multisystem dysfunction over time, resulting in structural and functional damage (commonly vascular remodeling, insulin resistance, and immune imbalance), is the basis of chronic diseases. People respond differently to stress: resilient phenotypes generate flexible, coordinated responses and return to the normal state rapidly, while vulnerable phenotypes exhibit prolonged or blunted responses, slow recovery, and greater biomarker drift towards dysregulation. Psychosocial resources (social support, effective coping), healthy behaviors (sleep, exercise, diet), and the early life environment set the course of these patterns, indicating that resilience is to some extent a characteristic that can be changed, and interventions-behavioral, social, or pharmacologic can lessen allostatic load and the health effects resulting from it (Lopresti et al., 2022; Song et al., 2025).

3.2.4 How adaptogens modulate stress physiology

Adaptogens are phytochemicals that help the body adapt to stress and restore homeostasis. They impact multiple mechanisms:

Regulation of the HPA axis: Adaptogens may help normalize the HPA axis response, preventing cortisol overproduction during chronic stress. Ashwagandha, for instance, has been shown to reduce cortisol levels and increase stress resilience.

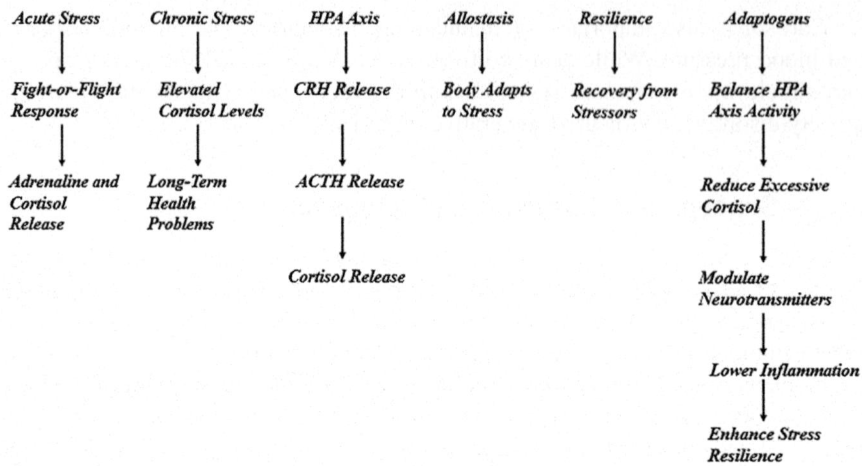

FIGURE 3.1 Overview of stress, hypothalamic-pituitary-adrenal (*HPA*) axis, allostasis, and adaptogen action.

Classification of stress.

Modulation of neurotransmitters: Adaptogens affect neurotransmitter levels, including serotonin and dopamine, which are key players in mood regulation. Adaptogenic modulation of these chemicals can reverse depressive and anxiety symptoms.

Antiinflammatory effects: Most adaptogens have antiinflammatory effects, helping to mitigate the harmful effects of chronic stress-induced inflammation. *R. rosea* and *S. chinensis*, for example, are reported to decrease neuroinflammation and thus potentially enhance cognitive function and emotional well-being.

Enhancing resilience: By supporting physiological balance and minimizing the effects of stressors, adaptogens can increase overall resilience, equipping people to handle better what life throws at them (see Fig.3.1) (Scudamore and Ciossek, 2018; Wróbel-Biedrawa and Podolak, 2024).

3.2.5 Stress, inflammation, and oxidative load

Chronic psychosocial and occupational stress continuously stimulates the HPA axis, increasing cortisol exposure over time; long-term cortisol dysregulation is linked to less physiological resilience and negative consequences in ageing and mental disorders, thus emphasizing the role of glucocorticoid signaling in stress-related pathologies. Continued high levels of cortisol facilitate oxidative stress through increased production of mitochondrial reactive oxygen species, reduced antioxidant defense, and changes in redox-sensitive signaling pathways; the resulting oxidative stress leads to the damage of lipids, proteins, and DNA, and further enhances inflammatory signaling cascades, thus establishing a vicious cycle between stress, inflammation, and cellular injury. The above mechanisms drive the development of metabolic dysregulation via stress-induced hyperglycemia, insulin resistance, and visceral fat accumulation, which are the pathological conditions of metabolic syndrome and cardiometabolic disease. Accordingly, occupational stress and cortisol elevation have been demonstrated to raise hypertension and related risks in the cohort studies mentioned. Neuronal dysfunction, DNA damage, and protein misfolding, which are pivotal to neurodegeneration, are exacerbated by chronic cortisol and oxidative/inflammatory stress in the brain, thereby leading to accelerated cognitive decline and increased susceptibility to neurotoxins. On the other hand, the immune response is weakened or disordered during prolonged stress: glucocorticoid exposure alters the immune system, making an individual more prone to infections; at the same time, it causes low-grade systemic inflammation that maintains the cycle of oxidative tissue damage (Ashraf et al., 2024; Er et al., 2025; George et al., 2025). Since adaptogens work on the main stress response systems (neuroendocrine, immune and cellular stress pathways) and regulate redox and inflammatory signaling instead of directly affecting a single symptom, they are capable of restoring the balance of various systems and breaking the stress → cortisol → oxidative/inflammatory chain; such a holistic action is the reason why adaptogenic herbs are often researched as general health enhancers for resilience, metabolic health and neuroprotection in integrative approaches (Kunicki et al., 2018; Malec, 2024).

3.3 Mechanisms of action of adaptogenic herbs

One of the most well-known qualities of adaptogenic herbs is their potential to support the organism in dealing with stressful situations and balancing itself. Their working principles are diverse as they influence different bodily systems. Next, we will discuss each of these systems in depth (see Fig. 3.2).

3.3.1 Neuroendocrine regulation (hypothalamic-pituitary-adrenal axis balance)

The HPA axis is basically the body mechanism that turns on when it senses stress. The situation is quite complicated, as it involves the interaction between the hypothalamus, pituitary gland, and adrenal glands that ultimately regulates cortisol release, the main stress hormone. The HPA axis is vital for the body to maintain balance during stressful situations, and it also uses different organs, such as the metabolism, immune system, and nervous system, to regulate mood. The HPA axis is a stress response system functioning through a CRH-ACTH-cortisol cascade: stress induces CRH release from the hypothalamus, which triggers pituitary ACTH secretion and adrenal cortisol production, and cortisol inhibits CRH/ACTH

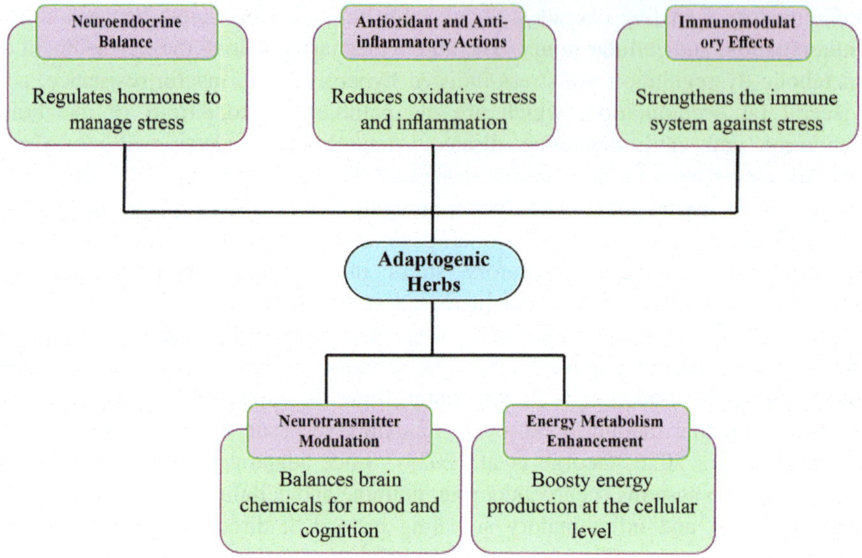

FIGURE 3.2 Key mechanisms of adaptogenic herbs.

Adaptogenic herbs mechanism.

secretion via negative feedback. Adaptogens (like ashwagandha and *Rhodiola*) are thought to reduce the overproduction of CRH and ACTH, thereby moderating cortisol peaks and restoring feedback control. By regulating glucocorticoid levels, adaptogens can lower cortisol levels, reduce associated neuronal damage, and facilitate neuroplasticity by increasing brain-derived neurotrophic factor (BDNF) expression, which, in turn, supports synaptic resilience and cognitive function. Adaptogens such as *Withania somnifera* (ashwagandha) and *R. rosea* have been known to change the conditions of the HPA axis by several different paths:

Cortisol modulation: Adaptogens can modulate cortisol levels, helping prevent excessive cortisol production even during prolonged periods of stress. For instance, the study on ashwagandha reveals a significant reduction in serum cortisol levels, accompanied by relief of stress-related symptoms.

CRH and ACTH regulation: Several studies support the idea that adaptogens may influence the release of CRH in the hypothalamus and ACTH in the pituitary gland. The effect of this modulation is a balanced physiological response to stress.

Neuroprotective effects: By regulating the HPA axis, adaptogens can act as neuroprotective agents, thereby enhancing cognitive function and psychological equilibrium. An example of this is research on ashwagandha, which not only promotes memory but also helps with the mental decline caused by stress.

One of the main ways adaptogenic herbs assist the body in dealing with stress is their capacity to regulate the HPA axis. By adjusting cortisol levels and strengthening neuroprotective effects, the herbs support both mental and physical health (Bledsoe et al., 2009; Kaushik et al., 2025; Ullagaddi, 2025).

3.3.2 Modulation of neurotransmitters (serotonin, dopamine, gamma-aminobutyric acid)

Neurotransmitters refer to substances that convey information and are necessary to the process of signal transduction in the brain and the rest of the nervous system. The most important neurotransmitters regulating the mood are serotonin, dopamine, and gamma-aminobutyric acid (GABA). Firstly, adaptogens influence neurotransmitters: they might raise levels of serotonin and dopamine (thereby helping mood and motivation) and also boost GABAergic tone (lessening anxiety and enhancing sleep). Secondly, enhanced serotonergic and GABA signaling may suppress hypothalamic CRH release, thereby further assisting the HPA axis in returning to normal. The neurotransmitter changes, combined with lower cortisol levels, will eventually lead to synaptic plasticity and stress resilience via BDNF-dependent mechanisms. Adaptogenic herbs can be illuminating to the brain by influencing the levels of these chemicals, along with their activity:

Serotonin: One of the major adaptogenic herbs, such as ashwagandha, is known to elevate serotonin levels, thereby brightening the mood and lessening the symptoms of anxiety and depression.
Dopamine: It is also possible that some adaptogenic herbs can stimulate the production of dopamine, which performs a triumphant role in human motivation, pleasure, and reward. As a result, this adjustment can lead not only to a better mood but also to enhanced cognitive function.
GABA: Adaptogens have the potential to increase GABA production, which is responsible for relaxation and anxiety alleviation. This can be extremely helpful during stressful situations and can also improve sleep quality.

Such neurotransmitter modulation is undoubtedly the major route by which adaptogenic herbs exert their calming and antistress effects. These medicinal herbs can indeed be a beauty secret for mental health and emotional well-being, thanks to their ability to balance serotonin, dopamine, and GABA levels (Chikkodi and Landge, 2025; Jadhav et al., 2025).

3.3.3 Antioxidant and antiinflammatory pathways

Oxidative stress and inflammation have been at the center of research that connects them to chronic diseases as well as mental health disorders. Antioxidants are agents that absorb free radicals, and antiinflammatory molecules are those that alleviate inflammation. Natural health products derived from plants, also known as adaptogenic herbs, have strong antioxidant and antiinflammatory characteristics:

Antioxidant activity: Several adaptogens, like ashwagandha and *Rhodiola*, have in their compositions bioactive compounds that neutralize free radicals, thus lowering oxidative stress and shielding cells from harm.
Antiinflammatory effects: Adaptogens have the potential to influence inflammatory pathways by lessening the production of pro-inflammatory cytokines. The suppression of these symptoms can be an eventual outcome of a chronic inflammatory process, including pain and fatigue.

The antioxidant and antiinflammatory properties of adaptogenic herbs are essential to their existence, as they help counter oxidative stress and inflammation, thereby supporting overall health and increasing the body's resistance to stress-related disorders (Lee et al., 2016).

3.3.4 Energy metabolism and mitochondrial support

Energy metabolism is necessary to keep cellular functions going and to maintain good health in general. Mitochondria are the organelles primarily responsible for energy production via oxidative phosphorylation. Adaptogenic herbs, in fact, are capable of boosting energy metabolism and supporting mitochondrial function:

Mitochondrial biogenesis: Some adaptogens stimulate mitochondrial biogenesis, thereby increasing energy production efficiency and endurance.
Enhanced ATP production: When mitochondrial function is optimized with adaptogens, ATP (adenosine triphosphate) production can be greatly elevated, providing cells with more energy for essential processes.

The adaptogenic herbs' proficiency in energizing metabolism and supporting mitochondria is a very important enabling factor that, in turn, leads to higher stamina, less fatigue, and a better overall sense of vitality (Catoni et al., 2022).

3.3.5 Immunomodulatory effects

The immune system is one of the major systems that keep the body safe from stress and diseases. Immunomodulation is the ability to either raise or lower immune responses.

Enhancing immune response: Medicinal herbs such as tulsi (*Ocimum sanctum*) and ashwagandha can boost the energy of immune cells, thereby increasing the body's capacity to resist viruses and other diseases.
Reducing immunosuppression: Adaptogens, by adjusting the HPA axis, can thus reverse the effects of stress-induced immunosuppression, leading to a more balanced immune response.

The immunomodulatory properties of adaptogenic herbs are instrumental in supporting immune health, especially when one is under pressure. By improving immune function and eliminating immunosuppression, these herbs become a source of positive health (Tan et al., 2023).

3.3.6 Epigenetic and gene expression modulation

Adaptogens may promote an individual's ability to withstand stress by changing stress-responsive transcriptional and epigenetic programs. Stressors use glucocorticoid signaling and other conserved pathways to reconfigure gene expression and chromatin states. These changes can be seen in altered methylation of glucocorticoid receptor loci and in long-term transcriptional reprogramming following early-life glucocorticoid exposure. Botanicals and small molecules affect chromatin modifiers, histone acetylation, and noncoding RNAs in cellular models, demonstrating the pathways through which phytochemicals may modify transcriptional outcomes under stress. The heat shock response, HSP70 in particular, is a core, inducible cytoprotective program that maintains proteostasis, limits the extent of oxidative damage, and facilitates recovery; therefore the mechanism by which adaptogens may exert their effects is to enhance HSP70 induction or stabilize chaperone networks. Adaptogens by altering chromatin enzyme activity or metabolic cofactors that feed epigenetic enzymes could (1) increase HSP70 and related chaperone gene expression, (2) correct maladaptive glucocorticoid-driven gene programs (e.g., by regulating NR3C1/FKBP5 dynamics), and (3) reduce proinflammatory transcriptional cascades that cause allostatic load. These molecular changes would go hand in hand with physiological effects, thus minimizing cellular damage and allowing cells to recover more quickly after reexperiencing stress (Guan et al., 2024; Sharma et al., 2018).

3.3.7 Gut–brain axis and adaptogens

Psychological and physiological stressors are the main drivers of disturbances in gut microbial communities, leading to dysbiosis that escalates systemic inflammation and alters metabolite profiles, which are linked to host signaling. Dysbiotic shifts diminish the production of beneficial metabolites, such as short-chain fatty acids and indoles, while increasing the formation of potentially harmful compounds, thereby weakening barrier integrity and stimulating low-grade inflammation that overlaps with cardiovascular, metabolic, and immune pathways. These peripheral disturbances signal back to the brain through vagal afferents, immune mediators, and derived neuromodulation, thereby altering HPA axis reactivity and behavioral responses. Polyphenols, flavonoids, fibers, and other bioactives in botanical adaptogens are usually metabolized by gut microbes, thereby modulating their community composition and functions; notably, the relationship between flavonoids and microbes is well documented for the production of health-promoting metabolites and the reshaping of ecological niches in the gut. By restoring microbial balance and increasing the production of beneficial metabolites, adaptogens might strengthen barrier function, decrease proinflammatory signaling, and normalize metabolite cues that affect central stress circuits. Phytochemical reviews of herbal medicine emphasize the potential of plant constituents to improve intestinal homeostasis, reduce neuroinflammation, and, in turn, confer indirect neuroprotective effects via the gut-brain axis. Collectively, these mechanisms imply that adaptogens might decrease allostatic load not merely through direct neuroendocrine actions, but also by nurturing a more robust microbiota, which, in turn, modulates immune, metabolic, and neural pathways to the brain (Pandian et al., 2020; Shinde et al., 2023).

3.4 Major adaptogenic herbs and their evidence

For hundreds of years, the use of adaptogenic herbs has been a common feature of traditional medicine systems like Ayurveda and TCM. The herbs are called so because they can soothe the body under various kinds of stress and help keep it healthy. Further, we will review some of the main adaptogenic herbs, their botanical characteristics, folk uses, physiological compounds, clinical trials, and safety.

3.4.1 *Withania somnifera* (ashwagandha)

Botanical profile: *W. somnifera* is the scientific name of ashwagandha or Indian ginseng. This plant belongs to the Solanaceae family. It is a small shrub native to India and North Africa.
Traditional uses: Ashwagandha has been used in Ayurvedic medicine for a very long time to support the body's energy, reduce stress, and promote general well-being.
Bioactive compounds: The herb is buffered with bioactive compounds, namely, withanolides, alkaloids, flavonoids, and phenolic compounds, which position it as a medicament.
Clinical evidence: Numerous modern studies have confirmed that ashwagandha can lower cortisol levels, improve stress tolerance, and enhance cognitive function. Apart from that, it also soothes anxiety and enhances the body's physical performance (Haber et al., 2024; Murthy et al., 2010).
Safety: Ashwagandha is almost free from side effects and hence safe for use. However, it should be noted that, in very large doses, this herb may cause discomfort in the gastrointestinal tract (Jia and Qian, 2011).
Molecular targets and signaling pathways: The main bioactive compounds in *W. somnifera* (withanolides and sitoindosides) are reported to interact with several molecular pathways involved in stress. In vivo and in vitro studies reveal that these compounds can regulate the activity of the HPA axis by dampening hypothalamic CRH secretion, pituitary ACTH secretion, and adrenal cortisol release. Consequently, feedback inhibition is restored during chronic stress conditions. Ashwagandha is also shown to possess antiinflammatory and antioxidant properties by downregulating NF-κB signaling and decreasing proinflammatory cytokine production. Thus the neuroprotective role of ashwagandha is highlighted, as it protects neurons from glucocorticoid-induced injury. Meanwhile, ashwagandha

increases neurotrophic factors (animal studies have shown increased levels of BDNF upon treatment with this herbal drug), thus enhancing synaptic plasticity and cognitive flexibility; other mechanisms include altering GABAergic neurotransmission and reducing oxidative injury, which eventually lead to mitochondrial stability (Chen et al., 2020).
Comparative clinical evidence: Standardized ashwagandha extracts prepared from the root, with different extraction methods and varying concentrations of withanolides, have been experimentally tested and systematically reviewed, confirming their ability to reduce subjective stress and lower serum cortisol levels. Clinical trials also show that these extracts elevate anxiety scores, improve sleep quality, and, in some cases, enhance memory performance. Reviews of such studies have demonstrated the point at which the supplement's effects on physical endurance capacity and athletic performance become random, and that some trials have shown gains. Safety records in the long-term for higher doses are scarce, but, in general, at normal doses, it is very safe; nevertheless, inconsistencies in dosage and extract composition remain the main issue in the studies (Natalie et al., 2022).

3.4.2 *Panax ginseng* (Asian ginseng)

Botanical profile: *P. ginseng* is Korean or Asian ginseng. It is a plant species in the Araliaceae family. The original habitat of this species is East Asia.
Traditional uses: Just as in a holistic medicine system known as TCM, *P. ginseng* is considered one of the most vital, mind-brightening, and immune-strengthening herbs.
Bioactive compounds: One of the main active parts is ginsenosides, which are saponins that exhibit adaptogenic and antifatigue effects.
Clinical evidence: Scientific studies have shown that ginseng is used primarily to improve physical performance, reduce fatigue, and enhance cognitive function. In addition, studies have also suggested that ginseng may have beneficial effects in the treatment of diabetes and cancer (Bai et al., 2018; Gan and Karmazyn, 2018).
Safety: *P. ginseng* is a very healthy product, but a few people may experience sleeplessness or stomach-related problems if they take it (Jiang et al., 2024).
Molecular targets and signaling pathways: The therapeutically active compounds of *P. ginseng*, ginsenosides (e. g., Rg1, Rb1), operate through several intracellular signaling pathways. They activate AMP-activated protein kinase, which helps maintain cellular energy balance and metabolic flexibility. Also, they promote nitric oxide (NO) signaling in the blood vessels, which may enhance brain blood flow and recovery from fatigue. Ginsenosides regulate neuroinflammatory pathways (lowering NFkB activation) and use PI3K/Akt and MAPK signaling to promote neuronal survival and synaptic plasticity; a few ginsenosides also increase BDNF expression in animal models, thus connecting metabolic and neurotrophic pathways. Furthermore, ginseng may alter monoaminergic neurotransmission, which may underlie mood and cognitive alterations (Hoho et al., 2002).

Comparative clinical evidence: Several randomized, placebo-controlled studies provide evidence that *P. ginseng* helps reduce fatigue and has the potential to improve cognitive performance, particularly during the medium-term follow-up of 412 weeks. Further, it can be expected to increase physical endurance and aid recovery. There is a good reason to anticipate that more extensive clinical trials will confirm the metabolic benefits (e.g., glycaemic control); however, this will require further studies to draw conclusive recommendations. Side effects are mild overall (insomnia, GI discomfort), and they are often dose-dependent; inconsistencies between ginseng species and preparations hinder direct comparisons (Ghiorghita et al., 2016).

3.4.3 *Rhodiola rosea*

Botanical profile: *R. rosea*, also named Arctic root or golden root, is a member of the Crassulaceae family and grows in the cold mountainous regions of Europe and Asia.

Traditional uses: The herb was used by traditional systems of healing to improve bodily and mental abilities, especially during times of distress.

Bioactive compounds: The main bioactive compounds of the plant are salidroside and rosavin, which have been widely reported as antioxidants and antifatigue agents.

Clinical evidence: Studies corroborate *Rhodiola's* use in enhancing mental performance, elevating mood, and reducing feelings of exhaustion. Symptoms related to stress have also been alleviated by *Rhodiola* administration (Limanaqi et al., 2020).

Safety: *Rhodiola* is considered a safer agent; however, in some cases, it may cause side effects such as dizziness and dry mouth (Murray, 2020; Schmidt et al., 2014).

Molecular Targets and Signaling Pathways: *R. rosea's* primary constituents (salidroside, rosavins) regulate stress responses, enzymes involved in neurotransmitter metabolism, and neurotransmitter signaling. From a mechanistic point of view, *Rhodiola* inhibits monoamine oxidase activity and limits catecholamine depletion under stress, thereby increasing the availability of serotonin and dopamine in the brain. Furthermore, it reduces stress-induced cortisol levels by suppressing CRH/ACTH signaling and exerts an antioxidant effect by activating endogenous antioxidant defenses. In animal studies, *Rhodiola* elevates BDNF levels and enhances mitochondrial function, thereby protecting neurons against stress and fatigue. These combined actions on monoamines, the glucocorticoid system, and neurotrophic factors explain its adaptogenic effects.

Comparative clinical evidence: Clinical trials have demonstrated that *Rhodiola* helps relieve perceived fatigue, improves cognitive functioning under stress, and shows some effectiveness against depressive symptoms in short-term studies (usually 2 to 12 weeks). There have also been reports on *Rhodiola's* effects on physical endurance and continuous concentration tasks, but these depend on the dose and the degree of extract standardization. *Rhodiola* has a good safety profile; the reported side effects are infrequent and mild (e.g., dizziness, dry mouth). As with other adaptogens, variation in study quality and extract composition hampers definitive comparative conclusions (Soejarto and Farnsworth, 1978).

3.4.4 Eleutherococcus senticosus (Siberian ginseng)

Botanical profile: Siberian ginseng, scientifically known as *E. senticosus*, is a member of the Araliaceae family and grows naturally in Eastern Siberia and some areas of Asia.
Traditional uses: The plant has been used in TCM and Russian herbal medicine to energize the body and reduce susceptibility to stress.
Bioactive compounds: The most notable components are eleutherosides, which contribute mainly to the plant's adaptogenic properties.
Clinical evidence: Studies indicate that the herb can enhance physical performance and alleviate tiredness, as well as stimulate the immune system (Rybnikář et al., 2019; Zhao et al., 2024).
Safety: While it is mostly safe, the herb may interact with some drugs and cause mild digestive system irritation (Gao and Wu, 2019).

3.4.5 *Schisandra chinensis*

Botanical profile: *S. chinensis*, the five-flavor berry, is a member of the Schisandraceae family and originates from China.
Traditional uses: It has been utilized in TCM for endurance boosting, liver function enhancement, and overall vitality promotion.
Bioactive compounds: The main energizing compounds in the work are dibenzo-cyclo-octadiene lignans, which have biological adaptogenic properties of the drug *Schisandra*.
Clinical evidence: Animal and human clinical trials have been conducted on its use to improve liver function, alleviate stress, and enhance physical performance (Singh et al., 2022; Singh et al., 2023).
Safety: *Schisandra* is safe and well-tolerated in most cases, but a few people may experience some gastrointestinal discomfort (Verma, 2016).

3.4.6 *Ocimum sanctum* (holy basil/tulsi)

Botanical profile: *Ocimum sanctum*, generally known as holy basil or tulsi, is a member of the Lamiaceae family and is a native of the Indian subcontinent.
Traditional uses: Among the medicaments of the ayurvedic system of medicine, it is the most venerable one, which is recognized for its calming, antistress, and strengthening the body's immunity -energizing the body's defense mechanism -properties.

Bioactive compounds: The main substances are eugenol and rosmarinic acid, which have been characterized as antiinflammatory and antioxidant in nature.
Clinical evidence: Studies confirm the plant's use for reducing stress and anxiety, elevating mood, and strengthening the immune system (Elkhateeb and Daba, 2020; Sundaram et al., 2012).
Safety: Mostly safe; nevertheless, an overdose may result in digestive tract irritation (Mazza et al., 1984; Wu et al., 2021).

3.4.7 Other emerging adaptogens

3.4.7.1 *Maca (*Lepidium meyenii*)*

Maca (*Lepidium meyenii*) is a root from Peru that has been traditionally used to increase energy, stamina, and fertility. Research indicates that maca might help with sexual function and the alleviation of tiredness; it is a widely used natural supplement to revitalize the body and promote general health (Wang et al., 2021).

3.4.7.2 *Licorice (*Glycyrrhiza glabra*)*

Glycyrrhiza glabra, commonly known as Licorice, is a medical herb that has been around for ages and is generally used to support the adrenal glands and decrease the side effects of stress. Research on the matter reveals that it possesses antiinflammatory properties and immune-modulating properties, making it a promising combination that helps keep the body fit and raise its defenses (Elkhateeb and Daba, 2020; Ghahremaninejad, Joharchi, Memariani, Behroozian, 2022).

3.4.7.3 *Cordyceps (*Cordyceps sinensis*)*

Cordyceps (*Cordyceps sinensis*) is a medicinal fungus known for increasing energy and improving athletic performance. Research has demonstrated that it can enhance physical exercise capacity and relieve fatigue; thus it can be used as a general support for physical endurance (Ghahremaninejad, Joharchi, Memariani, 2022; Jácome et al., 2021).

3.4.7.4 *Astragalus (*Astragalus membranaceus*)*

Astragalus (*Astragalus membranaceus*) is a popular traditional herbal medicine that is widely used to enhance the immune system and reduce nervousness. Research indicates that it is a strong antioxidant and has an antiinflammatory effect, so it contributes to the body's revitalization and makes it less vulnerable to different ailments (see Table 3.1) (Kholmogorova et al., 2020; Zhang et al., 1904).

Table 3.1 Key adaptogenic herbs.

Herb	Main use	Key compound	Safety
Ashwagandha	Stress, anxiety	Withanolides	Safe; mild GI upset
Panax ginseng	Energy, cognition	Ginsenosides	Insomnia in some
Rhodiola	Fatigue, mood	Salidroside	Dry mouth, dizziness
Siberian ginseng	Energy, immunity	Eleutherosides	Mild GI irritation
Schisandra	Endurance, liver	Lignans	GI discomfort
Holy basil	Stress, immunity	Eugenol	GI irritation
Maca	Energy, stamina	Macamides	Well tolerated
Licorice	Stress, adrenal	Glycyrrhizin	High dose affects BP
Cordyceps	Athletic performance	Cordycepin	Rare nausea
Astragalus	Immunity	Polysaccharides	Mild GI symptoms

Table 3.2 Clinical applications of major adaptogens.

Health area	Key adaptogens	Main benefit
Stress, anxiety and depression	*Rhodiola*, ashwagandha	Lowers cortisol, improves mood, reduces fatigue
Cognition (memory, attention, learning)	Ginseng, bacopa	Enhances memory, learning, and mental performance
Physical endurance and performance	Cordyceps, *Rhodiola*	Increases energy, reduces fatigue, improves exercise capacity
Immune support	*Echinacea*, ashwagandha	Strengthens immune response, protects against infections
Sleep and circadian rhythm	Valerian root, ashwagandha	Promotes relaxation, improves sleep quality

3.5 Clinical applications and health benefits

Adaptogenic herbs are organic materials that soothe the body and make it resistant to stress and, at the same time, promote general health. To a great extent, they have become popular because they are thought to provide a wide range of therapeutic benefits, primarily in the management of stress, improvement of mental abilities, enhancement of physical performance, strengthening of the immune system, and regulation of sleep. In each of the following sections, we will discuss how these substances can be used as medicines and their advantages for the human body (see Table 3.2).

3.5.1 Adaptogens in stress management (anxiety, depression, burnout)

Adaptogens are mostly known for their role in normalizing the stress response, which in turn makes the body more efficient in handling stress. In other words, they mitigate the symptoms of anxiety and depression that usually go along with chronic stress. Research indicates that treatment with such adaptogens as *R. rosea* and ashwagandha results in the lowering of cortisol levels, the stress-related hormone, and improvement of the mood and the capacity to resist exhaustion.

Rhodiola rosea: Research has shown that it can improve brain function and lessen tiredness, especially when under pressure.

Ashwagandha: This herb was associated with less anxiety and better general mental health, thus becoming a favorite among mentally exhausted people (Andargeery et al., 2024; Papi et al., 2023).

Adaptogen herbs can support standard treatments in cases of anxiety and depressive symptoms, especially when such symptoms are mild to moderate and mainly the result of chronic stress or burnout. Clinical and animal studies show that *R. rosea* and ashwagandha can help correct stress mechanisms (e.g., lowering cortisol), reduce tiredness, and improve mood and concentration, thus helping patients be more receptive to psychotherapy or pharmacotherapy rather than replacing these treatments. A considerable number of adaptogens have a non-sedative effect, meaning they are unlikely to cause heavy daytime sleepiness or mental dulling, which are typically associated with benzodiazepines or sedative-hypnotics; thus they can be used during the day to maintain energy and performance. Usually, the effects of the treatment become obvious after a few weeks and depend on the person's constitution. The main points to be remembered are doctor supervision for combined therapies, checking for potential interactions between herbs and drugs (especially with psychotropics), and not stopping prescribed medicines without a doctor's permission. In cases of severely depressed or anxious patients, psychiatric care according to the guidelines will be the first choice; adaptogens can only be considered under medical supervision and as one element of the overall treatment plan (Pincus et al., 2023).

3.5.2 Cognitive health: memory, attention, learning

Adaptogens are equally necessary for a mental health upgrade. In fact, they can improve memory, attention, and learning by neuroprotection as well as by enhancing general brain function. To illustrate, research on ginseng and *B. monnieri* indicates that they help cognitive performance and memory retention.

Ginseng: One of the major effects of the compound is that it can not only stimulate the brain but also make the mental function better and fatigue lighter.

Bacopa monnieri: Studies suggest that this herb can enhance memory and learning by increasing synaptic communication in the brain (Husna et al., 2025; Sahoo et al., 2021).

3.5.3 Physical endurance and athletic performance

Adaptogens have also been recognized for their potential to improve physical endurance and athletic performance. They can raise energy levels, reduce fatigue, and speed recovery after a workout. In particular, Cordyceps and *R. rosea* are the two substances that are most frequently mentioned for their supportive effects in this area.

Cordyceps: One study found that the fungus can increase ATP production, which in turn can elevate energy levels and improve exercise performance.

Rhodiola rosea: Besides the improvement of brain functions, the physical performance can be enhanced as well, through a decrease in the perception of effort during a workout (Schmid et al., 2021; Xiao, 2023).

3.5.4 Immune support and resilience

Adaptogens deliver a powerful immune system, and hence, they can be a good natural remedy against stress-related diseases. The health-promoting effects of *Echinacea* and ashwagandha have been confirmed through experiments.

Echinacea: One of the major uses of this product is to strengthen the immune system; thus it is a good preventive measure against colds and other infections.

Ashwagandha: Moreover, to be a good stress-reliever, the herb is also beneficial for the immune system as it helps regulate the immune response (Colombi, 2008; Jorquera Gutiérrez and Herrera Gallardo, 2020).

3.5.5 Role in sleep and circadian rhythm regulation

The majority of adaptogens also assist in deeper sleep and the normalization of circadian rhythms. Relaxation is often facilitated by valerian root and ashwagandha, and sleep is improved with their use.

Valerian root: This herb has long been recognized for its calming effects and is regularly used for insomnia treatment and sleep quality enhancement.

Ashwagandha: On top of its other advantages, it can be a great sleep regulator and a general sleep quality improver (Kalaivani et al., 2023; Meena et al., 2025).

3.5.6 Adaptogens in occupational and lifestyle stress

There has been a considerable increase in the stress people face due to high-demand professions or disrupted schedules. This stress negatively affects their mental functioning, moods, sleep, and physical health. Healthcare workers struggle with long shifts, emotional burdens, and broken sleep. All these things increase their risk of burnout and decrease their attention and decision-making abilities under pressure. Students, on the other hand, have to deal with their studies, peer pressure, and irregular sleep. Normally, they get stressed out during exams with acute stress and then suffer from performance anxiety for a long time. Employees in corporates are victims of chronic workload, job stress, and the "always-on" culture. This leads to emotional exhaustion and eventually leads to them getting away from their work. Shift workers (healthcare, transport, emergency services) have to face circadian misalignment, sleep deprivation, and an inability to recover properly, which leads to their metabolic, mood, and cognitive functions being impaired. According to several researches, adaptogenic herbs could be considered as adjunct interventions in these populations: By moderating the HPA, axis responses, aiding neurotransmitter balance, and facilitating BDNF as well as antioxidant pathways, adaptogens could be effective at: reducing the capitalized on perceived stress, improving the quality of sleep, and strengthening cognitive resilience during one, time and reoccurred stress exposures. Clinical studies, which are mostly of short to medium duration, show reductions in stress- and fatigue-related symptoms and improvements in sleep metrics in occupational cohorts and student samples. However, the variation in extracts, dosing, and outcome measures makes it difficult to generalize the results. In practical terms, the focus should be on the use of standardized preparations, correct dosing, consideration of interactions and contraindications, and combination with evidence-based workplace measures (shift scheduling, rest breaks, mental health support). As a whole, adaptogens can be a supportive, low-risk strategy for stress resilience in high-stress jobs when used in combination with organizational and behavioral interventions (Karole et al., 2019).

3.6 Formulations and delivery systems

The potency of adaptogenic herbs is greatly dependent on their composition and delivery systems. Different ways of preparing and administering these herbs can increase their bioavailability and enhance their therapeutic effects. We will discuss different formulations and delivery systems for adaptogens in the following subheadings.

3.6.1 Herbal teas, powders, capsules, tinctures

Adaptogenic herbs may be taken in different forms, such as:

Herbal teas: Herbal teas are prepared by steeping dried herbs in hot water. The method is widely known for its ease and for the extraction of water-soluble compounds. Herbal teas can offer a relaxing experience and are often used as calming agents.
Powders: Ground herbs are also great additions to any smoothie, juice, or food. Powder forms are more convenient for dosing and can be easily added to any daily meal plan.
Capsules: These are a convenient way to take adaptogens without experiencing the herbs' taste. Capsules may contain herbs in powder form or a standardized extract, ensuring a consistent dose.

Tinctures are concentrated liquid extracts obtained by soaking herbs in alcohol or vinegar. Tinctures are strong and can be taken up rapidly; thus they are great for short-term effects.

3.6.2 Standardized extracts and bioactive markers

Standardization plays a vital role in herbal medicine if we want to get the same strength and effect of the medicine.

Standardized extracts: The extracts contain a defined amount of the active compounds, which is the main factor for achieving predictable therapeutic results. For example, the adaptogen ashwagandha can be standardized to contain a specific percentage of withanolides, the active compounds responsible for its effects.

Bioactive markers: The detection and measurement of bioactive markers in plant-based formulations is an essential condition for the production of drugs. To find these markers, methods such as high-performance liquid chromatography and high-performance thin-layer chromatography are utilized, ensuring that herbal products are safe and efficient (Hazarika, 2025; Hoy, 2010).

3.6.3 Synergistic blends (polyherbal formulations)

Polyherbal formulations are made up of several herbs with the intent to amplify the healing effects by synergism. Polyherbal formulations are traditional medicines that rely on the principle of synergism to enhance efficacy and minimize side effects. One of the concepts in Ayurveda, rasayana, focuses on rejuvenation, immune system strengthening, and increasing lifespan through the use of herbs in a natural way. Herbs are combined in ways that potentiate therapy, moderate side effects, and support whole-system homeostasis. TCM also uses a mixture of various preparations where each herb plays a different role, such as the main herb, the supportive herb, etc., in a way that the principle of harmony is achieved and both the root causes and symptoms are thus dealt with. Contemporary phytopharmacology reveals that such combos may affect different targets simultaneously, increase bioavailability through the activity of complementary substances, and minimize the dosage of single actives; however, there is a need for clinical studies to identify the mechanisms and therefore optimize the safety and efficacy ratios.

Synergistic effects: Different adaptogens mixed can produce a higher impact than a single-herb product. For instance, the pair of *R. rosea* and ashwagandha may offer a broader range of benefits, such as improved stress resistance and enhanced cognitive function.

Traditional practices: Polyherbal compositions are the preferred way of treatment in the majority of cultural medical systems, like Ayurveda, to acquire the maximum therapeutic effect. These combinations are created by following the rule of harmonizing different herbs so that their effects can be utilized to the greatest extent (Ajgaonkar et al., 2022; Siwek et al., 2023).

3.6.4 Nutraceutical and functional food applications

Adaptogens are being progressively added to nutraceuticals and functional foods that are made to give health benefits over and above the basic nutrition.

Nutraceuticals: These products are derived from food sources and provide health benefits, for example, dietary supplements containing adaptogenic herbs. They are created to promote certain health effects, such as reducing stress or improving mental abilities.

Functional foods: Consuming these foods allows the intake of additional nutrients and promotes good health. For instance, energy bars or drinks with adaptogenic herbs may be promoted as stress-relief snacks or brain function enhancers.

3.6.5 Bioavailability and pharmacokinetics

Withanolides (isolated from *W. somnifera*) and ginsenosides (from Panax species) are typical representatives of biologically active classes whose oral bioavailability is severely limited by low aqueous solubility, extensive first-pass metabolism, and poor intestinal permeability. Withanolides are lipophilic steroidal lactones that are absorbed differently in each individual; they are metabolized via hepatic phase I/II enzymes and are susceptible to gut microbial action, which can both lead to the production of active metabolites and the inactivation of other metabolites. Ginsenosides are glycosylated saponins with high molecular weight and polarity; intestinal microbiota-mediated deglycosylation frequently yields smaller aglycones (e.g., compound K) with greater permeability and systemic exposure. Both classes thus exhibit variable pharmacokinetic profiles: relatively low C_{max}, delayed T_{max}, and limited oral bioavailability, with significant interindividual variations due to microbiome and hepatic enzyme activity.

Formulation strategies have quite a significant effect on absorption enhancement. Take Liposomal encapsulation, for example, it can increase the compound's solubility and shield sensitive ingredients from degradation by the digestive system, enabling lymphatic uptake and reducing first-pass loss. Various nanoparticle platforms, including polymeric nanoparticles, solid lipid nanoparticles, and nanoemulsions, help increase dissolution rate, enhance mucosal adhesion, and enable controlled release; meanwhile, molecular-level modifications (PEGylation, targeting ligands) can also be used to prolong circulation time and alter tissue distribution. Such drug delivery systems demonstrated increased plasma exposure and pharmacodynamic effects in animal and human studies, thereby providing a rationale for their potential to enhance the therapeutic outcomes of withanolide and ginsenoside formulations (Shaikh et al., 2020).

3.7 Safety, dosage, and precautions

Before using adaptogenic herbs, it is very important to know their safety features, possible interactions with other drugs, dosage, and contraindications. This part will

Table 3.3 Safety and precautions for adaptogens.

Category	Key points
General safety	Mostly safe when used correctly; mild side effects like GI upset, headache, or allergies may occur
Drug interactions	St. John's Wort → interacts with antidepressants; Ginseng and Garlic → may lower BP too much with antihypertensives; caution with anticoagulants and antidiabetics
Dosage guidelines	Start low and increase slowly; dosage varies by herb and form; seek medical guidance
Contraindications	Avoid in pregnancy/lactation; use caution in autoimmune diseases; consult a doctor for hormone-sensitive cancers or before surgery

explain these elements in detail to help users use adaptogens safely and effectively (see Table 3.3).

3.7.1 General safety profile of adaptogens

In general, adaptogens are safe for almost everyone when used properly. Nevertheless, the safety of these products may depend on the specific herb, the dose, and the person's health condition.

Common adaptogens: There are numerous scientific studies on the safety of short-term use of ashwagandha, *R. rosea*, and holy basil. Studies have confirmed very few side effects for the medicines used at the given dosages (Markowitz et al., 2025; Sakti et al., 2025).

Side effects: Most adaptogens are safe, but a few people may experience mild side effects such as digestive issues, headaches, or allergic reactions. It is very important to be alert for adverse symptoms, particularly when a new herb is taken (Palleria et al., 2020).

3.7.2 Drug-herb interactions (with antidepressants, antihypertensives, etc.)

Interactions between drugs and herbs are a matter of great concern, especially in the context of polypharmacy. An adaptogen can alter drug exposure by affecting cytochrome P450 enzymes, UGTs, or drug transporters (e.g., P-gp), and a few also have additive pharmacodynamic effects (e.g., on blood pressure, bleeding risk, or CNS activity). Enzyme induction or inhibition that alters plasma levels of anticoagulants, antidiabetics, antihypertensives, and psychotropics is one example that is clinically relevant. To lessen the risk, you should first do an exhaustive medication reconciliation before suggesting adaptogens, verify the known mechanisms of interaction, begin with a low dose, keep an eye on the relevant clinical and laboratory parameters, and if there are psychotropics, anticoagulants, or drugs with a narrow therapeutic index, you should consult the prescribing clinician. Using adaptogens together with prescription medications can cause major herb-drug interactions.

Antidepressants: One of the major concerns is the interaction of adaptogens like St. John's Wort with antidepressants that may change their effectiveness and give rise to side effects (Doke et al., 2023; Tamboli et al., 2025). For example, the hyperforin contained in St. John's Wort activates cytochrome P450 enzymes, thus causing lowered plasma levels of certain antidepressants (Asokkumar and Ramachandran, 2019).

Antihypertensives: On the other hand, some adaptogens, such as Ginseng and Garlic, might enhance the effectiveness of antihypertensive drugs, leading to a drop in blood pressure to dangerous levels. People taking blood pressure medications should not risk it with these herbs and must get their doctor's approval first (Ajgaonkar et al., 2022; Shaikh et al., 2020).

Other Interactions: Besides that, adaptogens may also affect the potency of anticoagulants, antidiabetics, and other drugs that require close observation and advice from medical practitioners (Siwek et al., 2023).

3.7.3 Recommended dosage ranges and guidelines

The proper consumption of adaptogens can vary significantly between different herbs, different individuals, and even different forms (e.g., extract, powder, or tincture). For this reason, there are no universal dosing guidelines that can be applied to everyone. In most cases, it is advised to start with a minimal amount and gradually increase it as you get accustomed to it. Each person's organism may respond differently to adaptogens, and factors such as age, health conditions, and medications should also be considered. Hence, it is best to seek advice from a qualified medical professional who will assess your individual health needs and recommend a treatment plan that is in line with the latest research (Heghes et al., 2022; Karttunen et al., 2019).

3.7.4 Contraindications (pregnancy, autoimmune conditions, etc.)

Certain groups of people must be very careful when using adaptogens or should refrain from using them completely.

Pregnancy and lactation: Most adaptogens lack sufficient safety data during pregnancy and breastfeeding. In consequence, ashwagandha, for instance, is typically not advised for pregnant women because it might cause changes in hormone levels.

Autoimmune conditions: Individuals with autoimmune diseases should take great care when using substances such as *Echinacea* and ashwagandha, as these compounds may overstimulate the immune system and worsen symptoms.

Other conditions: Individuals suffering from certain health conditions, like cancer, that are sensitive to hormones, or people who are going to have surgery, should talk to their doctors before starting the use of adaptogens (Ara et al., 2020; Newcomer et al., 2023).

3.7.5 Adaptogens in long-term use and special populations

Long-term use of adaptogens in elderly and chronically ill patients presents special considerations, as pharmacokinetics, comorbidity burden, and polypharmacy alter the benefit-risk profile. In seniors, age-related decreases in renal and hepatic clearance, changes in protein binding, and increased sensitivity to both therapeutic and adverse effects could lead to higher exposure and risk; thus dose titration from the lowest possible initial dose with regular review is recommended. Frailty, cognitive impairment, and comorbid cardiovascular disease require closer monitoring for falls, blood pressure changes, and sedation (if applicable), and simplified regimens to facilitate adherence. In patients with chronic disease (cardiovascular, hepatic, renal, diabetes, autoimmune disorders), adaptogens should be selected and dosed with consideration of disease-specific risks; for example, substances that alter blood glucose, coagulation, blood pressure, or immune activity necessitate coordination with treating specialists and more frequent laboratory or clinical monitoring. Differentiate between prevention and treatment aims: for stress prevention in the healthy, intermittent low-dose schedules (cyclical dosing, seasonal use) may limit cumulative exposure and interaction risk; for active stress-related syndromes, therapeutic dosing under clinician supervision with scheduled reassessment (412 weeks) is recommended. Medication reconciliation, renal/hepatic function tests, and evaluation of continuation should be performed at regular intervals to avoid adverse events and to enable early detection of interactions if long-term use is intended (Gao et al., 2024).

3.8 Future perspectives and research directions

The world of adaptogenic herbs keeps changing very fast as scientists keep on discovering new things about how they work, their effectiveness, and their uses. In this part, we will look at the next steps and changes in this field of research, including the absence of clinical trials, the use of omics technologies, personalized medicine, and the possible use of these substances to treat stress-related disorders and chronic diseases (see Fig. 3.3).

3.8.1 Clinical trial gaps and standardisation issues

Despite the increasing interest in adaptogens, there are still large gaps in clinical trial methodologies and standardization practices.

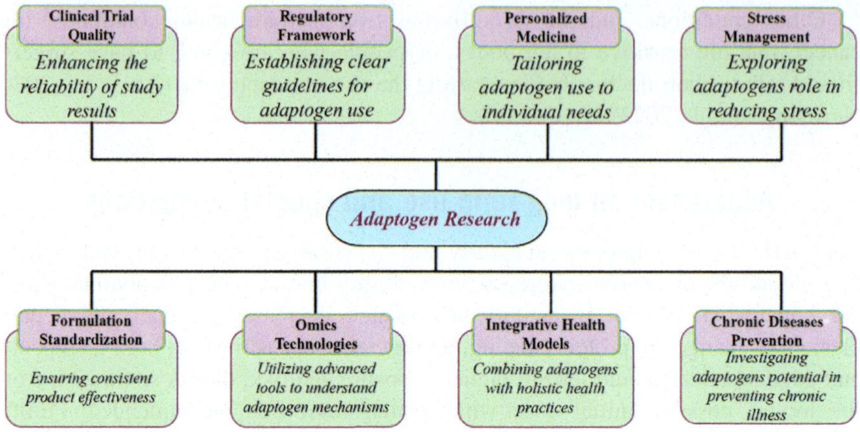

FIGURE 3.3 Future research directions for adaptogenic herbs.

Adaptogenic herbs future uses.

Enrollment challenges: According to recent studies, a very small percentage of eligible patients participate in clinical trials, even though most patients express willingness to do so (Zhang et al., 2023). This difference indicates a need for novel strategies to increase participant recruitment and retention in studies on adaptogens.

Standardization of extracts: The absence of standardized formulations and dosages makes it very difficult to compare results across studies. Subsequent investigation should devote its resources to establishing clear guidelines for the preparation and dosing of adaptogenic herbs to achieve uniformity and the reliability of clinical results.

Regulatory frameworks: It is necessary to establish robust regulatory systems to govern the approval and surveillance of adaptogenic products. Along with quality control, safety, and efficacy, this comprises establishing the standards that must be observed to facilitate their integration into the healthcare system of the future (Urgessa and Woldesemayat, 2023).

3.8.2 Omics technologies (genomics, metabolomics) in adaptogen research

The use of omics technologies, such as genomics and metabolomics, is a major step forward in our understanding of adaptogens.

Genomic insights: Studies in genomics can enable us to understand the genetic basis of individuals' responses to adaptogens, offering a pioneering approach to personalized approaches in herbal medicine. By understanding how genetic variations affect the effectiveness of adaptogens, it is possible to develop treatments customized to the individual, thereby maximizing benefits.

Metabolomic profiling: Through metabolomics, we can identify the physiological pathways influenced by adaptogens and thus prevent, at the molecular level, the breakdown of these mechanisms of action. Such facts can trigger the concepts of how these herbs work in stress response, the immune system, and general wellness.

Integration of multiomics: Furthermore, one could combine genomics, transcriptomics, proteomics, and metabolomics to understand fully how biological systems interact with adaptogens. The integrations approach facilitates the discovery of novel bioactive compounds and their therapeutic potential (Bleaney et al., 2024; Wang et al., 2025).

3.8.3 Role in personalised medicine and integrative health

Adaptogens are gradually being acknowledged as a potential source of personalized medicine as well as a part of integrative health approaches.

Tailored interventions: By understanding each person's reaction to adaptogens, medical practitioners can design individual treatment plans that take into account not only genetic but also environmental and lifestyle factors. Such a personalized approach can make the use of adaptogens as stress-relievers and general health promoters much more effective.

Integrative health models: Including adaptogens in holistic health models can be seen as a step towards a more comprehensive wellness plan. The benefits of adaptogens can be raised to a new level, and prolonged health can be achieved by combining adaptogens with other therapeutic modalities, for example, nutrition, exercise, and mindfulness practices (Yang et al., 2025).

3.8.4 Potential in stress-related disorders and chronic diseases

One of the major research areas of adaptogens is their use to relieve stress-related disorders and chronic diseases.

Stress management: As they can control the stress response, adaptogens can be very helpful in situations such as anxiety, depression, and low energy. Eventually, research should be limited to clinical trials involving patients with these disorders.

Chronic disease prevention: In addition, adaptogens may help prevent chronic diseases associated with stress, such as heart disease, diabetes, and autoimmune disorders. Studies should be conducted to understand the long-term effects of adaptogens on such diseases.

Mechanistic studies: More molecular-level studies are needed to identify the specific brain and body regions that interact with adaptogens in stress-related pathways and chronic disease processes. Knowing these mechanisms can help devise targeted treatments and amplify the therapeutic potential of adaptogens.

3.8.5 Artificial intelligence and network pharmacology in adaptogen research

AI-assisted network pharmacology can radically transform adaptogen research by integrating large-scale chemical, phenotypic, and omics data to map compound-target pathway networks. Artificial intelligence, graph algorithms, and machine learning are applied to identify potential bioactive substances, predict polypharmacology, assess drug combination synergism, and validate biomarker mechanisms. Before in vivo testing, generative and causal models may also suggest optimized extract compositions and warn of safety/ADMET liabilities. In silico network predictions, combined with high-throughput experimental perturbations (transcriptomics, phenotypic screens), can accelerate target validation and the rational design of clinical trials, thereby making multicomponent adaptogen formulations more feasible for precision phytomedicine (Durga Devi et al., 2014).

3.8.6 Clinical biomarkers for stress adaptation and resilience

Standardized biomarker panels across domains are necessary to assess adaptogen efficacy and stratify responders. Potential markers may include neuroendocrine (waking and diurnal cortisol, ACTH), autonomic (heart rate variability, catecholamines), inflammatory (IL-6, TNF, CRP), cytokines, metabolomic stress signatures (energy, tryptophan/kynurenine metabolites), and microbiome-derived metabolites. Sensitivity and clinical importance will be increased by multimodal composite indices that incorporate objective biomarkers and validated patient-reported outcomes. Resilient phenotypes can be characterized using high-throughput metabolomics combined with machine-learning classifiers, thereby enabling personalized dosing and endpoint selection in clinical trials (Kanakadurgadevi, 2010).

3.8.7 Regulatory challenges and need for standardisation

The regulatory translation of adaptogens depends on their thorough standardization: In addition, to authenticated botanical identity and reproducible extraction methods, chemical fingerprints (e.g., LC-MS profiles), quantitative marker assays, contaminant limits, and stability specifications are required. There is a pressing need for consensus monographs and broadly accepted criteria for bioequivalence of complex extracts. The use of AI-powered quality monitoring and multiomics batch characterization could help establish traceability and quality-by-design concepts. Global standardization of naming, reference standards, and clinical endpoints will enable study comparability and facilitate regulatory approval, paving the way for safer, more evidence-based use of adaptogens in healthcare.

3.9 Conclusion

Adaptogenic herbs offer a natural and effective way to support the body during stress by balancing hormones, improving resilience, and enhancing overall well-being. Backed by traditional knowledge and growing scientific evidence, adaptogens show benefits across mental, physical, and immune health. With proper use and guidance, they have strong potential as valuable tools in modern integrative healthcare.

References

Agarwal A. Adaptogenic plants within food supplement. Annals of Innovation in Medicine 2023;1(1). Available from https://doi.org/10.59652/aim.v1i1.29.

Ajgaonkar A, Jain M, Debnath K. Efficacy and safety of Ashwagandha (*Withania somnifera*) root extract for improvement of sexual health in healthy women: a prospective, randomized, placebo-controlled study. Cureus 2022;14(10). Available from https://doi.org/10.7759/cureus.30787.

Andargeery SY, Taani MH, Alhalwani RA, El-Gazar HE. Psychological distress, academic stress, and burnout among Saudi undergraduate nursing students. Journal of Clinical Medicine 2024;13:3357. ("Menke A. Glucocorticoid signalling in stress-related psychiatric disorders uncovered by dexamethasone-induced gene expression. ScienceOpen Res. 2017; Available from: http://doi.org/10.14293/s2199-1006.1.sor-uncat.eitgtn.v1," no date). http://doi.org/10.3390/jcm13123357.

Ara I, Maqbool M, Bukhari B, Ara N, Hajam TA. Present status, standardisation, and safety issues with herbal drugs. International Journal of Research in Pharmacy Science & Technology 2020;1(Santhanu and Senthil, 2021):95–101.. Available from https://doi.org/10.33974/ijrpst.v1i3.169 Available from:.

Ashraf MV, Khan S, Misri S, Gaira KS, Rawat S, Rawat B, et al. High-altitude medicinal plants as promising source of phytochemical antioxidants to combat lifestyle-associated oxidative stress-induced disorders. Pharmaceuticals 2024;17(8). Available from https://doi.org/10.3390/ph17080975 Available from: https://www.mdpi.com/journal/pharmaceuticals.

Asokkumar K, Ramachandran S. Herb-drug interactions: focus on adverse drug reactions and pharmacovigilance of herbal medicines. Herbal Medicine in India: Indigenous Knowledge, Practice, Innovation and its Value. Singapore, India: Springer; 2019. p. 547–71. Available from: https://doi.org/10.1007/978-981-13-7248-3 10.1007/978-981-13-7248-3_33.

Bai L, Gao J, Wei F, Zhao J, Wang D, Wei J. Therapeutic potential of ginsenosides as an adjuvant treatment for diabetes. Frontiers in Pharmacology 2018;9(May). Available from https://doi.org/10.3389/fphar.2018.00423.

Bhardwaj K, Kumar P. Ayurveda, panchakosha model and manomayakosh: its relevance in modern mental health care. Journal of Psychosocial Well-Being 2024;5(Panossian and Efferth, 2022):40–9. Available from https://doi.org/10.55242/jpsw.2024.5205.

Bleaney CW, Abdelaal H, Reardon M, Anandadas C, Hoskin P, Choudhury A, et al. Clinical biomarkers of tumour radiosensitivity and predicting benefit from radiotherapy: a systematic review. Cancers 2024;16(10):1942. Available from https://doi.org/10.3390/cancers16101942.

Bledsoe JM, Kimble CJ, Covey DP, Blaha CD, Agnesi F, Mohseni P, et al. Development of the Wireless Instantaneous Neurotransmitter Concentration System for intraoperative neurochemical monitoring using fast-scan cyclic voltammetry: Technical note. Journal of Neurosurgery 2009;111(4):712–23. Available from: http://thejns.org/doi/pdf/10.3171/2009.3.JNS081348 10.3171/2009.3.JNS081348.

Catoni M, Alvarez-Venegas R, Worrall D, Holroyd G, Barraza A, Luna E, et al. Genome-wide changes in DNA methylation mark long-lasting defence priming by β-aminobutyric acid in tomato. Frontiers in Plant Science 2022;13:836326. Available from https://doi.org/10.3389/fpls.2022.836326 Available from:.

Chen W, Balan P, Popovich DG. Comparison of ginsenoside components of various tissues of New Zealand forest-grown Asian ginseng (panax ginseng) and American ginseng (*Panax Quinquefolium* L.). Biomolecules 2020;10(3):372. Available from https://doi.org/10.3390/biom10030372.

Chikkodi PB, Landge DD. A review on role of polyherbal formulation as energy booster in traditional medicine: phytochemistry and mechanism of action. International Journal of Advanced Research in Science, Communication and Technology 2025:151–63. Available from https://doi.org/10.48175/ijarsct-28921.

Colombi A. The unbearable lightness of mental health. American Health & Drug Benefits 2008;1(Bhardwaj and Kumar, 2024):10–2.

Doke CO, Tiwari RS, Todekar VB, Mane AP, Sawant VS, Sawant GV. Herb drug interaction in antibiotic drug, antiulcer drug, antihypertensive drug, antidiabetic drug, anticancer drug. Research Journal of Science and Technology 2023:119–26. Available from https://doi.org/10.52711/2349-2988.2023.00020.

Durga Devi NK, Shaheena S, Dinesh Babu CH. Evaluation of adaptogenic activity of plant extracts by microorganisms. Der Pharmacia Lettre 2014;6(4):184–9. Available from: http://scholarsresearchlibrary.com/dpl-vol6-iss4/DPL-2014-6-4-184-189.pdf.

Elkhateeb W, Daba G. Review: the endless nutritional and pharmaceutical benefits of the Himalayan gold, Cordyceps: current knowledge and prospective potentials. Biofarmasi Journal of Natural Product Biochemistry 2020;18(2). Available from https://doi.org/10.13057/biofar/f180204.

Er B, Ozmen B, Sahin E, Orhan C, Sahin N, Morde AA, et al. A novel Ashwagandha (*Withania somnifera*) formulation mitigates sleep deprivation-induced cognitive impairment and oxidative stress in a rat model. Biomolecules 2025;15(5):710. Available from https://doi.org/10.3390/biom15050710.

Fan X, Li D, Zhang Y, Green TA. Differential phosphoproteome regulation of nucleus accumbens in environmentally enriched and isolated rats in response to acute stress. PLoS One 2013;8(11). Available from https://doi.org/10.1371/journal.pone.0079893 Available from: http://www.plosone.org/article/fetchObject.action?uri=info%3Adoi%2F10.1371%2Fjournal.pone.0079893&representation=PDF.

Gan XT, Karmazyn M. Cardioprotection by ginseng: experimental and clinical evidence and underlying mechanisms. Canadian Journal of Physiology and Pharmacology 2018;96(9):859–68. Available from: https://doi.org/10.1139/cjpp-2018-0192, www.nrc.ca/cgi-bin/cisti/journals/rp/rp_desy_e?cjpp.

Gao Z, Lu Y, Li M, Chong Y, Hong J, Wu J, et al. Application of pan-omics technologies in research on important economic traits for ruminants. International Journal of Molecular Sciences 2024;25(17):9271. Available from https://doi.org/10.3390/ijms25179271.

Gao Y, Wu S. Thermal and oxidation stability of functional oleogels formed by edible wax/ starch and *Schisandra chinensis* oil. Food & Function 2019;10:8056–68.. Available from https://doi.org/10.1039/c9fo01727e "Menke A. Glucocorticoid signalling in stress-related psychiatric disorders uncovered by dexamethasone-induced gene expression. ScienceOpen Res. 2017; Available from: http://doi.org/10.14293/s2199-1006.1.sor-uncat.eitgtn.v1," no date).

George MY, Abdel Mageed SS, Mansour DE, Fawzi SF. The cortisol axis and psychiatric disorders: an updated review. Pharmacological Reports 2025;77(6):1573–99.. Available from https://doi.org/10.1007/s43440-025-00782-x Available from: https://www.springer.com/journal/43440.

Ghahremaninejad F, Joharchi MR, Memariani F, Behroozian M. Astragalus nigriceps (Fabaceae, Papilionoideae): an emended description with new data on distribution and conservation status. Phytotaxa 2022;536(3):243–51. Available from https://doi.org/10.11646/phytotaxa.536.3.4.

Ghahremaninejad F, Joharchi MR, Memariani F. *Astragalus durandianus* (sect. Trachycercis) as an Iranian endemic species: emended description, distribution, and conservation. Phytotaxa 2022;550(1):32–44.. Available from https://doi.org/10.11646/phytotaxa.550.1.2 Available from: https://www.mapress.com/pt/article/view/phytotaxa.550.1.2.

Ghiorghita G, Maftei D, Maftei D, Rhodiola rosea l. – a valuable plant for traditional and modern medicine. Semanticscholar.org, 2016.

Guan Y, Tang G, Li L, Shu J, Zhao Y, Huang L, et al. Herbal medicine and gut microbiota: exploring untapped therapeutic potential in neurodegenerative disease management. Archives of Pharmacal Research 2024;47(2):146–64. Available from https://doi.org/10.1007/s12272-023-01484-9.

Haber M, Czachor A, Kula P, Juśkiewicz A, Grelewicz O, Kucy N, et al. Ashwagandha as an adaptogen: its influence on sleep patterns, stress response, and anxiety in modern life. Journal of Education, Health and Sport 2024;68:55327. Available from https://doi.org/10.12775/jehs.2024.68.55327.

Hazarika N. Quality standardization and statistical optimization of polyherbal blends targeting rheumatoid arthritis. International Journal of Environmental Sciences 2025:383–91. Available from https://doi.org/10.64252/3e37wx94.

Heghes SC, Vostinaru O, Mogosan C, Miere D, Iuga CA, Filip L. Safety profile of nutraceuticals rich in coumarins: an update. Frontiers in Pharmacology 2022;13. Available from https://doi.org/10.3389/fphar.2022.803338 Available from: http://www.frontiersin.org/Pharmacology.

Hoho OH. *Rhodiola rosea*. Monograph. Alternative Medicine Review: A Journal of Clinical Therapeutic 2002;7:421–3. Panossian et al., 2021.

Hoy MB. Firefox add-ons for medical reference. Medical Reference Services Quarterly 2010;29:269–74.. Available from https://doi.org/10.1080/02763869.2010.494514 Available from:.

Husna AB, Rahmi AS, Putri NINNI. Exploring the root causes of burnout syndrome among college students: a systematic literature review of contributing factors. Jurnal Promkes 2025;13(SI1):246–60. Available from https://doi.org/10.20473/jpk.v13.isi1.2025.246-260.

Inaba Y, Iwamoto S, Nakayama K. Genome-wide DNA methylation status of Mongolians exhibits signs of cellular stress response related to their nomadic lifestyle. Journal of

Physiological Anthropology 2022;41(1). Available from https://doi.org/10.1186/s40101-022-00305-0.

Jadhav MP. A review: study of Tulsi as an immune booster. International Journal for Research in Applied Science & Engineering Technology 2025;13:771–5. Panossian et al., 2021 http://doi.org/10.22214/ijraset.2025.70198.

Jiang L, Yang D, Zhang Z, Xu L, Jiang Q, Tong Y, et al. Elucidating the role of *Rhodiola rosea* L. in sepsis-induced acute lung injury via network pharmacology: emphasis on inflammatory response, oxidative stress, and the PI3K-AKT pathway. Pharmaceutical Biology 2024;62(1):272–84. Available from https://doi.org/10.1080/13880209.2024.2319117.

Jia L, Qian K. An Evidence-based Perspective of Panax Ginseng (Asian Ginseng) and Panax Quinquefolius (American Ginseng) as a Preventing or Supplementary Therapy for Cancer Patients. Springer Science and Business Media LLC; 2011. p. 85–96. Available from 10.1007/978-94-007-0526-5_4.

Jorquera Gutiérrez R, Herrera Gallardo F. Salud mental en funcionarios de una universidad chilena: desafíos en el contexto de la COVID-19. Revista Digital de Investigación en Docencia Universitaria 2020;14:1310. Available from https://doi.org/10.19083/ridu.2020.1310. Panossian and Efferth, 2022.

Jácome C, Seixas A, Serrão C, Teixeira A, Castro L, Duarte I. Burnout in Portuguese physiotherapists during the COVID-19 pandemic. Physiotherapy Research International: The Journal for Researchers and Clinicians in Physical Therapy 2021;26:e1915. Available from https://doi.org/10.1002/pri.1915. Santhanu and Senthil, 2021.

Kalaivani AM, Shamshalniha S, Anbu N. Standardization of Gowthamar chooranam – a polyherbal siddha formulations in the management of swasakasam (bronchial asthma). Journal of Pharmaceutical Research International 2023;35(25):47–54. Available from https://doi.org/10.9734/jpri/2023/v35i257432.

Kanakadurgadevi N. Holy basil – holistic lifestyle management. Pharmazine 2010:10–1.

Karole S, Shrivastava S, Thomas S, Soni B, Khan S, Dubey J, et al. Polyherbal formulation concept for synergic action: a review. Journal of Drug Delivery and Therapeutics 2019;9(1-s):453–66. Available from https://doi.org/10.22270/jddt.v9i1-s.2339.

Karttunen M, Sneck S, Jokelainen J, Männikkö N, Elo S. Safety checks, monitoring and documentation in medication process in long-term elderly care–Nurses' subjective perceptions. Journal of Nursing Education and Practice 2019;9(8):26. Available from https://doi.org/10.5430/jnep.v9n8p26.

Kaushik A, Rathia S, Thakur P, Misra S, Prasad K. Targeting brain health with medhya rasayana: natural approach to mental disorders. International Ayurvedic Medical Journal 2025;13(08):2351–63. Available from https://doi.org/10.46607/iamj3913082025.

Kholmogorova AB, Petrikov SS, Suroyegina AY, Mikita OY, Rakhmanina AA, Roy AP. Burnout and its factors among healthcare workers caring for patients with COVID-19 across different stages of the pandemic. Russian Sklifosovsky Journal of Emergency Medical Care 2020;9:321–37. Available from https://doi.org/10.23934/2223-9022-2020-9-3-321-337. Santhanu and Senthil, 2021.

Knezevic E, Nenic K, Milanovic V, Knezevic NN. The role of cortisol in chronic stress, neurodegenerative diseases, and psychological disorders. Cells 2023;12. Available from https://doi.org/10.3390/cells12232726 "Cui L, Zheng F, Zhang M, Wang Z, Meng X, Dong J, et al. Selenium suppressed LPS-induced oxidative stress in bovine endometrial stromal cells under a high-cortisol background via the Nrf2 pathway. J Anim Sci. 2024;102. Available from: http://doi.org/10.1093/jas/skae260," no date.

Kunicki MA, Amaya Hernández LC, Davis KL, Bacchetta R, Roncarolo M-G. Identity and diversity of human peripheral Th and T regulatory cells defined by single-cell mass cytometry. Journal of Immunology 2018;200:336–46.. Available from https://doi.org/10.4049/jimmunol.1701025 Available from:.

Lee G, Joo JC, Choi BY, Lindroth AM, Park SJ, Park YJ. Neuroprotective effects of *Paeonia lactiflora* extract against cell death of dopaminergic SH-SY5Y cells is mediated by epigenetic modulation. BMC Complementary and Alternative Medicine 2016;16(1). Available from https://doi.org/10.1186/s12906-016-1205-y Available from: http://www.biomedcentral.com/bmccomplementalternmed/.

Limanaqi F, Biagioni F, Busceti CL, Polzella M, Fabrizi C, Fornai F. Potential antidepressant effects of *Scutellaria baicalensis, Hericium erinaceus* and *Rhodiola rosea*. Antioxidants 2020;9(3):234. Available from https://doi.org/10.3390/antiox9030234.

Lopresti AL, Smith SJ, Drummond PD. Modulation of the hypothalamic–pituitary–adrenal (HPA) axis by plants and phytonutrients: a systematic review of human trials. Nutritional Neuroscience 2022;25(8):1704–30.. Available from https://doi.org/10.1080/1028415X.2021.1892253 Available from: http://www.tandfonline.com/loi/ynns20#.VvukQLdf1Hh.

Machín RP, Florido M, Chirino-Godoy R, López-Rios L. Adaptogenic botanicals with emphasis on *Rhodiola rosea* and *Withania somnifera*. European Journal of Medicinal Plants 2023;34(11):20–39. Available from https://doi.org/10.9734/ejmp/2023/v34i111168.

Malec K. The impact of ashwagandha on sleep quality, anxiety reduction, and stress lowering: comparative analysis of available studies. Journal of Education, Health and Sport 2024;52:205–14. Available from https://doi.org/10.12775/jehs.2024.52.015.

Markowitz JS, PharmD, DeVane ACL. PharmD. The emerging recognition of herb-drug interactions with a focus on St. John's wort (*Hypericum perforatum*). Psychopharmacol Bull 2025;35:53–64.. Available from https://doi.org/10.64719/pb.4270 Available from:.

Mazza F, Catzola A, Catalano M, De Lucia A. Liquorice as a possible cause of arterial hypertension. Description of a clinical case. Minerva Cardioangiol [Internet] 1984;32:463–5. Panossian, 2017, 2003.

Meena AK, Singh A, Sharma P, Khanduri S, Sharma BS, Rekha P, et al. Harnessing HPTLC and HPLC techniques for identification and quantitative estimation of bioactive constituents in Ayush D formulation and its three main ingredients viz: *Syzygium cumini, Withania coagulans* and *Caesalpinia bonducella*. Research Journal of Pharmacy and Technology 2025;18(5):2227–36. Available from: https://doi.org/10.52711/0974-360X.2025.00319, www.rjptonline.org.

Misiak B. Stress, allostatic load, and psychosis: one step forward in research but where to go next? Frontiers in Psychiatry 2020;10. Available from https://doi.org/10.3389/fpsyt.2019.00937.

Murray MT. *Eleutherococcus senticosus* (Siberian ginseng). Textbook of Natural Medicine. Elsevier; 2020. p. 574–7.

Murthy MRV, Ranjekar PK, Ramassamy C, Deshpande M. Scientific basis for the use of Indian ayurvedic medicinal plants in the treatment of neurodegenerative disorders: 1. Ashwagandha. Central Nervous System Agents in Medicinal Chemistry 2010;10(3):238–46.. Available from https://doi.org/10.2174/1871524911006030238 Available from: http://www.eurekaselect.com/613/journal/central-nervous-system-agents-medicinal-chemistry.

Natalie VP, Lay DS, Luisa J. Potency of Asian ginseng (*Panax ginseng*) and turmeric (*Curcumin*) extract compound in promoting cerebral perfusion in vascular dementia: a narrative review. International Journal of Research and Review 2022;9(2):40–52. Available from https://doi.org/10.52403/ijrr.20220208.

Newcomer K, Goodman A, Fenstermacher D. Colorectal cancer clinical trial summit: devising solutions and interventions. Journal of Clinical Oncology 2023;41(16_suppl):e15603. Available from https://doi.org/10.1200/jco.2023.41.16_suppl.e15603.

Palleria C, Roberti R, Iannone LF, Tallarico M, Barbieri MA, Vero A, et al. Clinically relevant drug interactions between statins and antidepressants. Journal of Clinical Pharmacy and Therapeutics 2020;45(2):227–39.. Available from https://doi.org/10.1111/jcpt.13058 Available from: http://onlinelibrary.wiley.com/journal/10.1111/(ISSN)1365-2710.

Pandian A, Ashokkumar K, Sekar S, Sivakumar P, Selvaraj KSV, Karthik M, et al. Botany and ethnopharmacological potential of Ashwagandha. Current Opinion in Crop Science 2020;1:35–40.. Available from https://doi.org/10.62773/jcocs.v1i1.16 Available from:.

Panossian A. Understanding adaptogenic activity: specificity of the pharmacological action of adaptogens and other phytochemicals. Annals of the New York Academy of Sciences 2017;1401(1):49–64. Available from https://doi.org/10.1111/nyas.13399.

Panossian A, Efferth T. Network pharmacology of adaptogens in the assessment of their pleiotropic therapeutic activity. Pharmaceuticals 2022;15(9):1051. Available from https://doi.org/10.3390/ph15091051.

Panossian AG, Efferth T, Shikov AN, Pozharitskaya ON, Kuchta K, Mukherjee PK, et al. Evolution of the adaptogenic concept from traditional use to medical systems: pharmacology of stress- and aging-related diseases. Medicinal Research Reviews 2021;41(1):630–703. Available from: https://doi.org/10.1002/med.21743, http://onlinelibrary.wiley.com/journal/10.1002/(ISSN)1098-1128.

Panossian A, Wikman G. Evidence-based efficacy of adaptogens in fatigue, and molecular mechanisms related to their stress-protective activity. Current Clinical Pharmacology 2009;4(3):198–219. Available from https://doi.org/10.2174/157488409789375311.

Panossian AG. Adaptogens: tonic herbs for fatigue and stress. Alternative and Complementary Therapies 2003;9:327–31. Machín et al., 2023 http://doi.org/10.1089/107628003322658610.

Papi S, Barmala A, Amiri Z, Vakili-Basir A, Cheraghi M, Nejadsadeghi E. Mediating roles of stress, anxiety, and depression in the relationship between constipation and sleep quality among the elderly: structural equation modeling (SEM). Sleep Science 2023;16(1):1–6.. Available from https://doi.org/10.1055/S-0043-1767750 Available from: http://www.thieme-connect.de/DOI/DOI?10.1055/s-0043-1767750.

Pincus T, Schmukler J, Li T, Block J. AB1373 are fibromyalgia and/or depression the most common diagnoses seen in routine rheumatology care, albeit as secondary diagnoses, but which may complicate clinical assessment, prognosis and management? Annals of the Rheumatic Diseases 2023;82:1916. Available from https://doi.org/10.1136/annrheumdis-2023-eular.3171.

Polick CS, Harris-Gersten ML, Dennis PA, Noonan D, Hastings SN, Calhoun PS, et al. Allostatic load, morbidity, and mortality among older adults: a multi-wave analysis from the national health and aging trends study. Journal of Applied Gerontology 2024;43(8):1052–9.. Available from https://doi.org/10.1177/07334648241230010 Available from: https://journals.sagepub.com/home/JAG.

Rybnikář M, Šmejkal K, Žemlička M. Schisandra chinensis and its phytotherapeutical applications. Ceska a Slovenska Farmacie: Casopis Ceske Farmaceuticke Spolecnosti a

Slovenske Farmaceuticke Spolecnosti 2019;68:95–118.. Available from https://doi.org/10.36290/csf.2019.012 Available from:.

Sahoo S, Naskar C, Singh A, Rijal R, Mehra A, Grover S. Sensory deprivation and psychiatric disorders: association, assessment and management strategies. Indian Journal of Psychological Medicine 2021;44(5):436–44.. Available from https://doi.org/10.1177/02537176211033920 Available from: https://journals.sagepub.com/home/szj.

Sakti AS, Nugroho GA, Wediasari F, Khan MA, Sibadu MSA. Herb–drug interactions between common Indonesian medicinal plants and cardiovascular medications: a narrative review. Journal of Health Sciences, Medicine, Biotechnology, and Pharmaceutical Research 2025;1(2):81–97. Available from https://doi.org/10.63142/jhsmbp.v1i2.263.

Santhanu K, Senthil K. Therapeutic potential of *Withania somnifera* (Linn) Dunal (Ashwagandha) in historical perspective and pharmacological evidence. Annals of Ayurvedic Medicine 2021;10(2):135. Available from https://doi.org/10.5455/aam.122229.

Schmidt M, Thomsen M, Kelber O, Kraft K. Myths and facts in herbal medicines: *Eleutherococcus senticosus* (Siberian ginseng) and its contraindication in hypertensive patients. Botanics: Targets and Therapy 2014:27. Available from https://doi.org/10.2147/BTAT.S60734.

Schmid B, Schulz SM, Schuler M, Göpfert D, Hein G, Heuschmann P, et al. Impaired psychological well-being of healthcare workers in a German department of anesthesiology is independent of immediate SARS-CoV-2 exposure – a longitudinal observational study. GMS German Medical Science 2021;19. Available from https://doi.org/10.3205/000298 Available from: https://www.egms.de/static/pdf/journals/gms/2021-19/000298.pdf.

Scudamore O, Ciossek T. Increased oxidative stress exacerbates α-synuclein aggregation in vivo. Journal of Neuropathology and Experimental Neurology 2018;77:443–53.. Available from https://doi.org/10.1093/jnen/nly024 Available from:.

Shaikh AS, Thomas AB, Chitlange SS. Herb–drug interaction studies of herbs used in treatment of cardiovascular disorders—a narrative review of preclinical and clinical studies. Phytotherapy Research 2020;34(5):1008–26.. Available from https://doi.org/10.1002/ptr.6585 Available from: http://onlinelibrary.wiley.com/journal/10.1002/(ISSN)1099-1573.

Sharma N, Cheemalapati VN, Mangal AK, Srikanth N, Murthy SN. Evaluation of the best procurement time for Ashwagandha [*Withania somnifera* (L.) Dunal] root by liquid chromatographic and pharmacognostical studies. Journal of Drug Research in Ayurvedic Sciences 2018;3(4):206–13. Available from https://doi.org/10.5005/jdras-10059-0053.

Shinde S, Balasubramaniam AK, Mulay V, Saste G, Girme A, Hingorani L. Recent advancements in extraction techniques of Ashwagandha (*Withania somnifera*) with insights on phytochemicals, structural significance, pharmacology, and current trends in food applications. ACS Omega 2023;8(44):40982–1003. Available from: https://doi.org/10.1021/acsomega.3c03491, pubs.acs.org/journal/acsodf.

Singh BR, Chandra M, Agri H, Karthikeyan R, Yadav A, Jayakumar V. The assessment of good and bad bacteria in holy basil (*Ocimum sanctum*) leaves. Infectious Diseases Research 2023;4(4):17. Available from https://doi.org/10.53388/idr2023017.

Singh BR, Kumar A, Agrawal RK, Karthikeyan R, Sinha DK, OR V, et al. Evaluation of Fifatrol, Septilin, giloy (Guduchi, *Tinospora cordifolia*) and oils of guggul (*Balsamodendron mukul*) and holy basil (Tulsi, *Ocimum sanctum*) for their antimicrobial potential. Acta Scientific Microbiology 2022:31–41. Available from https://doi.org/10.31080/asmi.2022.05.1149.

Siwek M, Woroń J, Wrzosek A, Gupało J, Chrobak AA. Harder, better, faster, stronger? Retrospective chart review of adverse events of interactions between adaptogens and antidepressant drugs. Frontiers in Pharmacology 2023;14. Available from https://doi.org/10.3389/fphar.2023.1271776 Available from: http://www.frontiersin.org/Pharmacology.

Soejarto DD, Farnsworth NR. The correct name for Siberian ginseng. The Botanical Museum Leaflets, Harvard University 1978;26:339–43.. Available from https://doi.org/10.5962/p.295218 Available from:.

Song G, Liu X, Lu Z, Guan J, Chen X, Li Y, et al. Relationship between stress hyperglycaemic ratio (SHR) and critical illness: a systematic review. Cardiovascular Diabetology 2025;24(1). Available from https://doi.org/10.1186/s12933-025-02751-3.

Sundaram RS, Ramanathan M, Rajesh R, Satheesh B, Saravanan D. LC-MS quantification of rosmarinic acid and ursolic acid in the *Ocimum sanctum* Linn. Leaf extract (Holy basil, Tulsi). Journal of Liquid Chromatography & Related Technologies 2012;35(5):634–50. Available from https://doi.org/10.1080/10826076.2011.606583.

Tamboli FA, Jogi PR, Waghmare R, Bhosale V, Jogi P, Gavhane D, et al. Herbal medicines and drug interactions: a growing concern in healthcare. International Journal of Pharmaceutical Chemistry and Analysis 2025;12:8–15.. Available from https://doi.org/10.18231/j.ijpca.2025.002 Available from:.

Tan J, Taitz J, Nanan R, Grau G, Macia L. Dysbiotic gut microbiota-derived metabolites and their role in non-communicable diseases. International Journal of Molecular Sciences 2023;24(20):15256. Available from https://doi.org/10.3390/ijms242015256.

Todorova V, Ivanov K, Delattre C, Nalbantova V, Karcheva-Bahchevanska D, Ivanova S. Plant adaptogens—history and future perspectives. Nutrients 2021;13(8):2861. Available from https://doi.org/10.3390/nu13082861.

Ullagaddi R. Nourishing the mind: role of functional foods in stress management. South Eastern European Journal of Public Health 2025:1623–34. Available from https://doi.org/10.70135/seejph.vi.5145.

Urgessa OE, Woldesemayat AA. OMICs approaches and technologies for understanding low-high feed efficiency traits in chicken: implication to breeding. Animal Biotechnology 2023;34(8):4147–66. Available from: https://doi.org/10.1080/10495398.2023.2187404, www.tandf.co.uk/journals/titles/10495398.asp.

Verma S. Chemical constituents and pharmacological action of *Ocimum sanctum* (Indian holy basil-Tulsi). The Journal of Phytopharmacology 2016;5(5):205–7. Available from https://doi.org/10.31254/phyto.2016.5507.

Wang Y, Dai YD, Yang ZL, Guo R, Wang YB, Yang ZL, et al. Morphological and molecular phylogenetic data of the chinese medicinal fungus cordyceps liangshanensis reveal its new systematic position in the family ophiocordycipitaceae. Mycobiology 2021;49(4):297–307.. Available from https://doi.org/10.1080/12298093.2021.1923388 Available from: https://www.tandfonline.com/loi/tmyb20.

Wang Z, Zhang T, Wang B, Li S. TCM network pharmacology: new perspective integrating network target with artificial intelligence and multi-modal multi-omics technologies. Chinese Journal of Natural Medicines 2025;23(11):1425–34. Available from https://doi.org/10.1016/s1875-5364(25)60986-1.

Wheaton MG, Prikhidko A, Messner GR. Is fear of COVID-19 contagious? The effects of emotion contagion and social media use on anxiety in response to the coronavirus pandemic. Frontiers in Psychology 2021;11. Available from https://doi.org/10.3389/fpsyg.2020.567379 Available from: http://www.frontiersin.org/Psychology.

Wróbel-Biedrawa D, Podolak I. Anti-neuroinflammatory effects of adaptogens: a mini-review. Molecules 2024;29(4):866. Available from https://doi.org/10.3390/molecules29040866.

Wu SY, Wang WJ, Dou JH, Gong LK. Research progress on the protective effects of licorice-derived 18β-glycyrrhetinic acid against liver injury. Acta Pharmacologica Sinica 2021;42(1):18–26. Available from: https://doi.org/10.1038/s41401-020-0383-9, http://www.nature.com.

Xiao B. Intervention effect of online and offline physical education model on anxiety disorder of undergraduates. CNS Spectrums 2023;28(S2):S114. Available from https://doi.org/10.1017/s1092852923005400.

Yang L, Wang H, Zhu Z, Yang Y, Xiong Y, Cui X, et al. Network pharmacology-driven sustainability: AI and multi-omics synergy for drug discovery in traditional Chinese medicine. Pharmaceuticals 2025;18(7):1074. Available from https://doi.org/10.3390/ph18071074.

Zhang T, Kishore V, Wu F, Weinberger KQ, Artzi Y. BERTScore: evaluating text generation with BERT. ArXiv, abs. 1904.

Zhang W, Zeng Y, Jiao M, Ye C, Li Y, Liu C, et al. Integration of high-throughput omics technologies in medicinal plant research: the new era of natural drug discovery. Frontiers in Plant Science 2023;14. Available from https://doi.org/10.3389/fpls.2023.1073848.

Zhao SL, Liu D, Ding LQ, Liu GK, Yao T, Wu LL, et al. *Schisandra chinensis* lignans improve insulin resistance by targeting TLR4 and activating IRS-1/PI3K/AKT and NF-κB signaling pathways. International Immunopharmacology 2024;142 Available from: https://doi.org/10.1016/j.intimp.2024.113069, https://www.sciencedirect.com/science/journal/15675769.

CHAPTER

Botanicals for immune system support

4

Manne Anupama Ammulu[1], Kanaka Durga Devi Nelluri[2], Minakshi Kommoju[2], Radha Madhavi Baluguri[2] and Surendra Kundurthi[3]

[1]*Department of Mechanical Engineering, NIT, Warangal, Telangana, India,*
[2]*Department of Pharmaceutics and Biotechnology, KVSR Siddhartha College of Pharmaceutical Sciences, Vijayawada, Andhra Pradesh, India,*
[3]*Medical School, University of Birmingham, Edgbaston, Birmingham, United Kingdom*

4.1 Introduction

The last several decades have seen extensive research on plants due to their therapeutic properties and profound influence on the human immune system. Compared to allopathic treatments, plants are among the best sources of traditional medicine, as they are less costly and nontoxic for treating a variety of illnesses, particularly those involving altered immunological processes. The use of botanicals in immune defense has long been supported by historical records of Indian Ayurvedic medicine and traditional Chinese medicine, which attest to their effectiveness and safety. Flavonoids, terpenoids, polysaccharides, lactones, alkaloids, glycosides, and saponins are among the many phytochemicals found in plants that have been shown to have immunomodulatory qualities. With this, the increased demand over the last couple of decades has greatly increased the use of phytomedicines as immunostimulants and/or immunosuppressants. It has also led to their application as prophylactic therapy for several diseases. Thus plant-based immunomodulators, which are less harmful and inexpensive, are gaining importance as an alternative to conventional therapies (Zebeaman et al., 2023).

4.1.1 Overview of the immune system

The initial line of defense that responds quickly to typical pathogen patterns is the innate immune system, also known as nonspecific immunity. It is composed of cells, including dendritic cells, neutrophils, natural killer (NK) cells, and macrophages, as well as physical barriers such as the skin and mucous membranes. To swiftly eradicate infections from the body, they combine with soluble components, including complement proteins and cytokines. In contrast, the adaptive immune system is antigen-specific and uses immunological memory to give long-term immunity. T cells and B cells, or lymphocytes, are the primary cells that mediate the adaptive immune response. T lymphocytes mediate cell-mediated immunity by destroying infected cells,

Healing Botanicals. DOI: https://doi.org/10.1016/B978-0-443-44295-7.00011-2

whereas B cells contribute to humoral immunity by secreting antibodies. Both mechanisms are complementary in ensuring that the body has sufficient defense and immune protection (Hooda et al., 2024; Kadiyska et al., 2023).

4.1.2 Significance of immune modulation in disease prevention

Since a healthy immune response is essential for preserving health, immunological modulation is required for both disease prevention and therapy. By identifying and eliminating dangerous germs and poisons, the immune system defends the organism. Due to compromised immune surveillance, immunological hypofunction raises the risk of infections and malignancies. Conversely, immune hyperfunction leads to autoimmune and chronic inflammatory diseases. Immune modulation is required to regulate the immune response and achieve homeostasis. Immunosuppression is used to control an overactive immune response, whereas immunostimulation is used to strengthen a weakened immune system. For immunological regulation, which is essential for preventing illness and improving treatment results, botanicals provide affordable and secure substitutes (Swaroop et al., 2022).

4.1.3 Historical use of botanicals in immune health

Botanicals have been used for thousands of years to modulate the immune system. Conventional medical systems, including Ayurveda and traditional Chinese medicine, have established frameworks to describe the therapeutic benefits of plants. The immune-boosting qualities of ginseng, astragalus, and reishi mushrooms have been utilized in traditional Chinese medicine. Ashwagandha, guduchi, and turmeric have been used in Ayurvedic medicine as rasayanas for their immune-enhancing properties. The use of these botanicals for centuries has demonstrated their efficacy in treating infections, reducing inflammation, and supporting recovery from diseases. Modern science has isolated phytochemicals like flavonoids, terpenoids, polysaccharides, and alkaloids for their immune-enhancing properties. The use of botanicals as immunomodulators has been justified by their low cost and safety (Shukla et al., 2022; Pathak et al., 2022).

4.2 Mechanisms of immune modulation by botanicals

Various botanicals have been employed to modulate the immune system through diverse mechanisms that involve complex interactions with cellular and molecular targets. These also include a wide array of bioactive compounds from plants that can act directly on immune cells, intracellular signaling pathways, and the production of immune mediators. Their mechanisms of action often involve antiinflammatory and antioxidant effects, which are essential for preserving immunological homeostasis and preventing tissue damage. Botanicals can modulate cytokine expression and activate distinct subpopulations of immune cells, thereby coordinating an effective immune response against pathogens and other potentially harmful agents. Through this interplay of mechanisms, botanicals have the potential to be powerful immunomodulators, with therapeutic potential for a wide array of immune-related diseases (Sharma et al., 2017).

4.2.1 Molecular targets and signaling pathways

While the major mechanism involved in the immunomodulatory roles of botanicals involves interaction with specific molecular targets and subsequent modulation of key intracellular signaling pathways within the immune cells, they may eventually lead to an activation or suppression of immune responses depending on various factors that may include the type of botanical compound, concentration, and physiological context. Targeting various immunological mediators and transcription factors, immunomodulators have been shown to modulate a variety of cellular processes, such as apoptosis, lipid and free radical peroxidation, and protein synthesis (Singh et al., 2016).

One of the most researched molecular targets is Nuclear Factor Kappa B (NF-κB). It is a crucial transcription factor that controls genes involved in immune responses, cellular inflammation, and cell viability. Numerous inflammatory conditions, including cancer, diabetes, atherosclerosis, and autoimmune illnesses, have been linked to NF-κB activation. According to reports, most herbs inhibit NF-κB activation. For example, curcumin, the naturally occurring yellow substance found in *Curcuma longa* (turmeric), significantly reduces inflammation by inhibiting NF-κB production. It works by reducing NF-κB activation and inhibiting protein kinase C, an upstream activator of NF-κB.

Additionally, it decreases NF-κB's DNA binding activity, which is essential for its transactivation function. Likewise, andrographolide, a prominent diterpenoid lactone found in *Andrographis paniculata*, appears to suppress proinflammatory proteins by blocking NF-κB's ability to bind DNA. Preventing cyclooxygenase (COX) enzymes, which normally modulate NF-κB, may prevent activation of the alternative NF-κB pathway, providing an additional antiinflammatory effect.

The signal transducer and activator of transcription 3 (STAT3) pathway is another mechanism by which botanicals work. The transcription factor STAT3 plays a critical role in immunological responses to inflammation and cancer, as well as cell growth and survival. Curcumin acts biphasically on STAT3 modulation: at high concentrations, it inhibits STAT3 binding to DNA and STAT3 phosphorylation, whereas at low concentrations it increases STAT3 phosphorylation, demonstrating a concentration-dependent effect on its antiinflammatory activity. Andrographolide also downmodulates the STAT3 pathway by affecting the signaling of cytokines 1 and 3, adding to its anticancer and antiinflammatory activities.

Mechanistically, they exert their effect by modulating several enzymes involved in inflammation. These include the potent inflammatory mediators cyclooxygenase-2

(COX-2) and inducible nitric oxide synthase (, which synthesize prostaglandins and nitric oxide, respectively. Curcumin was shown to lower COX-2 and iNOS synthesis, thereby reducing inflammation. By preventing Lipopolysaccharide (LPS)-induced expression in macrophage cell lines, andrographolide inhibited COX-2 and iNOS expression in microglia and macrophages. The Toll-like receptor 4 pathway, a crucial component of innate immunity that may recognize pathogen-associated molecular patterns, is another target. Andrographolide demonstrated its role in regulating innate immune responses by acting more effectively on pathways such as the Toll-Like Receptor 4 (TLR4)/NFκB pathway in insulinoma cancers (Tharakan et al., 2021).

More specifically, Activator Protein 1 (AP-1), another transcription factor that controls inflammatory and immunological responses, is thought to be impacted by botanicals. In fact, it has been suggested that curcumin's antiinflammatory properties stem from reduced AP-1 binding. Another mechanism is the modulation of Protein Kinase C (PKC). PKC mediates cell proliferation and survival and also activates NF-κB. Curcumin inhibits PKC, thereby weakening NF-κB activation.

Besides these pathways, the general mode of action of phytocompounds broadly involves modulation of molecular targets in immune cell signaling and the intracellular release of various soluble substances, such as transcription factors and interleukins. These broad molecular interactions underlie the complex, multi-targeted nature of botanical immunomodulation, thereby eliciting comprehensive effects on the immune system (Shah et al., 2007) (see Fig. 4.1).

4.2.2 Antioxidant and antiinflammatory mechanisms

Most botanicals possess important properties related to antioxidant and antiinflammatory activities, which, in fact, form the very basis for their immunomodulatory effects. It follows that such activities are significant for immune homeostasis, protection against oxidative cellular damage, and the resolution of inflammation, all of which help protect against chronic diseases. Indeed, antiinflammatory, antioxidant, and anticancer properties are associated with immunomodulatory mechanisms.

4.2.2.1 Antioxidant mechanisms

Reactive oxygen species production and the body's neutralizing capacity are out of balance, which leads to oxidative stress. In addition to causing cellular damage, this may also be connected to long-term inflammation and immunological compromise. This is mitigated by the abundance of antioxidants in botanicals. Phenols and flavonoids are two of the main antioxidants found in plants. A higher amount of total flavonoids with better radical scavenging activity has been reported in an aqueous extract of *A. paniculata* compared to an ethanol extract. However, the latter contained a higher amount of phenols. Because it scavenges free radicals and inhibits lipid peroxidation, andrographolide and its analogs from *A. paniculata* have been shown to have significant antioxidant activity. The latter is how free radicals harm cell membranes. According to experimental research, extracts from

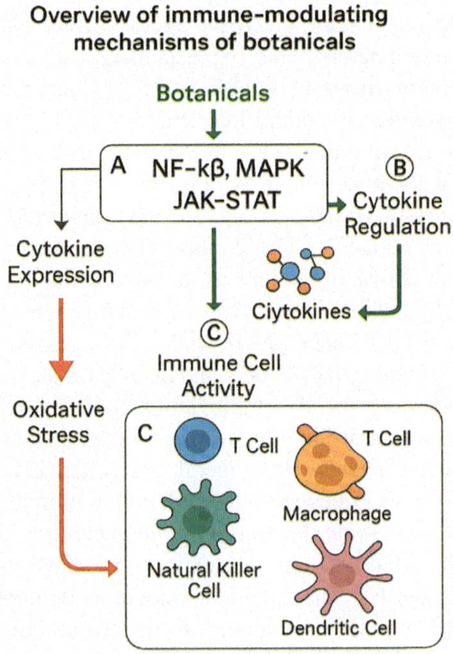

FIGURE 4.1 Overview of immune-modulating mechanisms of botanicals.

A. paniculata significantly increase levels of endogenous antioxidant enzymes, such as glutathione-S-transferase, catalase, and superoxide dismutase, while lowering oxidative damage markers, such as lactate dehydrogenase and reactive substances of thio barbituric acid (Aggarwal and Harikumar, 2009). Other contributors to antioxidant defense include curcumin, derived from *C. longa*. The antiinflammatory actions of this compound are implicit in its function of modulating pathways that reduce oxidative stress, as evidenced by its wide-ranging effects on cellular processes. Quercetin, ellagic acid, magniferin, and withanolide were also reported to reduce immune-related diseases through their antioxidant, antiinflammatory, and immunomodulatory actions. Most phytochemical antioxidants work either by directly neutralizing free radicals or by chelating metal ions; some upregulate the body's own antioxidant defense systems (Ren et al., 2021).

4.2.2.2 *Antiinflammatory mechanisms*

Although inflammation is a vital defense mechanism, prolonged, unchecked inflammation can cause tissue damage and a range of illnesses. Botanicals modulate inflammatory processes through various mechanisms. Curcumin suppresses NO, COX-2, NFκB, iNOS, and lipoxygenase production as part of its antiinflammatory activity. It inhibits the production of proinflammatory cytokines by activated macrophages, including Tumor Necrosis Factor Alpha (TNF-α) and Interferon Gamma (IFN-γ). Inflammatory gene expression is lost, and NF-κB activation is weakened when protein kinase C is inhibited. It prevents the release of additional proinflammatory cytokines, including Interleukin 1 (IL-1), Interleukin 6 (IL-6), and Interleukin 12 (IL-12), produced by monocytes, macrophages, dendritic cells, activated immune cells, and splenic lymphocytes in response to LPS or Phorbol 12-Myristate 13-Acetate (PMA) stimulation. By lowering the levels of IL-12, TNF-α, NO, prostaglandin E2, COX-2, and iNOS in microglia and macrophages, andrographolide, the active ingredient of *A. paniculata*, has also been shown to have potent antiinflammatory properties. In macrophage cell lines, it has been shown to suppress LPS-induced iNOS and COX-2 expression primarily. Neo-andrographolide, iso-andrographolide, and skullcap flavone are other active principles of *A. paniculata*, mostly employed for their antiinflammatory properties. They block the production of inflammatory cytokines such as IL-6, NO, and IL-1β in activated LPS-macrophages. In general, the alternative NF-κB pathway is inactivated when COX enzymes are inhibited, which ultimately reduces inflammation and its related symptoms. Numerous plant secondary metabolites, including flavonoids, tannins, alkaloids, and essential oils, have been linked to these antiinflammatory properties. The botanicals modulate oxidative stress and dampen excessive inflammatory responses, aiming to restore immune balance and protect the host from disease (Jin et al., 2014) (see Fig. 4.2).

4.2.3 Cytokine regulation and immune cell activation

By modulating cytokine production and action, botanicals can have strong effects on the immune system and thus directly influence the activation and function of various immune cell populations. This depth of involvement in immune signaling and cellular responses is indeed at the very core of their immunomodulatory capability.

4.2.3.1 Cytokine regulation

Low-molecular-weight proteins called cytokines mediate immune responses and serve as intercellular messengers. Their balanced production is required for effective immunity. Several botanicals have been demonstrated to modulate cytokine profiles.

Curcumin from *C. longa* has a profound effect on cytokine modulation. The pro-inflammatory cytokines TNF-α, IL-1, IL-6, and IL-12, which are generated by activated immune cells such as monocytes, macrophages, dendritic cells, and splenic lymphocytes, are inhibited when LPS or PMA induces their production. On the other hand, unrestored murine macrophages and splenocytes showed increased levels of IL-2, IL-6, IL-10, IL-12, IFN-γ, NO, MCP-1, and TNF-α in response to pure polysaccharides found in curcumin, which are not dependent on curcuminoids.

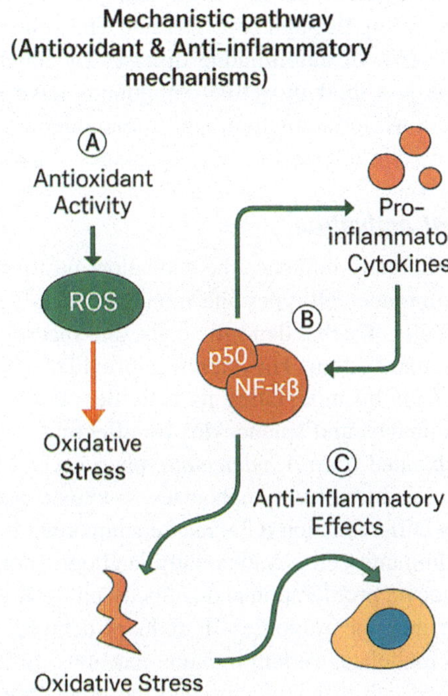

FIGURE 4.2 Antioxidant and antiinflammatory pathways.

Hence, different components of *C. longa* may cause different or opposite effects on cytokine production. These observations undeniably describe the complexity with which botanical extracts can execute their active phytochemical actions. Curcumin was also observed to perform a dual task in modulating STAT3, a cytokine signaling protein, whereby high concentrations repressed the expression of IL-1β and TNF-α. In contrast, low concentrations enhanced STAT3 phosphorylation, demonstrating concentration-dependent activity in cytokine-related pathways (Zielińska et al., 2021).

Andrographolide from *A. paniculata* has also been reported to have a potent activity on cytokine levels, reducing the production of IL-12, TNF-α, and NO in microglia and macrophages. Its treatment has also been suggested to improve the structure of TNF-α by lymphocytes; other studies have presented that it increases the production of various cytokines by human peripheral blood lymphocytes. Other substances that decrease inflammatory cytokines, including IL-6, NO, and IL-1β in activated LPS-macrophages, include neo-andrographolide, iso-andrographolide, and skullcapflavone. In general, plant chemicals have been shown to influence the release of immunoglobulins (Ig) and soluble factors from immune cells, including IL-2, IL-4, IL-6, IL-12, and IL-17. Phytochemicals tend to increase levels of a range of cytokines, acting through both Th1 and Th2 subsets, and have become useful for reducing the risk of autoimmune diseases by rebalancing the immune response. Plant extracts and their bioactive compounds have played an important role in managing infections, inflammation, and immunodeficiencies by modulating several cellular activities via interleukins and cytokines (Mondal et al., 2011).

4.2.3.2 *Immune cell activation*

Beyond that, botanicals directly influence the activation, proliferation, and functional capabilities of various immune cell types and modulate overall immune responses.

It interacts with B cells, T cells, dendritic cells, and macrophages, all important components of the immune system. This therefore provides a basis for it to exert its immunomodulatory effects by modulating its activities. For instance, it influences the functions of macrophages and splenocytes by affecting cytokine production.

Andrographolide, obtained from *A. paniculata*, plays several roles in immune cell activation. For instance, it increases lymphocytes' cytotoxic activity and raises the expression of Cluster of Differentiation (CD) markers important for the recognition and proper functioning of immune cells. Additionally, andrographolide increases human peripheral blood lymphocyte proliferation and antibody production, demonstrating that it strengthens adaptive immune responses. It reduces delayed-type hypersensitivity responses; thus, it also modulates certain immune reactions. In its anticancer studies, andrographolide increased cell differentiation and the production of IL-2 and IFN-γ, which are believed to be very important for T cell activation and antitumor immunity. It has also been reported to enhance CD4+ T cell responses against HIV-1 infection; hence, it has potential for antiviral immunity (Barreca et al., 2023). In general, a variety of immune cell functions, including those of dendritic cells, lymphocytes, neutrophils, monocytes, macrophages, basophils, mast cells, eosinophils, and NK cells, are modulated by plant chemicals. The majority of these target NK cells, which are crucial components of innate immunity against viral infections and malignancies, as well as phagocytic cells, including neutrophils, monocytes, macrophages, basophils, and eosinophils that produce inflammatory mediators. This is further underscored by the ability to directly target pathogens and enhance host immunity by modulating lymphocytes with plant-derived immunomodulators. General regulation of cytokines and direct activation/modulation of immune cells by botanicals are sophisticated ways of fine-tuning the immune system and, hence, are considered valued agents in immune health support against a wide range of diseases (Cherian and Vadivel, 2023).

4.3 Major bioactive phytochemicals with immunomodulatory properties

Plants are rich in various chemical compounds called phytochemicals or secondary metabolites. Most of these compounds possess considerable immunomodulatory properties. Active principles account for therapeutic efficacies observed in traditional medicine and are now under investigation against modern pharmaceutical uses. Indeed, among such phytochemicals include flavonoids, terpenoids, alkaloids, polysaccharides, lactones, glycosides, saponins, and polyphenols, which are major players in immune modulation. A basic understanding of major classes of bioactive compounds is useful for providing insights into how botanicals support immune health.

4.3.1 Flavonoids

One of the most significant groups of polyphenolic chemicals found across the kingdom of plants is flavonoids. Numerous biological activities, such as strong antioxidant, antiinflammatory, and immunomodulatory effects, have been extensively documented in the literature. Generally, they confer the bright colors of fruits, vegetables, and flowers. Flavonoids are major phytochemicals in plants such as *A. paniculata*. They have also been noted as primary antioxidant compounds in plants; aqueous extracts often exhibit higher concentrations of total flavonoids and superior radical-scavenging activity (Daley and Cordell, 2021).

The immunomodulatory effects of flavonoids are obviously multifaceted, acting directly through free radical scavenging, reducing oxidative stress, thereby limiting the injury to immune function and inflammation. Often, their antiinflammatory actions involve inhibiting pro-inflammatory enzymes and signaling pathways, including those of NF-κB and COX-2, which are critical regulators of inflammatory responses. Two of these flavonoids that interact with the immune system include quercetin and epigallocatechin gallate. Of these, quercetin's antiinflammatory, immunomodulatory, and antioxidant properties reduce the prevalence of immune-related illnesses (Vyshnavi et al., 2023) (see Table 4.1).

4.3.2 Terpenoids

Another major class of structurally varied natural products includes terpenoids or isoprenoids, which are built up from isoprene units containing five carbons. Everything from small monoterpenes to large triterpenes and diterpenes falls into this category; many possess major biological activities, including immunomodulatory effects. The main chemical constituent and the principal immunomodulatory terpenoid is andrographolide, a diterpene lactone that occurs abundantly in the leaves of *A. paniculata*. Andrographolide's antiinflammatory and immunostimulatory qualities have led to much research. It has antiviral, antibacterial, antiinflammatory, antitumor, and hepatoprotective properties via regulating both antigen-specific and nonspecific immune responses. Its actions include blocking key signaling pathways such as NF-κB and STAT3, inhibiting enzymes such as COX-2 and iNOS, and lowering levels of pro-inflammatory cytokines such as TNF-α and IL-12. The immune system is also affected by other terpenoids, such as oleanolic acid (Murphy et al., 2022; Moni et al., 2023).

Table 4.1 Immunomodulatory actions of the major phytochemical classes.

Phytochemical class	Key examples	Primary immunomodulatory actions	Mechanisms involved
Flavonoids	Quercetin, kaempferol, apigenin	Antioxidant, antiinflammatory, immune-balancing	NF-κB inhibition, ROS scavenging, cytokine suppression
Terpenoids	Curcuminoids, ginsenosides	Antiinflammatory, immune enhancement	Mitogen-activated protein kinase (MAPK) modulation, COX-2/iNOS inhibition
Alkaloids	Piperine, berberine	Immunosuppressive and antiinflammatory	Cytokine suppression (TNF-α, IL-6), NF-κB inhibition
Polysaccharides	β-glucans, arabinogalactans	Immunostimulation, macrophage and NK cell activation	TLR activation, enhanced phagocytosis
Polyphenols	Resveratrol, ellagic acid	Strong antioxidant; reduces inflammatory mediators	JAK-STAT modulation, Nuclear factor erythroid 2-related factor 2 (Nrf2) activation

4.3.3 Alkaloids

Alkaloids are a broad class of plant compounds that contain nitrogen and have pharmacologically active properties. They are one of the most important phytochemicals with substantial immunomodulatory properties. *Tinospora cordifolia* (guduchi) is an established immunomodulatory herb that contains alkaloids such as berberine, palmatine, and tinosporine. These alkaloids have been shown to enhance macrophage activity, regulate cytokine release, and activate the immune system.

Probably the most renowned immunomodulatory plant species, rich in a variety of alkaloids, is *T.cordifolia* or guduchi. Among these major alkaloids are berberine, palmatine, and tinosporine, which have been shown to contribute to the plant's therapeutic actions, including modulation of the immune response. Among the phytochemicals present in *T. cordifolia*, these alkaloids are believed to increase macrophage activity, modulate cytokine production, and improve overall immune function. Another plant considered an immunostimulant is *Glycyrrhiza glabra* (liquorice), which contains the alkaloid glycyrrhizin. The studies demonstrated that the liquorice extract enhanced immune parameters in serum and skin mucus, such as lysozyme activity, IgM levels, and complement activity, likely due to compounds such as glycyrrhizin. It also modulated the transcriptomic responses of immune-related genes, indicating a direct influence on immune cell function and gene expression (Liang et al., 2024; Das, 2022).

4.3.4 Polysaccharides

Algae, fungi, and plants all contain polysaccharides, which are complex carbohydrates made up of lengthy chains of monosaccharides. They are being increasingly recognized as important bioactive phytochemicals with strong immunomodulatory properties. Strong immunostimulatory effects that interact with multiple components of the immune system have been reported for plant and fungal polysaccharides. The most researched of these polysaccharides are β-glucans, which are present in seaweed, plants, and mushrooms. By improving phagocytosis, cytokine generation, and antitumor effects, β-glucans stimulate immune cells, such as NK cells and macrophages. Several receptors, including complement receptor-3, Toll receptors, and Dectin-1, mediate these activities. Immunomodulatory polysaccharides present in various medicinal plants have been shown to activate both innate and adaptive immune responses (Sarangi and Padhi, 2018).

Astragalus membranaceus is one such example where the Astragalus polysaccharides have been found to stimulate lymphocyte proliferation, NK cell activity, and cytokine secretion. Polysaccharides of *Ganoderma lucidum* (reishi mushroom) stimulate lymphocyte differentiation, macrophage activation, and cytokine regulation, which are generally involved in antiviral and antitumor activities. Acidic polysaccharides from *Gynostemma pentaphyllum* tea also activate macrophages and stimulate nitric oxide and cytokine secretion. Because of their ability to activate antigen-presenting cells and differentiate T-helper cells, polysaccharides are considered promising vaccine adjuvants. They enhance immune responses to coadministered antigens and improve vaccine efficacy. By integrating traditional knowledge with modern immunology, polysaccharides have a precious place in modern immune-based therapies (Singh et al., 2025).

4.3.5 Polyphenols

A broad class of secondary chemicals found in plants, characterized by many phenolic structures, is called polyphenols. Flavonoids, phenolic acids, tannins, and lignans are among them; some play a significant role in plants' therapeutic properties. Polyphenols have been demonstrated to have antidiabetic qualities in addition to their well-known antioxidant, antiinflammatory, and immunomodulatory effects. Polyphenols play a significant role in the therapeutic qualities of plants due to their existence and activity (Maji et al., 2025).

A large class of polyphenols, known as flavonoids, has been thoroughly investigated for its immunomodulatory properties. Scavenging free radicals and suppressing pro-inflammatory mediators, such as COX-2 and NF-κB, are the main mechanisms underlying the immunomodulatory impact. Because they suppress inflammation and oxidative stress, medications such as quercetin and epigallocatechin gallate (EGCG) may reduce the burden of disease caused by the immune system. Curcumin, a diarylheptanoid substance that is extracted from *C. longa* and is highly recognized for its immunomodulatory qualities, is another well-known polyphenol. Curcumin is a potent inhibitor of transcription factors such as NF-κB and STAT3, enzymes such as COX-2 and iNOS, and pro-inflammatory cytokines such as TNF-α, IL-1, IL-6, and IL-12. Additionally, it exhibits interactions with several immune cell types, including B cells, T cells, dendritic cells, and macrophages (Pedrosa and Fabi, 2024).

Several mechanisms underlie polyphenols' immunomodulatory effects. The body's natural antioxidant system is activated, and reactive oxygen species are scavenged by antioxidants, which shield immune cells from oxidative stress-induced oxidative damage. Pro-inflammatory mediators such as cytokines, chemokines, prostaglandins, and nitric oxide are suppressed by key signaling pathways as part of antiinflammatory effects. Furthermore, polyphenols affect both the innate and adaptive immune systems by modifying the growth, development, and activity of immune cells. The diverse structure and widespread distribution of polyphenols make them a very powerful class of natural immunomodulators (Yang et al., 2020).

4.4 Botanicals with established immunomodulatory effects

Traditionally, several botanicals are used for health promotion, and recently, their immunomodulatory properties have been supported by a rapidly growing body of scientific evidence. Plants represent complex mixtures of bioactive phytochemicals

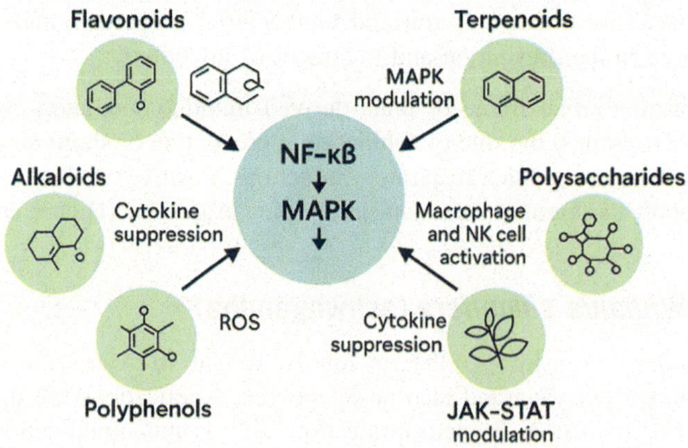

FIGURE 4.3 Cytokine regulation and immune cell activation.

that might act synergistically on various aspects of the immune system—a few key examples, along with their mechanisms and therapeutic potential (see Fig. 4.3).

4.4.1 *Echinacea purpurea*

One of the most recognizable and, without a doubt, most studied immunomodulatory botanical agents is *Echinacea purpurea*, commonly known as purple coneflower. For numerous centuries, this native plant of North America has been utilized in medicine to cure diseases, primarily upper respiratory infections.

Alkaloids, polysaccharides, glycoproteins, and derivatives of caffeic acid, such as chicoric acid, are examples of active phytochemical ingredients that have been linked to *Echinacea purpurea*'s immunomodulatory properties. These act synergistically to stimulate various components of the immune system (Hooda et al., 2024).

The key mechanisms of action include:

1. Enhanced phagocytic activity: Echinacea extracts have been documented to enhance the phagocytic activities of macrophages and neutrophils, representatives of highly involved cell types in the process of ingesting and pathogen elimination.
2. Modulation of cytokine production: It may modulate a defined set of cytokines, which are the pro-inflammatory cytokines called TNF-α and IL-1, and the antiinflammatory ones, resulting in a balancing of the immune response.
3. Stimulation of NK Cells: There have been indications that Echinacea reinforces the cytotoxic effect of NK cells, which is an important component in innate immunity against viral infection and tumor cells.

4. Antiviral and antibacterial properties: Besides direct modulation of immunity, Echinacea possesses antiviral and antibacterial properties, and thus it can be involved in the prevention and treatment of infections.

Within this broad frame of plant-derived immunomodulators, several clinical trials have examined the role of echinacea in preventing or shortening the course of the common cold and other respiratory infections. Positive results from such clinical trials support its immune-enhancing properties in general (Hieber et al., 2023).

4.4.2 *Withania somnifera* (ashwagandha)

W. somnifera, or ashwagandha, is highly valued in Ayurveda and is usually classified as a rasayana and also an adaptogen. Applications are diverse, ranging from stress reduction to neuroprotection and immunomodulation. The major bioactive principles of ashwagandha are the class of steroidal lactones called withanolides, amongst which the principal representative is withaferin A. Other phytoconstituents are alkaloids, saponins, and flavonoids (Sivakumar et al., 2011).

Ashwagandha exerts many different immunomodulatory actions:

1. Counteracting stress-induced immunosuppression: As an adaptogen, ashwagandha improves the body's adaptation to stress, which is usually associated with immunosuppression and indirectly supports immune function by modulating the response to stress.
2. Enhancement of immune cell activity: It has been documented that ashwagandha enhances the activities of immune cells, including macrophages, neutrophils, and NK cells. It even enhances phagocytic activity and promotes lymphocyte proliferation.
3. Modulation of cytokine production: Ashwagandha extracts act to alter the balance between pro- and antiinflammatory cytokines, with activities implicated in immune regulation. In certain instances, it has been able to decrease inflammation while enhancing protective immunity (Jantan et al., 2015).
4. Antioxidant properties: Withanolides have a strong antioxidant effect. They protect immune cells from oxidative damage, allowing them to perform optimally. In a randomized, double-masked, placebo-controlled trial with an open-label extension in healthy subjects, the immunomodulatory effect of an extract of *Withania somnifera* was shown. Ashwagandha supplementation significantly increased neutrophil, lymphocyte, and NK cell counts, as well as Ig levels, confirming its potential to enhance immune function in humans (Panossian et al., 2021).

4.4.3 *Tinospora cordifolia* (guduchi)

T. cordifolia, commonly known as guduchi or giloy, has been conventionally considered in Ayurveda an extremely valuable herbal remedy with immunomodulatory, adaptogenic, antiinflammatory, and antipyretic properties. It is more commonly known as "Amrita," which is a Sanskrit word meaning the root of immortality, since it has wide-spectrum therapeutic applications. The phytochemicals present in *T. cordifolia* include alkaloids, berberine, palmatine, tinosporine, diterpenoid lactones, glycosides, polysaccharides, and phenolic compounds; hence, its immunomodulatory effect can be attributed to it (Mintah et al., 2022).

Key mechanisms of action include:

1. Macrophage activation: The extracts of *T. cordifolia* have been found to activate macrophages, which enhance their phagocytic capacity and the ability of such cells to produce ROS as well as NO, important in the clearance of pathogens. (Nya and Austin, 2009).
2. Lymphocyte proliferation: It can induce the proliferation of lymphocytes like T and B cells, therefore enhancing cell-mediated and humoral immunity.
3. Cytokine modulation: Guduchi can modulate various cytokines like IL-1, IL-6, TNF-α, and IFN-γ involved with inflammatory and immune responses.
4. Antioxidant activity: The antioxidant compounds in it protect the immune cells against oxidative stressors, ensuring their integrity and functionality.
5. Adaptogenic properties: The adaptogenic nature of guduchi, just like ashwagandha, helps the body deal with different kinds of stressors and may contribute indirectly to immune resilience.

The medicinal properties of *T. cordifolia* have been studied for the prevention of a variety of infectious diseases, including COVID-19, in home quarantine settings. It enhances host immunity and is thereby considered an important botanical for immune system support (Lopresti et al., 2019).

4.4.4 *Curcuma longa* (turmeric)

C. longa, the botanical name for turmeric, is among the most commonly used herbs and spices in Chinese and Ayurvedic medicine. Its vivid yellow hue and several medicinal properties have often been attributed to a family of substances known as curcuminoids, among which curcumin is the most active and widely researched. Curcumin is a polyphenol with exceptional antioxidant, antiinflammatory, and immunomodulatory qualities. Numerous factors and several molecular targets are involved in its methods of action (Hooda et al., 2024):

1. Antiinflammatory effects: Curcumin has remarkable antiinflammatory qualities. By preventing the production of NO, COX-2, iNOS, and lipoxygenase, it reduces inflammation. Additionally, it prevents activated macrophages and other immune cells from expressing the pro-inflammatory cytokines TNF-α, IL-1, IL-6, and IL-12. This is accomplished by inhibiting protein kinase C, thereby reducing NF-κB's DNA-binding activity and lowering NF-κB activation (Darvishi et al., 2022).

2. Antioxidant activity: By scavenging free radicals, curcumin demonstrates potent antioxidant activity, shielding cells from oxidative damage—a crucial component necessary for healthy immune cell function.
3. Immune cell modulation: Curcumin's extensive modulatory activities allow it to affect a large range of immune cells, such as B cells, T cells, dendritic cells, and macrophages. It could alter how these cells differentiate and proliferate, potentially affecting both innate and adaptive immune responses.
4. Modulation of the STAT3 pathway: Curcumin also affects the STAT3 signaling system, which is linked to immunological responses, cell survival, and proliferation. Curcumin can increase STAT3 phosphorylation at low doses while decreasing its ability to bind to DNA and phosphorylate this transcription factor at high concentrations, indicating that curcumin's antiinflammatory effects are concentration-dependent.

Its immunomodulatory activity, along with that of its bioactive compounds, has been widely reviewed for possible therapeutic use in a range of immune-related conditions: neurodegenerative, cardiovascular, pulmonary, metabolic, autoimmune, and neoplastic diseases. Besides, nanoformulations have been prepared to overcome poor water solubility and low bioavailability; in fact, these formulations show enhanced anticancer effects due to the stimulation of autophagy and apoptosis and a decrease in metastasis in preclinical models (Tavares et al., 2022).

4.4.5 *Ocimum sanctum* (tulsi)

O. sanctum, commonly known as holy basil or tulsi, is an herb widely used in Ayurvedic medicine and has been referred to as the “Queen of Herbs” for the depth of its health benefits. Traditional uses of this adaptogen have included stress reduction, respiratory ailments, and immune support.

This plant has immunomodulatory properties due to its high levels of phytochemicals, including eugenol, rosmarinic acid, ursolic acid, carvacrol, linalool, and various flavonoids and polyphenols (Lou et al., 2024).

The key mechanisms of action include:

1. Immune cell activation: It has been noticed that extracts of tulsi increase the activity of different immune cells, such as T helper cells and NK cells. They can enhance antibody production and cell-mediated immunity.
2. Modulation of production of cytokines: It helps to maintain the balance between the levels of cytokines for a balanced immune response. While antiinflammatory agents blunt uncontrolled inflammation, the immunostimulatory components further heighten protective immunity.
3. Antioxidant and antiinflammatory effects: The various types of polyphenols and other antioxidant components in tulsi help to combat oxidative stress, reduce further inflammation, and generally promote immune health.

4. Adaptogenic properties: Tulsi is an adaptogen that permits the body to deal with physical and psychological stress, which normally suppresses immune functions. Indirectly, therefore, by reducing stress, it supports a robust immune system.

The present study was a double-masked, randomized controlled trial conducted to assess the immunomodulatory role of *O. sanctum* leaf extract in healthy volunteers. Supplementation with tulsi extract increased the percentage of T helper cells and NK cells with improved IFN-γ levels, hence indicating a positive influence on cellular immunity. It therefore provides clinical evidence to support the role of tulsi in immune system function (Ge et al., 2024).

4.4.6 Other emerging botanicals

In addition to these established botanicals, numerous other plant species have been investigated, which demonstrate promising immunomodulatory activity. Many of the newer, emerging botanicals exhibit diverse phytochemical profiles and mechanisms of action, contributing to an increasing understanding of plant-derived immunomodulators.

1. *A. paniculata*: The herb is composed of diterpenoid lactones, wherein andrographolide is the major constituent and exhibits potent immunomodulation activity. Andrographolide demonstrated antiviral, antibacterial, antiinflammatory, and antitumor properties, in addition to regulating both antigen-specific and nonspecific immune responses. It inhibits signaling pathways, including NF-κB and STAT3, decreases pro-inflammatory cytokines such as IL-12 and TNF-α, and inhibits enzymes such as COX-2 and iNOS (Yuandani et al., 2021). It enhances lymphocyte cytotoxicity, improves CD marker expression, and increases antibody production. Other active constituents, such as neoandrographolide, isoandrographolide, and skullcap flavone, reduce inflammation by blocking inflammatory cytokines (Arreola et al., 2015).
2. *Allium sativum* (garlic): Garlic exerts immunomodulatory and antiinflammatory activities. Generally, the activities have been related to the plant's bioactive compounds. (Firdous et al., 2025). Specifically, the organosulfur compounds contain allicin, diallyl disulfide, and ajoene. Garlic may activate NK cells, lymphocytes, and cytokine production. In addition, the antimicrobial activity of garlic protects against various pathogens by enhancing immune responses (Al-Obaidi et al., 2021).
3. *Glycyrrhiza glabra* (liquorice): The immunostimulant effect of liquorice extract was attributed to substances like glycyrrhizin, as was discussed in the preceding sections (Harun and Mohamad, 2022). In fact, it has been shown to alter immune-related gene expression and to improve immunological markers in serum and skin mucus, such as lysozyme activity, IgM levels, and complement activity (Ren et al., 2019).

4. *Zingiber officinale* (ginger): Since ginger contains substances like gingerols and shogaols, it has long been known to have antiinflammatory and antioxidant properties. Since then, it has also been investigated as an immunostimulant, for example, in aquaculture to manage Aeromonas hydrophila infection in rainbow trout, demonstrating its ability to boost the host's immune system (Shi et al., 2025).
5. *Astragalus membranaceus*: The main active ingredients in this traditional Chinese herbal remedy are astragalus polysaccharides, which have strong immunomodulatory properties. Both the innate and adaptive immune responses may become more active as a result of APS's ability to promote lymphocyte proliferation, boost NK cell activity, and regulate cytokine release.
6. *Ganoderma spp.* (reishi Mushroom) polysaccharides: Polysaccharides from *Ganoderma spp.* are typically characterized by impressive immunomodulatory effects on immune effector cells, inducing lymphocyte proliferation and differentiation. Activated macrophages, which modulate cytokine production, contribute to their antitumor and antiviral properties.
7. *Gynostemma pentaphyllum*: Acidic polysaccharide fraction from the herb exerted an immunomodulatory effect by the activation of macrophages with production of nitric oxide and pro-inflammatory cytokines. H. Hippophaerhamnoides: Sea buckthorn is a plant with a suite of bioactive phytoconstituents, including flavones, which confer its immunomodulatory profile. *Carica papaya* and *Chrysanthemum indicum*, both plants, are also reported to contain flavonoids that provide health benefits and may modulate immune responses. These, along with many other botanicals, represent a vast, largely underexploited resource for the identification of novel immunomodulatory agents. The complex phytochemical composition tends to confer synergistic effects, affording a holistic approach to supporting immune system functioning (Pan et al., 2019) (see Table 4.2).

4.5 Clinical evidence supporting botanical immunotherapy

Natural goods have been used medicinally for more than 5000 years, predating the development of Western medicine. The World Health Organization estimates that up to 80% of people worldwide use herbal remedies to prevent and treat nearly all diseases, including immunological ones, primarily in developing nations. More than 85,000 plant species are used medicinally worldwide. Additionally, it has been estimated that about 30% of all Food and Drug Administration (FDA)-approved medicines have botanical origins. This degree of historical and current usage underscores the need to investigate the chemical composition of traditional phytomedicines for their potential as immunomodulatory agents for immune-related diseases.

Table 4.2 Botanicals and their molecular immunological targets.

Botanical	Active compounds	Molecular targets	Immune pathway affected
Echinacea purpurea	Alkylamides	CB2 receptors, NF-κB	Cytokine modulation, immune stimulation
Withania somnifera (ashwagandha)	Withanolides	NF-κB, MAPK, IL-1β, TNF-α	Antiinflammatory; immune homeostasis
Tinospora cordifolia (guduchi)	Diterpenoids, polysaccharides	TLR-4, Macrophage receptors	Enhanced macrophage/NK activation
Curcuma longa (turmeric)	Curcumin	NF-κB, COX-2, iNOS, STAT3	Antiinflammatory and antioxidant pathways
Ocimum sanctum (tulsi)	Eugenol, rosmarinic acid	LOX, COX, IL-6, TNF-α	Cytokine suppression, ROS reduction

4.5.1 Preclinical studies

Studies conducted in vivo, in vitro, and ex vivo have validated the immunomodulatory characteristics of several phytochemicals, which have long been acknowledged in conventional medicine. Potential bioactive substances with immune-modulating qualities can be found in medicinal plants. Polyphenols (resveratrol, curcumin), flavonoids (EGCG, quercetin), alkaloids (berberine, colchicine), terpenoids (andrographolide, oleanolic acid), polysaccharides (pectin), and saponins are the main phytoconstituents that modulate the immune system. Dendritic cells, lymphocytes, macrophages, neutrophils, monocytes, mast cells, eosinophils, basophils, and NK cells are among the immune cells affected by these substances.

They play a significant role in controlling NK cells and phagocytic activity. In the immune signaling pathways, phytochemicals interact with a variety of molecular targets. They control the release of Ig, transcription factors, and interleukins (IL-2, IL-4, IL-6, IL-12, and IL-17). Because of their antiinflammatory and antioxidant properties, substances such as withanolides, ellagic acid, quercetin, and mangiferin help alleviate immune-related conditions. By altering the balance of Th1 and Th2 cytokines, these substances can lessen autoimmunity. The polyherbal drug Septilin is one such example that has been used in preclinical models. It showed immunomodulatory effects in cyclophosphamide-induced immunosuppressed mice. Cytotoxicity assays revealed selective anticancer effects without toxicity to normal cells. Such observations, therefore, indicate the safety and efficacy of plant-derived immunomodulators (Liu et al., 2022).

4.5.2 Human clinical trials

The idea of immunostimulation therapy was first proposed in clinical practice in 1857, when Ferdinand Ho used milk protein or blood protein injections as a treatment for malaria and lues. Since then, plant-derived biologically active compounds with immunomodulatory properties have completed clinical trials and been approved. Among the immunostimulators, curcumin from *C. longa* and genistein from soy have shown potent immune-stimulating properties. On the other hand, some botanical compounds have been found to have immunosuppressive properties. These compounds include resveratrol, which is found in grapevines, red wine, and peanuts; epigallocatechin-3-gallate, also known as EGCG, which is found in *Camellia sinensis*, or green tea; and quercetin. Other botanical compounds with immunosuppressive properties include colchicine from *Colchicum autumnale*, capsaicin from Capsicum species, and andrographolide from *A.paniculata*, highlighting the significance of botanical immunomodulators in clinical practice.

In addition to single compounds, polyherbal medicines have also demonstrated clinical efficacy. Many polyherbal formulations from the Indian traditional system of medicine, such as Vivartana, Chyawanprash, Brahma rasayana, IM-133, and Septilin, are clinically approved and marketed. Further research is continually discovering new immunomodulatory agents. One study examined the efficacy of a combination of four medicinal plants for their in vivo immunomodulatory activity in cyclophosphamide-induced immunosuppressed mice. The polyherbal medicine proved efficacious, showing a marked enhancement of immune cells: NK cells by 60%, B cells by 18%, CD4 cells by 14%, and CD8 cells by 7% ($P < .01$). These results suggest the potential of complex botanical formulations in modulating immune responses.

4.5.3 Botanicals as vaccine adjuvants

Although plant-based immunomodulators have been widely reviewed for their use in clinical settings, their application as vaccine adjuvants has not been investigated in depth. Adjuvants in vaccines enhance the immune system's response to antigens. This is a basic principle that is central to immunostimulation therapy. Most approved adjuvants are either bacterial or synthetic, not botanical. Bacterial toxins and toxoids of Escherichia coli and Vibrio cholerae have been used as mucosal adjuvants. Bacterial cell wall components, such as peptidoglycans and lipopolysaccharides, activate Toll-like receptors, thereby conferring immunity. Lipopolysaccharides with lipid A are highly effective adjuvants but are also highly toxic. To counter this, AS04, a detoxified monophosphoryl lipid A derivative, has been developed and is approved for use in humans. AS04 has been formulated with hepatitis B and human papillomavirus vaccines to bring about long-term antibody responses. Another adjuvant, AS03, formulated with squalene, has been used in influenza vaccines.

Parenteral and mucosal antigen delivery may be improved with biodegradable polymer microspheres and lamellar particles prepared from Polylactic Acid (PLA) and Poly(lactic-co-glycolic acid) (PLGA). Soluble biopolymers, including chitosan, may be used to deliver antigens and DNA nasally. More specifically, chitosan microspheres are superior adjuvants for Hepatitis B Surface Antigen (HBsAg) compared with PLGA and PLA. While the botanicals possess immunomodulatory properties, no clear-cut relationship has been established regarding the specific roles they may play as vaccine adjuvants, as with bacterial derivatives or synthetic polymers. Their broad immunomodulatory phytochemical effects do suggest potential areas for further study in this domain (Singh et al., 2021) (see Table 4.3).

4.6 Formulation strategies and quality control

The ever-increasing interest in plant-derived immunomodulators requires rigorous formulation strategies and quality control measures to ensure efficacy, safety, and consistent performance. Although the use of herbs has traditionally been widespread, their inclusion in current pharmaceutical practice poses several problems, mainly related to standardization and delivery.

4.6.1 Standardization and phytochemical characterisation

In many cases, the pathway from traditional medicines to Western pharmaceutical practices is tortuous, not in the least because of the variability of responses seen with phyto-medical practices. Much of this variability may stem from the fact that profiles of plant secondary metabolites are strongly influenced by environmental signals, which disrupt the reproducibility of findings across extracts. For such approaches to be workable, the principles behind the standardization of extracts and fractions must be scrupulously followed.

Some salient features of standardization and phytochemical characterization include:

1. Analyzing the chemical makeup of traditional phytomedicines: This is a crucial step in determining the potential processes underlying their historical use as immunomodulatory drugs. Determining the chemical makeup, molecular targets, and associated disorders of representative phytocompounds will be crucial.
2. Reproducibility: Inherent variability in plant secondary metabolites is a basic consequence of the environmental factors influencing their biosynthesis, which makes the reproducibility of quality and efficacy in various batches of plant products difficult.
3. Quantities and techniques of isolation: One major limitation with natural products is the often low available quantities for development and clinical use. Consequently, there is a need to develop new isolation techniques to improve the yield of active compounds in pharmaceutical applications.
4. Quality control procedures: Standard testing and quality control methods for ensuring the effectiveness and quality of medicinal compounds of plant origin

Table 4.3 Preclinical and clinical evidence supporting immunomodulatory botanicals.

Botanical	Study model	Dosage/extract used	Observed immune effects	Study type
Echinacea purpurea	Human volunteers	300–500 mg extract/day	Increased NK cell activity, reduced cold incidence	Clinical trial
Withania somnifera	Mice, human subjects	100–300 mg/day (root extract)	↓ IL-6, ↓ TNF-α; enhanced immune resilience	Preclinical + clinical
Tinospora cordifolia	Rodent models	Aqueous extract 100 mg/kg	↑ Macrophage phagocytosis, ↑ NO production	Preclinical
Curcuma longa	In vitro, human studies	Curcumin 500–1500 mg/day	↓ NF-κB, ↓ COX-2, ↓ oxidative stress markers	Clinical + cell studies
Ocimum sanctum	Rodent models	50–200 mg/kg	↓ Corticosterone; improved immune cell counts	Preclinical

for potential pharmaceutical uses are known to be lacking. This is particularly difficult for substances like polysaccharides, whose molecular weights and structures are so diverse that it is challenging to achieve uniform quality across batches.

4.6.2 Safety, toxicity, and regulatory aspects

Since the majority of currently used immunodrugs, including cytotoxic drugs, have severe side effects, one of the key areas of pharmacological research today is the search for natural products of plant origin for new leads of potent and safe immunosuppressant and immunostimulant agents. This underlines the relevance of safety and toxicity studies of botanical immunomodulators.

Key considerations about safety, toxicity, and regulatory aspects are:

A. Cytotoxicity testing: The safety profile of the bioactive phytoconstituents needs to be preceded by preclinical studies. The polyherbal drug Septilin exhibited no cytotoxic activity against normal human breast epithelial cells, MCF-10A, but showed cytotoxicity against cisplatin-induced human breast adenocarcinoma, MCF-7 cells, at 5 μg/mL. Based on that, the mentioned compound may act as a safe therapeutic agent.

B. Microbial contamination control: One of the most important factors is microbial contamination control, which has not been considered in many studies on the effects of plant extracts on the immune system. Immune system parameters can be significantly altered by microbial endotoxins, leading to unrepresentative results. For this, necessary precautions should be taken to guard against microbial contamination during research and production.

C. Risk classification: One other important task involves the classification of immunomodulatory herbal agents according to specific categories of risk. A widely accepted body of information derived from national registries, metaanalyzes of clinical studies, and expert opinions from medical professionals themselves can support this.

D. Clinical evidence and standardization of dosage: More empirical evidence is required to completely establish the immunomodulatory role of phytochemicals in the prevention and treatment of immune system-related diseases, even in the face of an abundance of preclinical and ex vivo studies. In addition to raising some characteristics, standard dose and duration of intervention are still necessary for the development of the best phytochemical immunomodulators.

E. Regulatory frameworks: Greater demand for an array of plant-derived products will encourage improvements in navigating market barriers through robust regulatory frameworks that ensure efficacy and quality. Safety and toxicity management of botanical immunomodulators, coupled with adherence to strict regulatory guidelines, are critical for their successful translation in the clinic.

4.6.3 Novel delivery systems (nanocarriers, phytosomes)

One of the major clinical constraints with most natural products is low bioavailability. Nanotechnology, along with other advanced drug-delivery strategies, is being utilized to enhance the in vivo efficacy of plant-origin immunomodulators upon administration to humans. The objective of such new drug delivery systems is to enhance the therapeutic profile of herbal active ingredients by overcoming their poor aqueous solubility, degradation, and inefficient targeting.

Polymeric nanoparticles, nanocapsules, liposomes, phytosomes, animations, microspheres, transfersomes, and ethosomes are among the innovative herbal preparations. Compared to traditional herbal extracts and actives, the new herbal formulations provide several benefits. The advantages include increased solubility and bioavailability, which increase systemic absorption and effectiveness. The novel herbal formulations also prevent toxicity and physical or chemical degradation, while improving pharmacological potency and stability. The new herbal formulations also allow better tissue and macrophage distribution, facilitate sustained and controlled drug delivery, and ensure consistency and reliability of therapeutic results compared to traditional herbal formulations.

4.6.3.1 *Phytosomes*

Phytosomes are a patented technology that encapsulates standardized plant extracts or water-soluble phytoconstituents, mainly polyphenols, within phospholipids. These phospholipids are generally phosphatidylcholine from soy that form lipid-compatible molecular complexes. The phytosomes exhibit a bioavailability much higher than that of the simple herbal extracts. This is attributed to their superior ability to cross lipid-rich biomembranes, enabling them to reach target sites more effectively.

Compared with their noncomplexed herbal extracts, phytosomal complexes exhibit enhanced pharmacokinetic and therapeutic profiles by greatly increasing the bioavailability of selected phytochemicals. While initially investigated for cosmetic applications, growing evidence over the past few years has highlighted their potential for drug delivery across various therapeutic areas, including cardiovascular, antiinflammatory, hepatoprotective, and anticancer applications. This is particularly advantageous for the bioactive constituents of phytomedicines, such as flavonoids, which often have poor oral bioavailability.

4.6.3.2 *Vesicular systems*

Targeted delivery and sustained release rates are considered critical parameters for phytochemical drug carriers to work effectively. Various kinds of nanocarriers have been used up to this point, mostly vesicular drug delivery systems, which encapsulate active ingredients inside a spherical shape. Liposomes, niosomes, transfersomes, and ethosomes are a few examples.

Liposomes are spherical, 0.05–5.0 micrometer-diameter vesicles made of cholesterol and phospholipids. They are highly promising drug-delivery vehicles due to their unique hydrophobic and lipophilic properties, which enable them to encapsulate both lipid-soluble and water-soluble drugs. Liposomes are biocompatible, biodegradable, and stable; their design aims to target drugs directly to the intended site of action (Falcinelli et al., 2025).

It will be necessary to develop and apply new delivery systems so that the full potential of botanicals in therapy can be realized and some of their inherent pharmaceutical shortcomings overcome, thereby allowing their application in conventional medicine.

4.7 Integration of botanicals in modern immunotherapy

The 21st century brought additional challenges to global health, including antibiotic, antiviral, and anticancer resistance, as well as drug side effects. This has led to increased dependence on traditional medicine and natural products because these are believed to be multivalent component agents in contrast with monovalent commercial drugs. Many plant-based medicines have less serious side effects and are cheaper than commercially available drugs. This shift in reliance underscores a potential avenue for incorporating botanicals into modern immunotherapy methods.

4.7.1 Combination therapies with conventional drugs

The integration of natural products and synthetic drugs is an exciting new trend in natural product and medicinal chemistry. This is because it has been realized that most synthetic immunomodulatory drugs, despite their high efficacy, are highly toxic and have adverse effects. Plant-based immunomodulatory drugs are safer due to their low toxicity and few side effects. One such example is the polyherbal medicine Septilin, which is derived from Indian traditional medicine. Its immunomodulatory effects were established in preclinical trials using cyclophosphamide-induced immunosuppressed mice. Cytotoxicity tests revealed selective toxicity against breast cancer cells, which is safe because it does not affect normal cells. These findings open the door to the formulation of botanicals in combination with chemotherapeutic drugs such as cisplatin. The adaptability of natural products enables targeting multiple pathways, making them excellent candidates for combination therapies.

4.7.2 Personalized botanical medicine

The transition from traditional to Western medicine via pharmaceuticals faces challenges due to the variability of phytochemical composition, which is influenced by environmental factors. Variability in phytochemical composition poses a challenge to reproducibility but can also be used for personalized botanical medicine. Standardization of extracts and phytochemically enriched fractions is an important step in ensuring consistency in quality, safety, and efficacy. Optimal dosage and duration of intervention are crucial for the best phytochemical immunomodulators. There is a need for a better understanding of the chemical nature, mechanisms, and disease associations of phytomedicines. The reverse pharmacology approach focuses on mechanism-based validation, safety optimization, and clinical acceptability of herbal medicines. Risk evaluation of plant-derived immunomodulators based on clinical data and physician knowledge is required for safe individualization.

4.7.3 Translational and public health perspectives

The incorporation of botanicals into modern immunotherapy has significant translational and public health implications, given their widespread use globally. Natural products have been in use in medicine for more than 5000 years, and many modern medicines have been derived from botanicals. The World Health Organization estimates that up to 80% of people in underdeveloped nations utilize herbal therapy, especially for conditions relating to the immune system. The growing antibiotic resistance and illness load have made the use of traditional medical systems to strengthen the immune system more crucial. Clinical studies are required to demonstrate efficacy and safety in broader groups, despite the abundance of preclinical data. There is a growing need for better standardization and regulation of botanicals. Global collaboration and information sharing are necessary to facilitate translational research and global acceptance. Overall, botanicals are poised to make a major contribution to future public health initiatives and immunotherapy (Awuchi, 2023).

4.8 Future directions and research challenges

The field of botanical immunomodulation is dynamic and holds great future potential; however, numerous issues remain to be investigated for its successful integration into modern medicine (see Fig. 4.4).

4.8.1 Molecular docking and network pharmacology approaches

Plant-based immunomodulators require a better understanding of their systemic, cellular, and molecular mechanisms to enable translation into a therapeutic modality. Sophisticated computational and systems biology approaches, such as molecular docking and network pharmacology, facilitate target-based identification of bioactive phytochemicals. Molecular docking calculates the binding affinity and pose of phytocompounds with immune receptors, enzymes, and transcription factors. These compounds interact with immune signaling pathways and cytokines,

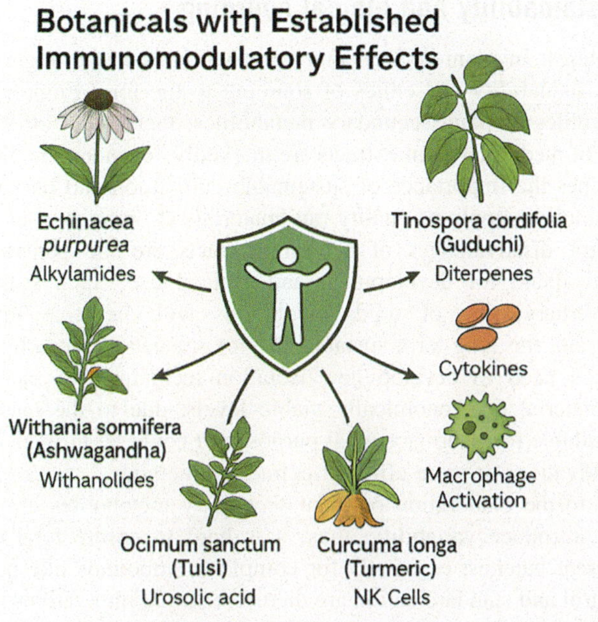

FIGURE 4.4 Major immunomodulatory botanicals and their active phytochemicals.

and Ig. Network pharmacology effectively models the polypharmacological and multitarget properties of herbal medicines. These approaches support reverse pharmacology by combining traditional and molecular knowledge. These approaches together help in designing a drug, optimizing natural compounds, and analyzing synergistic botanical-synthetic combinations (Awuchi, 2023).

4.8.2 Clinical validation and global acceptance

However, due to their long history of traditional use and the availability of preclinical data, botanical immunomodulators have yet to gain widespread clinical acceptance. More advanced clinical study designs are required to validate safety and efficacy. Standardization of dosage, duration, and intervention protocols is necessary for clinical relevance. The lack of universally accepted standards for herbal clinical trials is a major drawback. International collaboration is required to develop a balanced database. Appropriate regulatory policies and risk stratification are required for quality and efficacy. Further rigorous clinical validation will be required to establish the successful integration of approved plant-derived immunomodulators into conventional medicine (Saenger et al., 2025).

4.8.3 Sustainability and ethical sourcing

Growing interest in plant-derived immunomodulators has brought key questions regarding sustainability and ethics of sourcing to the fore. Environmental signals modulate profiles of plant secondary metabolites; therefore, both the quality and consistency of medicinal plant extracts are markedly dependent on the environment. This underlines the importance of sustainable cultivation and harvesting practices for the continuous supply of quality raw materials (Komariah et al., 2023).

The major disadvantages of natural products are the generally inadequate quantities available for development and clinical use. Such scarcity may also promote overharvesting of wild populations with negative implications for biodiversity and the long-term sustainability of species of medicinal value. There is therefore a need to develop new isolation techniques to increase yields of biological material to economically viable levels, due to the inadequacy of the amounts available for pharmaceutical purposes. It could be through biotechnology, artificial cultivation, or more efficient extraction methods.

Intrinsic to the elaboration of plant secondary metabolites are environmental factors that introduce variability; thus, it is hard to ensure total quality control across different batches, especially for complex compounds like polysaccharides. Quality control and standardization are therefore strong imperatives if plant-derived medicinal products are to ensure efficacy and safety, thereby contributing to sustainable practices through reduced waste and the efficient use of resources. However, the context of traditional medicine and the global reliance on herbal remedies raise the need for ethical sourcing practices that include nonexploitative relations with indigenous communities and ecosystems, so that the collection and cultivation of medicinal plants do not reach exploitative levels. This immense biodiversity demands protection through sustainable, ethical practices. Successful control of the challenges mentioned above will make the quest for botanical immunomodulators not only scientifically feasible but also ecologically friendly and socially just, thereby securing a long-term supply of this valuable natural resource (Barani et al., 2021).

4.9 Conclusion

Botanicals are the meeting point of traditional knowledge in medicine and modern biomedical innovation in the modulation of the immune system. Phytochemicals such as flavonoids, terpenoids, alkaloids, polysaccharides, and polyphenols modulate cytokines, transcription factors (NF-κB, MAPK, STAT3), and immune effector cells. Preclinical and emerging clinical evidence support their use in restoring immune homeostasis in immune dysregulation. Nanocarriers, phytosomes, and liposomes have enhanced bioavailability and targeting specificity. However, there are challenges in standardization, safety assessment, dose optimization, and acceptance by regulatory authorities worldwide. There is an urgent need to integrate molecular docking analysis, network pharmacology, and systems biology to understand the polypharmacological mechanisms of botanicals. By harmonizing traditional knowledge with scientific validation and sustainable development, botanicals will find their place as holistic, personalized immunotherapeutic agents.

References

Aggarwal BB, Harikumar KB. Potential therapeutic effects of curcumin, the anti-inflammatory agent, against neurodegenerative, cardiovascular, pulmonary, metabolic, autoimmune and neoplastic diseases. The International Journal of Biochemistry & Cell Biology 2009;41(1):40–59. Available from https://doi.org/10.1016/j.biocel.2008.06.010.

Al-Obaidi JR, Jambari NN, Ahmad-Kamil EI. Mycopharmaceuticals and nutraceuticals: promising agents to improve human well-being and life quality. Journal of Fungi 2021;7(7):503. Available from https://doi.org/10.3390/jof7070503.

Arreola R, Quintero-Fabián S, López-Roa RI, Flores-Gutiérrez EO, Reyes-Grajeda JP, Carrera-Quintanar L, et al. Immunomodulation and anti-inflammatory effects of garlic compounds. Journal of Immunology Research 2015;2015:1–13. Available from https://doi.org/10.1155/2015/401630.

Awuchi CG. Plants, phytochemicals, and natural practices in complementary and alternative system of medicine for treatment of central nervous system disorders. International Journal of Food Properties 2023;26(1):1190–213. Available from https://doi.org/10.1080/10942912.2023.2205039, https://www.tandfonline.com/loi/ljfp20.

Barani M, Sangiovanni E, Angarano M, Rajizadeh MA, Mehrabani M, Piazza S, et al. Phytosomes as innovative delivery systems for phytochemicals: a comprehensive review of literature. International Journal of Nanomedicine 2021;16:6983–7022. Available from https://doi.org/10.2147/ijn.s318416.

Barreca MM, Alessandro R, Corrado C. Effects of flavonoids on cancer, cardiovascular and neurodegenerative diseases: role of NF-κB signaling pathway. International Journal of Molecular Sciences 2023;24(11):9236. Available from https://doi.org/10.3390/ijms24119236.

Cherian A, Vadivel V. Perspective Chapter: Phytocompounds as Immunomodulators. IntechOpen; 2023. Available from http://doi.org/10.5772/intechopen.108858.

Daley SK, Cordell GA. Alkaloids in contemporary drug discovery to meet global disease needs. Molecules (Basel, Switzerland) 2021;26(13) Available from https://doi.org/10.3390/molecules26133800, https://www.mdpi.com/1420-3049/26/12/3758/pdf.

Darvishi M, Shamsaie Mehrgan M, Khajehrahimi AE. Effect of licorice (*Glycyrrhiza glabra*) extract as an immunostimulant on serum and skin mucus immune parameters, transcriptomic responses of immune-related gene, and disease resistance against *Yersinia ruckeri* in rainbow trout (*Oncorhynchus mykiss*). Frontiers in Veterinary Science 2022;9 Available from https://doi.org/10.3389/fvets.2022.811684, http://journal.frontiersin.org/journal/veterinary-science.

Das K. Herbal plants as immunity modulators against COVID-19: a primary preventive measure during home quarantine. Journal of Herbal Medicine 2022;32:100501. Available from https://doi.org/10.1016/j.hermed.2021.100501.

Falcinelli B, Benincasa P, Riahi J, Bulgari R. Micro-herbs: a valuable source of phytochemicals from aromatic and medicinal plants. Journal of Agriculture and Food Research 2025;24:102418. Available from https://doi.org/10.1016/j.jafr.2025.102418.

Firdous SM, Irfan Z, Giri S, Asaba NS. Unravelling the role of natural herbs for the management of diabetes mellitus. Diabetology and Metabolic Syndrome 2025;17(1) Available from https://doi.org/10.1186/s13098-025-01904-4, https://dmsjournal.biomedcentral.com/.

Ge W, Gao Y, He L, Jiang Z, Zeng Y, Yu Y, et al. Developing Chinese herbal-based functional biomaterials for tissue engineering. Heliyon 2024;10(6):e27451. Available from https://doi.org/10.1016/j.heliyon.2024.e27451.

Harun NH, Mohamad MF. The immunomodulatory effects of *Zingiber officinale* (Ginger): a systematic review. Research Journal of Pharmacy and Technology 2022;15(8):3776–81. Available from https://doi.org/10.52711/0974-360X.2022.00634, http://www.rjptonline.org, http://www.anvpublication.org.

Hieber C, Grabbe S, Bros M. Counteracting immunosenescence—which therapeutic strategies are promising? Biomolecules 2023;13(7):1085. Available from https://doi.org/10.3390/biom13071085.

Hooda P, Malik R, Bhatia S, Al-Harrasi A, Najmi A, Zoghebi K, et al. Phytoimmunomodulators: a review of natural modulators for complex immune system. Heliyon 2024;10(1):e23790. Available from https://doi.org/10.1016/j.heliyon.2023.e23790.

Jantan I, Ahmad W, Bukhari SNA. Plant-derived immunomodulators: an insight on their preclinical evaluation and clinical trials. Front iers in Plant Science 2015;6 Available from https://doi.org/10.3389/fpls.2015.00655, http://journal.frontiersin.org/article/10.3389/fpls.2015.00655/pdf.

Jin M, Zhao K, Huang Q, Shang P. Structural features and biological activities of the polysaccharides from *Astragalus membranaceus*. International Journal of Biological Macromolecules 2014;64:257–66. Available from https://doi.org/10.1016/j.ijbiomac.2013.12.002.

Kadiyska T, Tourtourikov I, Dabchev K, Zlatarova A, Stoynev N, Hadjiolova R, et al. Herbs and plants in immunomodulation (Review). International Journal of Functional Nutrition 2023;4(1). Available from https://doi.org/10.3892/ijfn.2023.31.

Komariah M, Amirah S, Maulana S, Abdurrahman MF, Ibrahim K, Platini H, et al. The efficacy of herbs as complementary and alternative therapy in recovery and clinical outcome among people with COVID-19: a systematic review, meta-analysis, and meta-regression. Therapeutics and Clinical Risk Management 2023;19:611–27. Available from https://doi.org/10.2147/tcrm.s405507.

Liang X, Zhou J, Wang M, Wang J, Song H, Xu Y, et al. Progress and prospect of polysaccharides as adjuvants in vaccine development. Virulence 2024;15(1). Available from https://doi.org/10.1080/21505594.2024.2435373.

Liu Y, Wang J, Wu C. Modulation of gut microbiota and immune system by probiotics, pre-biotics, and post-biotics. Frontiers in Nutrition 2022;8 Available from https://doi.org/10.3389/fnut.2021.634897, http://journal.frontiersin.org/journal/nutrition.

Lopresti AL, Drummond PD, Smith SJ. A randomized, double-blind, placebo-controlled, crossover study examining the hormonal and vitality effects of Ashwagandha (*Withania somnifera*) in aging, overweight males. American Journal of Men's Health 2019;13(2) Available from https://doi.org/10.1177/1557988319835985, http://www.sagepub.com.

Lou K, Luo H, Jiang X, Feng S. Applications of emerging extracellular vesicles technologies in the treatment of inflammatory diseases. Frontiers in Immunology 2024;15. Available from https://doi.org/10.3389/fimmu.2024.1364401.

Maji S, Dasgupta S, Das R, Smaoui S. Plant-based bio-actives encapsulated nanocarriers for gastrointestinal inflammations management. Gastrointestinal Inflammations and Gut Microbiota: Pathophysiology, Management, and Emerging Therapies. India: CRC Press; 2025. p. 322–45. Available from https://www.taylorfrancis.com/books/edit/10.1201/9781003493143/gastrointestinal-inflammations-gut-microbiota-neeraj-mishra-sumel-ashique-anoop-kumar, http://doi.org/10.1201/9781003493143-19.

Mintah SO, Archer M-A, Asafo-Agyei T, Ayertey F, Jnr PA-A, Boamah D, et al. Medicinal plant use in Ghana: advancement and challenges. American Journal of Plant Sciences 2022;13(03):316–58. Available from https://doi.org/10.4236/ajps.2022.133020.

Mondal S, Varma S, Bamola VD, Naik SN, Mirdha BR, Padhi MM, et al. Double-blinded randomized controlled trial for immunomodulatory effects of Tulsi (*Ocimum sanctum* Linn.) leaf extract on healthy volunteers. Journal of Ethnopharmacology 2011;136(3):452–6. Available from https://doi.org/10.1016/j.jep.2011.05.012.

Moni S, Abdelwahab S, Jabeen A, Elmobark M, Aqaili D, Gohal G, et al. Advancements in vaccine adjuvants: the journey from alum to nano formulations. Vaccines 2023;11(11):1704. Available from https://doi.org/10.3390/vaccines11111704.

Murphy EJ, Rezoagli E, Pogue R, Simonassi-Paiva B, Abidin IIZ, Fehrenbach GW, et al. Immunomodulatory activity of β-glucan polysaccharides isolated from different species of mushroom – a potential treatment for inflammatory lung conditions. Science of the Total Environment 2022;809 Available from https://doi.org/10.1016/j.scitotenv.2021.152177, http://www.elsevier.com/locate/scitotenv.

Nya EJ, Austin B. Use of dietary ginger, Zingiber officinale Roscoe, as an immunostimulant to control *Aeromonas hydrophila* infections in rainbow trout, *Oncorhynchus mykiss* (Walbaum). Journal of Fish Diseases 2009;32(11):971–7. Available from https://doi.org/10.1111/j.1365-2761.2009.01101.x.

Pan P, Huang YW, Oshima K, Yearsley M, Zhang J, Arnold M, et al. The immunomodulatory potential of natural compounds in tumor-bearing mice and humans. Critical Reviews in Food Science and Nutrition 2019;59(6):992–1007. Available from https://doi.org/10.1080/10408398.2018.1537237, https://www.tandfonline.com/loi/bfsn20.

Panossian AG, Efferth T, Shikov AN, Pozharitskaya ON, Kuchta K, Mukherjee PK, et al. Evolution of the adaptogenic concept from traditional use to medical systems: pharmacology of stress- and aging-related diseases. Medicinal Research Reviews 2021;41(1):630–703. Available from https://doi.org/10.1002/med.21743, http://onlinelibrary.wiley.com/journal/10.1002/(ISSN)1098-1128.

Pathak S, Fialho J, Nandi D. Plant-based immunomodulators and their potential therapeutic actions. Journal of Exploratory Research in Pharmacology 2022. Available from https://doi.org/10.14218/jerp.2022.00033.

Pedrosa LDF, Fabi JP. Polysaccharides from medicinal plants: bridging ancestral knowledge with contemporary science. Plants 2024;13(13):1721. Available from https://doi.org/10.3390/plants13131721.

Ren D, Zhao Y, Zheng Q, Alim A, Yang X. Immunomodulatory effects of an acidic polysaccharide fraction from herbal *Gynostemma pentaphyllum* tea in RAW264.7 cells. Food & Function 2019;10(4):2186–97. Available from https://doi.org/10.1039/c9fo00219g.

Ren L, Zhang J, Zhang T. Immunomodulatory activities of polysaccharides from *Ganoderma* on immune effector cells. Food Chemistry 2021;340:127933. Available from https://doi.org/10.1016/j.foodchem.2020.127933.

Saenger J, Happe J, Maier C, Kerber B, Uenal E, Bos D, et al. Mammographic calcifications in lung transplant recipients: prevalence and evolution. Biomedicines 2025;13(9):2318. Available from https://doi.org/10.3390/biomedicines13092318.

Sarangi MK, Padhi S. Novel herbal drug delivery system: an overview. Archives of Medicine and Health Sciences 2018;6(1):171. Available from https://doi.org/10.4103/amhs.amhs_88_17.

Shah SA, Sander S, White CM, Rinaldi M, Coleman CI. Evaluation of echinacea for the prevention and treatment of the common cold: a meta-analysis. Lancet Infectious Diseases 2007;7(7):473–80. Available from https://doi.org/10.1016/S1473-3099(07)70160-3, http://www.journals.elsevier.com/the-lancet-infectious-diseases.

Sharma P, Kumar P, Sharma R, Gupta G, Chaudhary A. Immunomodulators: role of medicinal plants in immune system. National Journal of Physiology, Pharmacy and Pharmacology 2017;7(6):552–6. Available from https://doi.org/10.5455/njppp.2017.7.0203808032017, http://www.ejmanager.com/mnstemps/28/28-1487571016.pdf?t=1496343555.

Shi YY, Li X, Li Z, Sun J, Gao T, Wei G, et al. Nano-formulations in disease therapy: designs, advances, challenges, and future directions. Journal of Nanobiotechnology 2025;23(1). Available from https://doi.org/10.1186/s12951-025-03442-7.

Shukla MK, Singh SK, Pandey S, Gupta PK, Choudhary A, Jindal DK, et al. Potential immunomodulatory activities of plant products. South African Journal of Botany 2022;149:937–43. Available from https://doi.org/10.1016/j.sajb.2022.04.055, https://www.sciencedirect.com/science/journal/02546299.

Singh R, Goel S, Bourgeade P, Aleya L, Tewari D. Ayurveda Rasayana as antivirals and immunomodulators: potential applications in COVID-19. Environmental Science and Pollution Research 2021;28(40):55925–51. Available from https://doi.org/10.1007/s11356-021-16280-5.

Singh S, Choudhury MD, Thakur A, Kiran V, Dulta K, Sharma AJ. Safety, toxicity, and regulation in use of medicinal plants. Medicinal Plants and Their Bioactives in Human Diseases. India: Springer Science+Business Media; 2025. p. 41–57. Available from http://doi.org/10.1007/978-3-032-01356-9_2.

Singh N, Tailang M, Mehta S. A review on herbal plants as immunomodulators. International Journal of Pharmaceutical Sciences and Research 2016;7(9) Available from https://doi.org/10.13040/ijpsr.0975-8232.7(9).3602-10.

Sivakumar SM, Safhi MM, Kannadasan M, Sukumaran N. Vaccine adjuvants - current status and prospects on controlled release adjuvancity. Saudi Pharmaceutical Journal 2011;19(4):197–206. Available from https://doi.org/10.1016/j.jsps.2011.06.003.

Swaroop AK, Lalitha CMVN, Shanmugam M, Subramanian G, Natarajan J, Selvaraj J. Plant derived immunomodulators: a critical review. Advanced Pharmaceutical Bulletin 2022;12(4):712–29. Available from https://doi.org/10.34172/apb.2022.074, https://apb.tbzmed.ac.ir/Article/apb-32199.

Tavares LP, Galvão I, Ferrero MR. Novel immunomodulatory therapies for respiratory pathologies. Comprehensive Pharmacology 2022;5:554–94. Available from https://doi.org/10.1016/B978-0-12-820472-6.00073-6, https://www.sciencedirect.com/referencework/9780128208762/.

Tharakan A, Shukla H, Benny IR, Tharakan M, George L, Koshy S. Immunomodulatory effect of *Withania somnifera* (Ashwagandha) extract—a randomized, double-blind, placebo controlled trial with an open label extension on healthy participants. Journal of Clinical Medicine 2021;10(16):3644. Available from https://doi.org/10.3390/jcm10163644.

Vyshnavi DV, Sivananda Swaroop S, Sudheer A, Kanala SR, Muralidhar Naik R, Varalakshmi O. Plant-derived selected bioactive saponins and tannins: an overview of their multi-target mechanisms and diverse biological activities. Pharmacognosy Research 2023;15(4):623–35. Available from https://doi.org/10.5530/pres.15.4.066.

Yang Q, Liang Q, Balakrishnan B, Belobrajdic DP, Feng Q-J, Zhang W. Role of dietary nutrients in the modulation of gut microbiota: a narrative review. Nutrients 2020;12(2):381. Available from https://doi.org/10.3390/nu12020381.

Yuandani, Jantan I, Rohani AS, Sumantri IB. Immunomodulatory effects and mechanisms of *Curcuma* species and their bioactive compounds: a review. Frontiers in Pharmacology 2021;12 Available from https://doi.org/10.3389/fphar.2021.643119, http://www.frontiersin.org/Pharmacology.

Zebeaman M, Tadesse MG, Bachheti RK, Bachheti A, Gebeyhu R, Chaubey KK. Plants and plant-derived molecules as natural immunomodulators. BioMed Research International 2023;2023 Available from https://doi.org/10.1155/2023/7711297, http://www.hindawi.com/journals/biomed/.

Zielińska M, Dereń K, Polak-Szczybyło E, Stępień AE. The role of bioactive compounds of *Nigella sativa* in rheumatoid arthritis therapy—current reports. Nutrients 2021;13(10):3369. Available from https://doi.org/10.3390/nu13103369.

CHAPTER

Herbal remedies for digestive health

5

Abdul Rahaman Shaik[1], P. Rao Tatineni[2], Bhavana Alaparthi[2], Kanaka Durga Devi Nelluri[2] and Sai Harsha Veganti[3]

[1]*Department of Pharmaceutical Sciences, Vignan's Foundation for Science, Technology and Research, Guntur, Andhra Pradesh, India,*

[2]*Department of Pharmaceutics and Biotechnology, KVSR Siddhartha College of Pharmaceutical Sciences, Vijayawada, Andhra Pradesh, India,*

[3]*University of Wisconsin, Milwaukee, WI, United States*

5.1 An overview of digestive health

Digestive health is one of the cornerstones of good health, as it reflects how the gastrointestinal (GI) tract and its associated organs function. The GI system has four major functions: the digestion of food, absorption of nutrients, immune function, and elimination of waste. Digestive disorders can be acute or chronic, resulting from dietary issues, lifestyle factors, stress, infection, and/or disease processes (Curtin et al., 2017).

Digestive disorders involve more than bodily symptoms. Chronic gut disorders are associated with significant psychosocial burdens, including anxiety, depression, stress-related symptom exacerbations, and impaired activities of daily living, causing reduced work performance and increased medical office visits, along with economic burden due to diagnostic testing and prescription costs (Strate and Gralnek, 2016).

Epidemiological studies indicate that a significant proportion of the world's population experiences digestive issues during their lifetime, and many patients seek complementary modalities when conventional treatments have not been effective or have resulted in negative side effects. The rising rates of digestive disorders show the need for affordable, safe, accessible, and more holistic treatment options (Black and Ford, 2020).

5.1.1 Burden of digestive disorders

Digestive disorders are the most common health complaint the world over, affecting individuals of every age group and culture. Conditions like indigestion or dyspepsia, constipation, diarrhea, acid reflux, and functional gastrointestinal disorders (FGIDs), including irritable bowel syndrome (IBS), are a major source of discomfort and disability. Digestive disorders can be acute or chronic, resulting from dietary issues, lifestyle factors, stress, infection, and/or disease processes (Dekker et al., 2020).

Healing Botanicals. DOI: https://doi.org/10.1016/B978-0-443-44295-7.00004-5

Digestive disorders involve more than bodily symptoms. Chronic gut disorders have significant psychosocial burdens related to anxiety, depression, stress-related symptom exacerbations, and affected activities of daily living, causing reduced work performance and increased medical office visits, along with economic burden for diagnostic testing and prescriptions.

Epidemiological studies indicate that a significant proportion of the world's population experiences digestive issues during their lifetime, and many patients seek complementary modalities when conventional treatments have not been effective or have resulted in negative side effects. The rising rates of digestive disorders show the need for affordable, safe, accessible, and more holistic treatment options (Peyrin-Biroulet, 2020).

5.1.2 The function of herbal medicine in promoting gastrointestinal health

The tradition of maintaining digestive health with herbal medicine spans thousands of years in Ayurvedic medicine, traditional Chinese medicine (TCM), Unani medicine, and Western herbalism. Many of the acute symptoms and chronic conditions people experience reflect an imbalance of the GI tract. Many herbs have carminative actions that relieve gas, as well as antiinflammatory, antispasmodic, antimicrobial, and mucosal protective properties that are particularly effective at supporting GI health (Biswas et al., 2017).

Some of the most well-known herbs used for generations to alleviate symptoms of bloating, cramping, nausea, and indigestion include ginger, peppermint, chamomile, licorice, fennel, and turmeric. Many of these uses have been supported by modern data, which reveal bioactive compounds consistent with their therapeutic claims, such as gingerols in ginger and menthol in peppermint (Lete and Allué, 2016).

A major reason people tend to choose herbal solutions is that they are thought to be safer, more accessible, and provide a more holistic approach to health care. The use of herbs as a complementary resource in integrative care has shifted in recent years, offering another dimension of care alongside traditional care modalities. When one thinks about herbs in terms of their ability to address so many things—from symptoms and physical ailments to inflammation, microbiome disruption, or gut issues related to stress—herbs can be a valuable way to approach gut health (Bent, 2008).

5.1.3 Aim and scope of the chapter

The chapter intends to provide a thorough overview of herbal therapies for digestive health through the lens of both traditional medicine and modern scientific rationale. The objectives of this chapter are to:

- Provide a straightforward summary of basic human digestive physiology.
- Discuss common digestive disorders and their basic physiological mechanisms.

- Discuss specific representative herbs and their traditional uses for digestive disorders.
- Discuss the available evidence regarding the effectiveness, safety, and mechanisms of action of the specific herbs noted.
- Discuss preparation, dosage forms, and clinical considerations.

Approaches will be considered from both historical medical disciplines and modern herbalist approaches. The chapter represents a synthesis of historical knowledge, clinical application, and evidence-based practice. It is intended to be a useful resource for practitioners, students, and health and wellness-oriented individuals interested in responsibly using herbal therapies. This chapter will highlight a practical, evidence-based service to improve GI health and wellness within an integrative model of care (Posadzki et al., 2013).

5.1.4 Epidemiology of digestive disorders

Digestive disorders such as gastroesophageal reflux disease (GERD), peptic ulcer disease, IBS, inflammatory bowel disease (IBD), constipation, dyspepsia, and functional bowel disorders are major contributors to worldwide morbidity and the use of health services. On a global scale, the pattern of digestive disease varies: lower- and middle-income countries generally have high levels of infectious GI illnesses and complications due to untreated *Helicobacter pylori* infection, while high-income regions have mostly chronic, noninfectious conditions such as GERD, IBS, and IBD. The presence of older people in the community also increases the prevalence of diseases such as diverticular disease and colorectal cancer. About India, the epidemiological profile is a mixture of both: infectious diarrheal and enteric diseases continue to be the main causes of morbidity and death in some parts, whereas changes in the economy and lifestyle have brought about an increase in the incidence of reflux disease, nonalcoholic fatty liver disease (NAFLD), constipation, and functional disorders. Moreover, India has significant regional disparities due to an imbalance in sanitation, healthcare, and diet. Urbanization and dietary changes are the main drivers of this new situation. Migrants to cities frequently replace traditional high-fiber, plant-based diets with processed, energy-dense, low-fiber foods. The overconsumption of fatty foods, foods made from refined cereals, and sugary drinks, besides the decreased physical activity, produces obesity and metabolic disorders that are risk factors for Nonalcoholic fatty liver disease (NAFLD), GERD, and gallstone disease. Stress, irregular sleep due to changes in habits, and high exposure to antibiotics can all affect the gut microbiota and, consequently, lead to the development of functional and inflammatory disorders, which are associated with city life. This combination of circumstances accounts for the fact that infectious GI diseases continue to be a problem among minority groups. At the same time, there is an increase in chronic digestive diseases among populations that are urbanizing (Delfino et al., 2013).

5.1.5 Role of diet and lifestyle in digestive health

Diet and lifestyle are major factors that determine how well the digestive system functions and are linked to symptoms, disease risk, and the composition of the gut microbiome. Dietary fiber, abundant in whole grains, legumes, fruits, and vegetables, plays a vital role in maintaining regular bowel movements. It also enables colonic microbes to ferment fiber into beneficial short-chain fatty acids, helping prevent constipation, diverticular disease, and colorectal cancer. Drinking enough water, along with fiber, helps make stool softer and easier to pass. Frequent physical exercise acts as a gut stimulant, increasing intestinal motility and lowering transit time, thereby reducing the risk of constipation and functional bowel symptoms. On the other hand, eating more processed foods, saturated fats, and refined sugars seems to contribute to dysbiosis, low-grade inflammation, and diseases such as NAFLD and GERD. Besides what you eat, how you eat is also important: irregular meals, large late-night meals, and rapid eating may cause reflux and dyspeptic symptoms. Furthermore, lifestyle factors such as psychosocial stress, sleep loss, and disturbances of the biological clock have a significant impact on GI physiology. One way stress can make IBS and functional dyspepsia worse is by activating the brain-gut axis, thereby altering gut motility, permeability, and visceral sensitivity. Activities like sleep deprivation and shift work interfere with circadian rhythms that govern not only GI motility and secretion but also the microbiome's diurnal cycles, and persistent circadian disruption has been linked to exacerbated GI symptoms and metabolic derangements. A blend of a fiber-rich, dense diet, plenty of fluids, regular physical activity, stress management (e.g., mindfulness, cognitive behavioural techniques), and sleep hygiene usually makes the greatest difference in symptom alleviation and long-term maintenance of good digestive health. For certain ailments, herbal medicines may serve as an adjunct to lifestyle strategies, but they should be carefully incorporated alongside diet and behavior changes (Langley-Evans, 2013).

5.2 Basic principles of digestive physiology

5.2.1 Composition and function of the digestive tract

The GI tract is an interdependent system of organs that work together to provide the necessary stages of food preparation, digestion, and absorption. Digestion begins in the mouth, where food is broken down into manageable particle sizes through chewing, and salivary enzymes begin to break it down chemically. The piece of food then passes through the esophagus and into the stomach, where it is further mixed with acidic gastric secretions and undergoes partial chemical breakdown and dissolution (Ananthakrishnan et al., 2014).

Once in the stomach, the partially digested food particles pass through the pyloric sphincter and enter the small intestine, which is primarily the area of nutrient absorption into circulation. The small intestine is a long, coiled organ with several specialized surface structures that assist with nutrient absorption, primarily villi and microvilli, which increase its surface area (Litke-Wager et al., 2021). The remaining material then passes into the large intestine, where fluids and electrolytes are absorbed, and waste is compacted and prepared for elimination at the rectum. Each of these areas of the GI tract has a specialized function, and any failure in a given area may lead to digestive health issues (Camilleri, 2021).

5.2.2 Digestive processes: motility and absorption

Digestion involves both mechanical and chemical processes. Mechanical digestion is simply the physical breakdown and disintegration of food. Chemical digestion is the act of enzymatically breaking down food into absorbable nutrients (Hopkins, 2020). Absorption predominantly occurs in the small intestine, where nutrients are absorbed through the intestinal lining into the bloodstream. Motility is the coordinated contractions of the GI tract that move food through the digestive system. Disturbances in either of these processes will trigger conditions such as diarrhea or constipation (Gupta et al., 2020).

5.2.3 Digestive enzymes and secretions

The digestive system relies on a coordinated series of secretions, including saliva, gastric juice, bile, and pancreatic enzymes, which chemically breakdown the food we eat into nutrients that can be absorbed. Saliva, mainly produced by the parotid, submandibular, and sublingual glands, plays a major role in carbohydrate digestion through the action of salivary alpha-amylase (ptyalin). It contains lingual lipase that starts a pinpoint lipid hydrolysis, which is especially significant in neonates. Saliva also lubricates the food bolus, contains mucins and bicarbonate that serve as protective and buffering mechanisms for the oral mucosa, and provides antimicrobial proteins (lysozyme, lactoferrin, immunoglobulin A) that support oral and upper GI immunity. Gastric juice, which is secretion of gastric glands located in the fundus and body, contains hydrochloric acid (HCl), pepsinogen (which is converted to pepsin in the presence of acid), intrinsic factor, mucus, and electrolytes. HCl facilitates protein unfolding, provides an acidic environment (pH ~1.53) for pepsin activity, enhances iron absorption, and acts as a barrier against many ingested pathogens. Intrinsic factor is necessary for vitamin B12 absorption at the terminal ileum; if it is lacking (e.g., due to autoimmune gastritis), pernicious anemia will develop. Gastric secretions are closely regulated through cephalic, gastric, and intestinal phases through neural (vagal) and hormonal (gastrin, somatostatin) pathways.

Bile is produced by liver cells, stored in the gallbladder, and consists of bile salts, cholesterol, phospholipids, bilirubin, and electrolytes. Bile salts are amphipathic molecules with detergent properties that break dietary fats into small droplets, thus increasing the surface area on which pancreatic lipase can act and allowing the formation of micelles for the absorption of monoglycerides, free fatty acids, and fat-soluble vitamins (A, D, E, K). About 90% of bile salts secreted into the GI tract are reabsorbed in the distal ileum, so it is through this that the enterohepatic circulation is maintained and the pool of bile salts is conserved for the next meal. Pancreatic secretions consist of a wide variety of enzymes and a fluid rich in bicarbonate. Among the digestive enzymes secreted by the exocrine pancreas are trypsinogen, chymotrypsinogen (both activated in the duodenum by enterokinase/trypsin cascade), carboxypeptidases for complete protein digestion, lipase and colipase for breaking down triglycerides, and pancreatic amylase for the degradation of starch. The large amount of bicarbonate neutralizes the acidic gastric chyme, thereby raising the pH to an optimum level for pancreatic enzymes to be effective. The major secretory regulators of the pancreatic duct are secretin (which increases the secretion of alkaline fluid) and cholecystokinin (which promotes the release of enzymes and contraction of the gallbladder), both of which are stimulated by luminal acid and nutrients. If any secretory component is disturbed, such as saliva, HCl, bile salts, or the exocrine pancreas, this will result in the characteristic patterns of maldigestion (e.g., steatorrhea, fat-soluble vitamin deficiencies, malabsorption of proteins) and symptoms that will direct the clinical evaluation and therapeutic substitution strategies (Vardakou et al., 2011).

5.2.4 Gut-brain axis

A bidirectional communication network, known as the gut–brain axis, connects the enteric nervous system (ENS), the central nervous system (CNS), the immune system, endocrine signals, and the gut microbiota. The ENS, a large, semiautonomous neural network located in the gut wall (myenteric and submucosal plexuses), controls motility, secretion, local blood flow, and reflexes independently but is influenced by the CNS. Enteric neurons and glia utilize typical neurotransmitters (acetylcholine, nitric oxide) and a great variety of neuropeptides to regulate smooth muscle and epithelial functions. Gut-produced neurotransmitters and neuromodulators are pivotal for gut–brain communication. Enterochromaffin cells in the intestinal mucosa produce most of the body's serotonin (5-HT), which regulates motility, secretion, and visceral sensitivity through enteric and extrinsic afferent pathways. The vagus nerve constitutes the main afferent and efferent route for intercommunication between the gut and the brain, allowing for very rapid signaling. Vagal afferents transmit to brainstem nuclei that regulate autonomic output, appetite, mood, and systemic inflammatory responses, as well as mechanical and chemical stimuli. Sympathetic pathways and spinal afferents also take part, especially in pain signaling.

Metabolites derived from microbiota (such as short-chain fatty acids, secondary bile acids, and tryptophan metabolites) and immune mediators influence neural signaling and the integrity of the blood-brain barrier, hence connecting diet, microbial ecology, and neural function. The dysfunctional gut–brain axis is a major contributor to many FGIDs, including IBS and functional dyspepsia. The symptoms of altered motility, visceral hypersensitivity, maladaptive central pain processing, low-grade mucosal inflammation, and microbial perturbations are intertwined, and result in chronic symptoms. The psychological stressors induce the hypothalamic pituitary adrenal axis and autonomic circuits, which subsequently cause a worsening of gut symptoms via motility, permeability, and immune changes; on the other hand, gut dysfunction influences mood and cognition, thus there is a strong association of anxiety/depression with FGIDs. There are several therapeutic strategies for treating the gut–brain axis that encompass dietary changes, microbiome-based treatment (probiotics, prebiotics, antibiotics only in selected cases), neuromodulators (low-dose antidepressants), behavioral therapies (cognitive behavioral therapy, gut-directed hypnotherapy), and interventions that influence vagal tone. It is critical to have a good understanding of the intricate, multilayered communication between the gut and the brain when dealing with functional and some inflammatory GI disorders, where symptoms are mainly caused by altered signaling rather than structural pathology (Tilg and Grander, 2018).

5.3 Common digestive disorders

Digestive disorders are some of the most common health conditions worldwide, affecting people of all ages, and they largely affect the quality of life. Nutrition and lifestyle, infection, stress, or chronic illness may be contributing factors to such disorders. Understanding the nature and mechanisms underlying common GI complaints provides a basis for choosing appropriate herbal remedies and integrative treatment approaches (Pai et al., 2020).

5.3.1 Indigestion (dyspepsia)

Dyspepsia, or indigestion, is defined as an uncomfortable or painful feeling in the upper abdomen that usually occurs soon after eating. Dyspepsia may also occur with early satiety, bloating, nausea, belching, fullness, or a burning sensation in the upper abdomen. Dyspepsia may be organic due to conditions including gastritis, peptic ulcer disease, or gallbladder disease. In contrast, dyspepsia may be classified as functional dyspepsia and lacks an identifiable structural cause. Other contributing lifestyle factors are overeating, fatty/spicy foods, stress, alcohol, caffeine, and irregular eating (Moayyedi et al., 2017). Stress and anxiety might further accentuate symptoms by increasing gastric acid secretion and altering gut motility through the gut-brain axis.

Herbal remedies, such as fennel, peppermint, chamomile, and ginger, have been used for dyspepsia based on their carminative, antiinflammatory, and prokinetic effects. Ginger may increase gastric emptying, and peppermint relaxes the smooth muscles of the GI tract, which together reduce cramping and discomfort (Yoon and Kim, 2015).

Pathophysiology: Dyspepsia is caused by gastric motor dysfunction, such as delayed gastric emptying and impaired fundic relaxation; hypersensitivity to visceral stimuli; acid or pepsin damage; bile reflux; or low-grade mucosal inflammation. Functional dyspepsia is characterized by the absence of structural abnormalities but is thought to result from altered gut–brain communication and increased afferent sensitivity.

Limitations of conventional treatment: Available therapies include PPIs, H2 blockers, prokinetics, *H. pylori* eradication when applicable, and neuromodulators. Some limitations include that many patients do not achieve complete symptom control; long-term use of PPIs may result in adverse effects (such as nutrient deficiencies and an increased risk of infections); some prokinetics may not be available or may be associated with safety concerns; and neuromodulators may produce varied responses.

5.3.2 Constipation

Constipation is generally described as passing difficult, infrequent, or incomplete bowel movements, very often accompanied by abdominal discomfort, straining, or the sensation of incomplete evacuation. It can be acute or chronic, and contributing factors range from dietary fiber and fluid intake insufficiency to physical inactivity, underlying metabolic conditions, for instance, hypothyroidism, or even side effects of medication, including opioids and iron supplements (Straumann and Schoepfer, 2016). From a functional point of view, constipation could mean either a disturbance in slow transit, pelvic-floor dysfunction, or an alteration in the gut microbiota. Chronic constipation can lead to impairment in quality of life and may further result in hemorrhoids or anal fissures (Jacobs and Richter, 2016).

Herbal therapies include bulk-forming laxatives such as psyllium husk, stimulant laxatives like senna and cascara, and demulcent herbs, used in appropriate amounts under supervision, such as aloe vera latex. Psyllium adds bulk to the stool and softens it with water, enhancing motility; senna stimulates intestinal contractions. Supporting herbal therapies, such as licorice and slippery elm, may soothe the lining of the intestines and support normal bowel elimination patterns (Suares and Ford, 2011).

Pathophysiology: Constipation results from **one or more of the following factors**: slow colonic transit, pelvic-floor dyssynergia, insufficient intake of dietary fiber and fluids, effects of medications, and metabolic disorders. Furthermore, an altered microbiome may play a role in perpetuating symptoms of chronic constipation.

Conventional treatment limitations: Bulk-forming agents, osmotics agents, stimulants, secretagogues, and biofeedback for dyssynergia are beneficial to the majority of patients. However, nonadherence, side effects (bloating, cramping), tolerance, and partial relief are common issues. Stimulant laxatives can also be abused.

5.3.3 Diarrhea

Diarrhea is defined as the passage of repetitive, loose, or watery stools; it may be acute or chronic. Common causes of diarrhea include viral and bacterial infections, food intolerances, drinking contaminated water, and dysbiosis induced by antibiotics or disease processes, including IBD or IBS with diarrhea. Symptoms of diarrhea can include abdominal cramping, urgency, bloating, or possibly dehydration; chronic diarrhea also presents risks for nutrient malabsorption and electrolyte imbalances (Thielman and Guerrant, 2004).

Antispasmodic, antiinflammatory, and soothing to the gut, herbs are chamomile, peppermint, marshmallow root, and blackberry root. Chamomile contains antiinflammatory phytochemicals (e.g., apigenin), which can help heal the gut mucosa, while peppermint reduces intestinal smooth muscle spasm. There have been historical uses of some herbs, including "goldenseal" for infectious causes of diarrhea, but they may pose challenges in terms of efficacy/safety due to their potency (In the Literature, 2016).

Pathophysiology: Acute infectious diarrhea is attributable to enteric pathogens; chronic diarrhea may be due to bile salt malabsorption, chronic infections, microscopic colitis, IBD, or functional diarrhea with rapid transit and secretion.

Conventional treatment limitations: Rehydration and pathogen-directed therapy are the mainstay; antimotility agents and antibiotics can be used wisely. The limitations are that these methods can mask a serious infection, promote resistance, and fail to address dysbiosis or inflammatory processes.

5.3.4 Acid reflux and gastroesophageal reflux disease

Acid reflux into the esophagus occurs when there is relaxation or incompetence of the lower esophageal sphincter. If symptoms persist, this may represent GERD, a chronic disorder that may lead to esophagitis, cough, hoarseness, and, in its most severe form, Barrett's esophagus. Common triggering foods of acid reflux include caffeine, chocolate, alcohol, citrus, high-fat meals, and lying down after a meal. Recent factors that have been identified with acid reflux include stress, obesity, and medications for other disease processes (Hwang et al., 2018).

In general, herbal therapy may help reduce dysbiosis and inflammation, soothe irritated mucosa, and support digestion. One of the most popular ingredients is deglycyrrhizinated licorice (DGL), which acts as a demulcent and may also enhance the mucosal barrier with mucosal protective action. Slippery elm, marshmallow root, and aloe vera gel can contain substantial amounts of mucilaginous compounds, which coat and soothe the esophageal and gastric linings. Some herbs may also normalize gastric acid output and enhance the lower esophageal sphincter (Spechler and Souza, 2014).

Pathophysiology: The primary mechanisms of acid reflux and esophagitis include transient lower esophageal sphincter relaxations, hiatal hernia, impaired esophageal clearance, and increased intra-abdominal pressure. Being overweight and the consumption of certain foods and beverages aggravate acid reflux.

Conventional treatment limitations: Dietary and behavioral changes, antacids, H2 blockers, and PPIs can control many cases of reflux disease. However, the treatment of refractory symptoms, PPI side effects, and nonacid reflux/functional heartburn is quite challenging. Prolonged use of PPIs requires careful consideration.

5.3.5 Bloating and gas bloating

Excessive gas is a very common complaint, the causes of which range from rapid eating, carbonated beverages, and high-fiber or fermentable foods to GI conditions like IBS, celiac disease, or small intestinal bacterial overgrowth. Abdominal distension, discomfort, and flatulence are some of the symptoms associated with gas (Chao and Vazquez, 2021). Carminative herbs are those that expel gas and reduce intestinal spasms. Commonly, fennel, ginger, peppermint, coriander, and cardamom facilitate digestion by reducing fermentation and helping prevent gut muscle spasms. Traditionally, fennel seeds are chewed at the end of meals in many cultures to aid digestion and reduce bloating (Wang et al., 2018).

Pathophysiology: At its core, too much gas results from fermentation of unabsorbed carbohydrates (like FODMAPs) in the intestine, SIBO, impaired movement of gas, or hypersensitivity of the intestines. Abdominal distension can result from altered function of the abdominal wall or excessive gas.

Conventional treatment limitations: The methods for eating (low FODMAP), simethicone, antibiotics for diagnosed SIBO, and prokinetics all help to varying degrees; however, the diet is hard to follow, and diagnosing/treating SIBO is not always reliable.

5.3.6 Functional gastrointestinal disorders (e.g., irritable bowel syndrome)

FGIDs include, but are not limited to, IBS, functional constipation, and functional dyspepsia, and present with chronic symptoms with no evidence of structural change. Features of FGIDs often coexist, including impaired gut–brain interaction, visceral hypersensitivity, motility issues, disturbances in the microbiome, and psychosocial components. IBS most commonly presents with recurrent abdominal pain associated with changes in bowel functioning (constipation, diarrhea, or both), bloating, and discomfort (Drossman and Hasler, 2016). All these disorders have a significant effect on daily functioning and mood. Herbal therapies are used in integrative treatments for multiple FGIDs. Peppermint oil is the most studied botanical agent for IBS. When taken in an enteric-coated capsule, peppermint oil has antispasmodic activity on the smooth muscle of the intestines. Other potentially appropriate botanicals include chamomile, turmeric, ginger, and artichoke leaf extract, all with potential antiinflammatory effects and the potential to improve motility by increasing bile flow. In some instances, where stress is a trigger for worsened symptoms of FGIDs, an adaptogenic herb may be utilized to modify the stress response, such as ashwagandha (Sayiner et al., 2016). Fig. 5.1 illustrates the major categories of common digestive disorders.

Pathophysiology: FGIDs result from the interplay of several factors, including impairment of the gut–brain axis, visceral hypersensitivity, dysmotility, low-grade inflammation, and microbial dysbiosis. Often, an individual's psychological state is a significant factor in determining symptom severity.

Conventional treatment limitations: Symptom-based treatments (fiber, antispasmodics, laxatives/antidiarrheals, antidepressants) can give relief, but only to a limited extent; many patients continue to suffer from the symptoms and the additional problem of anxiety/depression.

5.3.7 Inflammatory bowel disease

Herbal medicine as adjunctive therapy: Distinguish from IBS (pathophysiology): IBD (Crohn's disease, ulcerative colitis) comprises immune-mediated intestinal inflammation substantiated by objective tests and supported by the theory of dysregulated immune responses; hence, it manifests with cellular infiltrates and structural damage shown by biomarkers (elevated CRP, fecal calprotectin), endoscopic examination, and the imaging of the intestine contrarily to IBS that does not have these inflammatory markers and the structural pathology. Conventional treatment limitations: the use of aminosalicylates, corticosteroids, immunomodulators, biologics, and small molecules helps reduce inflammation; however, they come with risks (immunosuppression, secondary infections, and

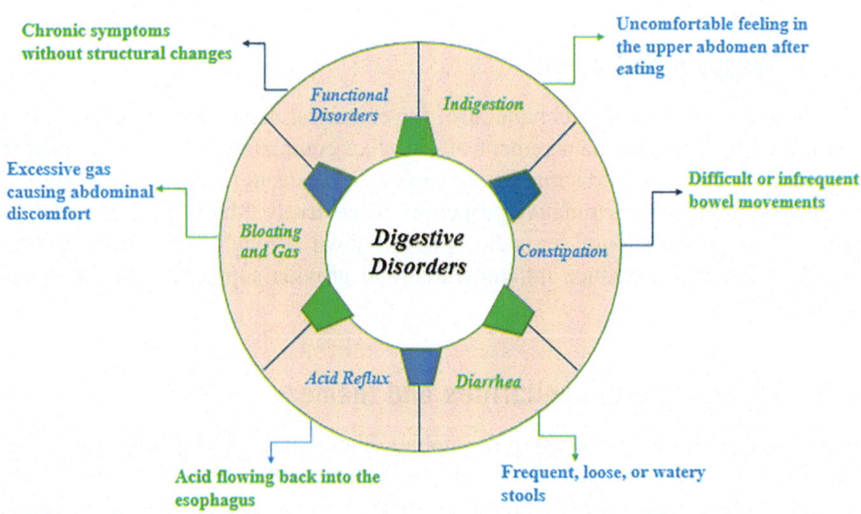

FIGURE 5.1

Common digestive disorders.

potential malignancy), high cost, and very limited long-term efficacy, and some patients are refractory (Asiri et al., 2023; Walker et al., 2020).

5.4 Historical and traditional perspective on digestive herbs

5.4.1 Ayurveda

Ayurveda, the ancient Indian system of healing, strongly advocates maintaining equilibrium among the body's three doshas: vata, pitta, and kapha. Proper digestion, or agni, is strongly linked to good health in this system, and herbal remedies to modulate the digestion processes are recommended in abundance. Traditionally, ginger, turmeric, and fennel have been an important part of Ayurvedic therapy, thought to kindle the fire of digestion, relieve bloating and indigestion, and normalize physiology throughout the GI tract (Ahmed et al., 2005).

5.4.2 Traditional Chinese medicine

TCM views digestion as the body's energetic center for transforming food into vital energy, or Qi. In practice, TCM also commonly uses licorice root and ginger to tonify the spleen and stomach, facilitating the easy movement of Qi and helping to rectify the deeper imbalance that causes symptoms of digestive disturbance. These herbs work together in formulas designed to restore harmony, rather than alleviate single symptoms (Candan et al., 2003).

5.4.3 Western herbalism

The western herbal traditions have also, for centuries, depended on various plant-based treatments in the management of digestive complaints. Chamomile, peppermint, and caraway are herbs that have enjoyed high esteem through many centuries for their soothing and carminative properties, respectively. Many Western herbalists specialize in using natural remedies passed down through generations to ease digestive discomfort, reduce inflammation, and provide support to the GI system (Hart, 2009).

5.4.4 Cross-cultural similarities and themes

While there may be divergent philosophies, the shared healing traditions often converge on the important issue of digestive health. Most digestive herbs, regardless of tradition, promote digestion, soothe the stomach, relieve inflammation, and support gut health. In nearly every traditional system, periods of strong emphasis are placed on the relevant connections between diet, lifestyle, and digestive function (Simões et al., 2020).

5.5 Prominent herbal remedies for aiding digestive health

5.5.1 Ginger (*Zingiber officinale*)

Ginger contains many important compounds that have shown antinausea, antiinflammatory, and antioxidant properties, as well as promoting digestion. Ginger enhances gastric emptying, reduces nausea, and decreases sensations of fullness after a meal by stimulating the release of saliva and digestive enzymes. Clinical studies have provided the most substantial scientific evidence regarding the antinausea effects of ginger, which are particularly relevant for pregnant women and chemotherapy patients. Ginger is generally deemed safe for most people, and although high-dose ginger may mildly irritate the GI tract, side effects are very rare. Pregnant women should seek the advice of an appropriately trained prescriber before using any concentrated preparation of ginger (Hu, 2010).

Botanical description: Ginger is a perennial herb of the family Zingiberaceae cultivated for its aromatic, fleshy rhizome; it has narrow green leaves and yellowish flowers and may be used fresh, dried, or as an extract.

Key phytochemicals: The major constituents that facilitate ginger's medicinal properties are gingerols, shogaols, paradols, zingerone, and the volatile essential oils (primarily zingiberene).

Pharmacological actions: Ginger has been shown to have gastroprokinetic effects (enhances gastric emptying), antiemetic activity (by inhibiting serotonin and other pathways), antiinflammatory and antioxidant properties, and slight antispasmodic effects.

Clinical evidence: Several randomized controlled trials and metaanalyzes have demonstrated ginger's efficacy in the treatment of pregnancy-related nausea, postsurgical, and chemotherapy-induced nausea, and some studies have identified effects on dyspeptic symptoms and gastric emptying; doses and formulations differ between studies.

Traditional usage: In Asian and Middle Eastern usage, ginger is used to treat nausea, indigestion, and colic, as well as to "warm" the digestive system. It can be prepared as tea, tincture, powder, or encapsulated extract.

Safety profile: At typical culinary and commonly supplemented doses, ginger is generally safe; taking large amounts may result in minor GI symptoms, heartburn, or a heightened bleeding risk if taken with anticoagulants, so be cautious in pregnancy and when using anticoagulants (Aimad et al., 2024).

5.5.2 Peppermint (*Mentha piperita*)

Peppermint (*M. piperita*) contains menthol as its main active ingredient, which gives it its well-known cooling and soothing properties in the GI tract. It produces effects by relaxing smooth muscle in the gut, which can help reduce spasms and be particularly helpful for functional disorders like IBS and generalized dyspepsia. Clinical trials have shown peppermint oil, especially in an enteric-coated capsule form, may help alleviate IBS-related pain, bloating, and discomfort. Overall, safe for most people, peppermint can also relax the lower esophageal sphincter and therefore could make symptoms worse for those with GERD (Weerts et al., 2020).

Botanical description: *M. piperita*, a hybrid mint, has aromatic, toothed leaves and purple flowers; it is a source of menthol oil and is cultivated worldwide.

Key phytochemicals: The essential oil mainly consists of menthol, menthone, menthyl acetate, and other volatile terpenoids; rosmarinic acid and flavonoids are also found.

Pharmacological actions: Peppermint oil induces relaxation of intestinal smooth muscle probably through blockade of calcium channels; it also possesses antispasmodic, carminative, and mildly antimicrobial properties. Besides, menthol influences sensory ion channels.

Clinical evidence: Randomized controlled trials have demonstrated the benefit of enteric-coated peppermint oil for IBS pain and global symptoms, with metaanalyzes indicating moderate efficacy over placebo; dose and formulation affect tolerability.

Traditional usage: Peppermint is used as a tea or oil mainly for the treatment of indigestion, colic, and flatulence and has also been used to flavor foods. Besides that, topical and aromatherapy uses are quite popular.

Safety profile: Usually safe when enteric-coated to prevent heartburn; can relax the lower esophageal sphincter and worsen GERD; avoid concentrated oil in young children and with G6PD deficiency concerns in rare cases (Mahendran and Rahman, 2020).

5.5.3 Licorice (*Glycyrrhiza glabra*)

Licorice root has, over centuries of folk medicine, earned a reputation as an herbal remedy that calms inflamed gut tissue. The active agent in licorice, glycyrrhizin, has a wide range of antiinflammatory and mucosal protective effects, which are useful for acid reflux, gastritis, ulcers, and so on. DGL, is a novel formulation, deemed safer for long-term ingestion (Isbrucker and Burdock, 2006).

Botanical description: Licorice is a legume that grows perennially, and the sweet roots of this plant are mainly used medicinally and for flavoring purposes; the roots are rich in glycosides and flavonoids.

Key phytochemicals: Glycyrrhizin/glycyrrhetinic acid, flavonoids (liquiritin), coumarins, and chalcones are the main components that contribute to the antiinflammatory and mucosal protective properties of the plant.

Pharmacological actions: DGL as well as whole root extracts exhibit mucosal protective, antiinflammatory, and mild antimicrobial properties; glycyrrhizin affects cortisol metabolism, and there are also antiviral claims.

Clinical evidence: DGL has been used in the past, and small clinical studies indicate its efficacy for peptic ulcer symptom relief and mucosal support; however, there are very few larger modern RCTs, and these are quite diverse.

Traditional usage: It has been used for gastritis, peptic ulcers, and sore throat, and as a demulcent, mainly in the form of lozenges and powders.

Safety profile: Long-term excessive consumption of whole root glycyrrhizin can lead to pseudoaldosteronism (hypertension, hypokalemia, edema). DGL extracts eliminate glycyrrhizin and are therefore safer for prolonged use, but product quality is variable (Pastorino et al., 2018).

5.5.4 Chamomile (*Matricaria chamomilla*)

While chamomile is often thought of as a sedative, it actually calms inflammation and muscle tension in the GI tract to aid digestion. Main uses include indigestion, gas, and bloating; even today, chamomile tea is the a gentle remedy for mild GI distress (Menghini et al., 2016).

Botanical description: Chamomile is a daisy-like herb with flowers that belong to the Asteraceae family; the flowers are used for tea and extracts.

Key phytochemicals: Apigenin, bisabolol, chamazulene, flavonoids, and volatile oils are the key ingredients responsible for the plant's antiinflammatory and antispasmodic effects.

Pharmacological actions: Chamomile possesses antispasmodic, antiinflammatory, mild sedative/anxiolytic, and mucosal-soothing effects that are beneficial for GI cramps and mild inflammation.

Clinical evidence: There are only a limited number of clinical trials, but they support the use of chamomile for indigestion and mild GI spasm, and as a sleep/anxiety adjunct; results of studies on standardized extracts vary.

Traditional usage: The tea has been a major part of the traditional use of chamomile for digestion, colic, and sleep; it is also used externally for inflammation.

Safety profile: Generally safe; very few cases of allergy to Asteraceae **have been reported, with** possible interaction with sedatives (Dai et al., 2023; El Mihyaoui et al., 2022).

5.5.5 Turmeric (*Curcuma longa*)

Turmeric contains a high curcumin content, a potent antioxidant and antiinflammatory substance that can improve digestive health by reducing intestinal inflammation, supporting liver function, and encouraging bile release to assist in the digestion of dietary fats (Zhang et al., 2016).

Botanical description: Turmeric is a perennial, rhizome-bearing plant of the ginger family, whose striking yellow rhizomes serve as a spice and medicine.

Key phytochemicals: The main active ingredients are the curcuminoids (curcumin, demethoxycurcumin) and the volatile oils (turmerones).

Pharmacological actions: Curcumin is an antiinflammatory, antioxidant, and bile-stimulating agent; it affects nuclear factor kappa B (NF-κB), cytokines, and other inflammatory pathways.

Clinical evidence: RCTs and metaanalyzes indicate that curcumin can be beneficial for mildly inflamed conditions and can be used as an adjuvant in ulcerative colitis therapy; however, formulations need to be optimized due to bioavailability issues.

Traditional usage: In Ayurvedic and traditional medicine, it has been used to enhance digestion, relieve bile stagnation, and treat inflammatory disorders.

Safety profile: In general, culinary amounts are safe; high-dose extracts may cause GI discomfort and interact with anticoagulant drugs, so the formulation and dose should be considered carefully.

5.5.6 Fennel (*Foeniculum vulgare*)

Historically, people have used fennel seeds for relief of gas, bloating, and abdominal cramping. Its carminative properties may help relieve intestinal gas and improve digestion. Fennel is usually eaten after meals. It can provide comforting support to the digestive system (Badgujar et al., 2014).

Botanical description: Fennel is an herbaceous plant of the Apiaceae family, which produces flowers and has leaves resembling feathers. Fennel seeds are aromatic and are used not only as a spice but also as medicine.

Key phytochemicals: The carminative and antispasmodic effects are derived from major constituents: anethole, fenchone, estragole, and flavonoids.

Pharmacological actions: Fennel seeds have antispasmodic properties that help reduce intestinal spasms. They also prevent the formation of gas and possess very slight antimicrobial and antiinflammatory activity; in addition, they stimulate digestion and relieve colic.

Clinical evidence: Although large RCTs are limited, small studies have demonstrated the benefit of fennel for infantile colic and for symptomatic relief of bloating/gas in adults.

Traditional usage: In many cultures, chewing the seeds right after meals or drinking them in tea is known to help reduce bloating and facilitate digestion.

Safety profile: In general, it is safe when used as a food; however, the concentrated essential oils may be poisonous if taken in large quantities, and there are theoretical carcinogenicity concerns with estragole. Therefore do not use it excessively over a long period (Hop and Son, 2023).

5.5.7 Aloe Vera (*Aloe barbadensis* miller)

The aloe vera gel contains constituents known to soothe the digestive tract, reduce intestinal inflammation, and promote regularity. When used correctly, aloe vera gel should relieve symptoms of acid reflux and positively impact gut health. The forms of aloe used as laxatives (i.e., aloin-containing extracts of aloe) require a thoughtful approach (Shi et al., 2021).

Botanical description: Aloe vera is a succulent from the Asphodelaceae family; the gel inside the leaf is used for medicinal purposes, while the latex (outer leaf) contains anthraquinone laxatives.

Key phytochemicals: Polysaccharides (acemannan), anthraquinones (aloin/aloesin in latex), sterols, and glycoproteins contribute to the impact on mucosa and motility.

Pharmacological actions: Aloe gel is mucosal-soothing, antiinflammatory, and has wound-healing properties; aloe latex is a very powerful stimulant laxative that acts through colonic irritation.

Clinical evidence: Topical and limited oral studies show mucosal benefits and symptomatic relief (e.g., in some ulcerative colitis trials with gel extracts); however, the quality of the evidence varies, and the formulations differ.

Traditional usage: Gel for soothing GI inflammations and topical wounds; latex was a laxative in the old days.

Safety profile: Oral latex (aloin) causes cramping and electrolyte loss and should not be employed for long-term use; the inner gel is better tolerated, but product purity and contamination risks require standardized preparations and medical supervision in cases of pregnancy and renal disease (Matara et al., 2021). Table 5.1 summarizes key herbal remedies commonly used to support digestive health, highlighting their essential actions and benefits.

5.6 Mechanisms of action of digestive herbs

5.6.1 Antiinflammatory effects

Antiinflammatory action is an important feature of many digestive herbs, and these herbs play a critical role in the management of gastritis, IBD, and various other irritation-based gut disorders. The bioactive phytochemicals of such plants reduce proinflammatory mediators and enzymes; for example, turmeric (curcumin) and ginger (gingerols). Such herbs not only reduce discomfort by decreasing inflammation in the digestive tract but also support the regeneration of mucosal tissues, contributing to better digestive resilience and long-term gut health (Zhou et al., 2022).

5.6.2 Antispasmodic activity

Various herbs that have historically been used to support GI function also have effects that can relieve smooth muscle activity in the GI system. Peppermint oil is one of the better-known examples; it mostly relieves abdominal spasms and cramping. This is an example of an herb with antispasmodic activity that can be quite helpful for people with issues such as IBS, as contraction of intestinal muscles can cause pain and irregular motility for many. By relieving smooth muscle spasm, antispasmodic herbs promote more comfortable, effective digestive function (Kelber et al., 2018).

5.6.3 Carminative properties

Carminative herbs, such as chamomile and fennel, reduce bloating, gas accumulation, and feelings of fullness. These plants allow the release and passage of gas that is usually trapped in the digestive system, easing intestinal muscles and thereby

Table 5.1 Prominent herbal remedies for aiding digestive health.

Herbal remedy	Key components	Primary actions on digestion	Clinical benefits/uses	Precautions
Ginger (*Zingiber officinale*)	Gingerols, shogaols	Enhances gastric emptying; stimulates saliva and digestive enzyme release; reduces nausea	Useful for nausea, indigestion, fullness, strong evidence in pregnancy-related nausea and chemotherapy-induced nausea	High doses may irritate gastrointestinal (GI) tract; pregnant women should seek guidance before concentrated use
Peppermint (*Mentha piperita*)	Menthol	Smooth muscle relaxation in GI tract; reduces spasms	Helpful for irritable bowel syndrome, dyspepsia; enteric-coated capsules reduce pain, bloating, discomfort	May worsen gastroesophageal reflux disease due to relaxation of lower esophageal sphincter
Licorice (*Glycyrrhiza glabra*)	Glycyrrhizin; deglycyrrhizinated liquorice (DGL)	Antiinflammatory; mucosal protective; supports healing of gut lining	Useful for acid reflux, gastritis, ulcers	Long-term use of regular licorice may affect blood pressure; DGL is safer
Chamomile (*Matricaria chamomilla*)	Apigenin, flavonoids	Reduces GI inflammation; relieves muscle tension	Helpful for indigestion, gas, bloating, chamomile tea is used for mild GI distress	Very safe; caution with allergies to Asteraceae family
Turmeric (*Curcuma longa*)	Curcumin	Antiinflammatory; antioxidant; supports bile release; improves liver-assisted digestion	Beneficial for intestinal inflammation and fat digestion	Excessive doses may irritate stomach or interact with anticoagulants
Fennel (*Foeniculum vulgare*)	Anethole, fenchone	Carminative effect; reduces gas; relieves cramping	Useful for gas, bloating, often eaten after meals to support digestion	Generally safe; excessive intake may cause photosensitivity
Aloe Vera (*Aloe barbadensis miller*)	Polysaccharides, vitamins, enzymes	Soothes intestinal lining; reduces inflammation; promotes bowel regularity	Helps with acid reflux, gut irritation, mild constipation	Laxative forms containing aloin must be used cautiously

improving digestion as a whole. These herbs are commonly used for indigestion and, for that reason, are very important allies for those who frequently experience digestive discomfort after meals (Wynn and Fougère, 2007).

5.6.4 Management of motility of the digestive tract

Some herbs can directly influence bowel movement coordination and speed. Ginger, for example, is an herb that can increase gastric emptying and facilitate regular intestinal motility, making it beneficial in situations of nausea and slow digestion and bowel transit. There are also herbs like senna that stimulate peristalsis directly and can be used safely and effectively in the short term to treat constipation. Knowing the different herbs that influence motility allows for a more individualized and appropriate use of herbs depending on the digestive complaint presenting (Hu, 2010).

5.6.5 Manipulate the microbiome

The gut microbiome is at the crux of digestion, immunity, and metabolic health. Some herbs are a useful means of contributing to a positive microbiome balance by introducing prebiotics, or substances that serve as a food source for beneficial bacteria. For example, chicory root and garlic contain beneficial fibers, such as inulin, which feed specific probiotic species. These herbs also improve digestive function and bioavailability/nutrient absorption, restore balance to the microbiome, and improve immunity (Hughes et al., 2022).

5.6.6 Mucosal protection and repair

Certain herbs directly protect and repair the intestinal lining. Aloe vera gel is soothing for inflamed tissues and is therefore utilized in protocols addressing the discomfort related to reflux or ulcers. Licorice root (an active ingredient derived from licorice root, glycyrrhizin) stimulates mucosal regeneration and creates a physical barrier that protects against injury. This pertains to digestive integrity and minimizes chronic irritation (Uygun et al., 2004) Fig. 5.2 illustrates the various mechanisms through which digestive herbs exert their therapeutic effects.

5.6.7 Antioxidant pathways in digestive protection

Oxidative stress, characterized by an imbalance between reactive oxygen species (ROS) and antioxidant defenses, is a major factor in mucosal damage, barrier dysfunction, and delayed wound healing, resulting from the increased inflammatory signaling typical of many digestive disorders such as gastritis, IBD, ischemia-reperfusion, and medication-induced injury. Phytochemicals with antioxidant properties, such as polyphenols, curcuminoids, gingerols, and flavonoids, are abundant in most digestive herbs and act synergistically through a dual mechanism

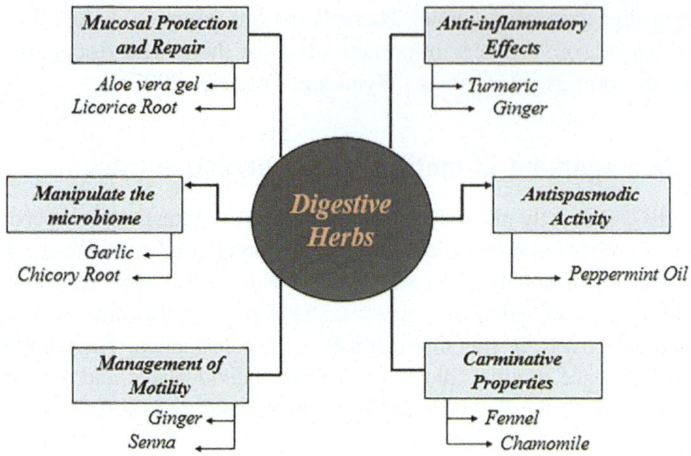

FIGURE 5.2 Mechanisms of action of digestive herbs.

Various digestive herbs action.

to safeguard the gut. In addition to direct ROS scavenging, they also trigger the body's natural cytoprotective pathways. On a molecular scale, multiple phytochemicals induce the nuclear factor erythroid2-related factor 2 (Nrf2) controlled gene network, thereby elevating the transcription of phase II detoxification enzymes and antioxidant proteins such as glutathione S-transferases, heme oxygenase-1, catalase, and superoxide dismutase. Nrf2-activated responses dampen oxidative damage to lipids and proteins in epithelial cells, help preserve tight junction proteins (claudin, occludin, ZO1), thereby upholding the barrier's integrity and reducing the passage of bacteria and antigens from the lumen.

Antioxidant effects also lessen proinflammatory signaling via ROS, which are the upstream activators of NF-κB. Antioxidant phytochemicals lower the transcription of cyclooxygenase-2 (COX-2) and various proinflammatory cytokines (TNF, IL-1, IL-6) by reducing ROS and also directly inhibiting NF-κB activation. A decrease in COX-2 expression reduces prostaglandin-mediated pain and swelling, while a reduction in cytokine production prevents immune overactivation and tissue damage. These synergistic antioxidant and antiinflammatory effects provide a logical explanation for the mucosal protective effects of Nrf2, as they activate phytochemicals (such as resveratrol, curcumin, and some flavonoids) in preclinical intestinal models and in limited clinical trials. Still, formulation and bioavailability are major factors that determine clinical efficacy.

5.6.8 Immunomodulatory effects of digestive herbs

Herbs regulate gut immunity through a variety of mechanisms, sometimes overlapping, that alter innate and adaptive responses, thereby supporting mucosal tolerance and controlled inflammation. At the signaling level, a large number of botanicals block NF-kB and mitogen activated protein kinase (MAPK) pathways in epithelial and immune cells, thus preventing COX-2 expression and the subsequent chain of proinflammatory cytokines (TNF, IFN, IL-17). At the same time, some phytochemicals induce the production of antiinflammatory cytokines (IL-10, TGF), as well as promote regulatory T cell (Treg) differentiation or tolerogenic dendritic cell phenotypes which are pathways by which the gut immune system becomes tolerant to dietary antigens and commensal microbes and prevents chronic, excessive immune activation typical for IBD and some food-triggered disorders.

Herbmicrobiome interactions form the core of this story: prebiotic fibers and polyphenols not only alter microbial composition and metabolite production (most notably short-chain fatty acids like butyrate) but also promote epithelial health, Treg induction, and barrier function. Herbal immunomodulation, by reducing cytokine release mediated by NF-kB and the expression of COX-2, prevents neutrophil and macrophage recruitment, thus limiting the extent of tissue damage; however, the initiating pathogen and directed immune defense remain unaffected. Still, herbs with a broad immunosuppressive action could theoretically impair host defense or have drug interactions with immunosuppressants; hence, in severe inflammatory disease, the use of herbal immunomodulators should be limited to adjuncts under strict medical supervision with monitoring of biomarkers (CRP, fecal calprotectin) and clinical status (Zhuang et al., 2019).

5.7 Dosage forms and preparation methods

5.7.1 Teas and infusions

Teas and infusions are among the most accessible and popular ways of extracting useful compounds from gentle, aromatic herbs. Steeping leaves, flowers, or soft plant materials in hot water is a method of releasing the therapeutic constituents of herbs. Examples of such herbs include peppermint, chamomile, and lemon balm. These preparations can be gentle and, when taken regularly, support the digestive system (Bone and Mills, 2013).

5.7.2 Decoctions

Decoctions are made by boiling the woody parts of the plant—in this instance, roots, barks, and seeds—for the extraction of the active principles. Ginger, licorice, and dandelion root are excellent examples of plants that can be decocted. It should be noted that the decoctions and teas made from these plants may be stronger and contain more concentrated active principles; they are often utilized in traditional herbal practices for their assertive effects (Jackson, 2002).

5.7.3 Tinctures and extracts

A tincture is simply a concentrated liquid herbal extract obtained when herbs are soaked in alcohol, glycerin, or vinegar. The preservation method retains the herb's active components and is excellent—it has a long shelf life and is easy to dose. Tinctures are warranted, especially if a person requires accurate dosing or a standardized strength of herbal support (Iwuji et al., 2021).

5.7.4 Capsules and powders

Capsules and powders are convenient forms that offer tasteless alternatives to teas or tinctures. Capsules may contain either dried, powdered herbs or standardized extracts, ensuring uniform potency. Capsules are good for people who want their herbal preparations to be straightforward to take on the go (Hossain et al., 2022).

5.7.5 Topical and external use (when applicable)

Although most digestive herbs are used internally, the preparations described in this section may be applied topically to provide symptomatic support in certain cases. For example, aloe vera gel is sometimes applied topically to soothe irritated skin, and an herbal poultice or compress may be used to help relieve abdominal tension or discomfort. Such topical routes provide local relief where needed (Samarh et al., 2017).

5.7.6 Standardization and quality considerations

Quality cannot be assumed when working with herbal medicine. Standardized extracts ensure a consistent amount of key active constituents is delivered per dose. Reliable products must be sourced from manufacturers that follow GMP and test for contaminants such as heavy metals, pesticides, and microbial impurities. Quality herbs ensure safer, more predictable therapeutic responses (Chaachouay and Zidane, 2024). Fig. 5.3 illustrates the major methods used in herbal medicine preparation.

5.8 Connections among herbs

5.8.1 Historical blends

Most forms of traditional healing are based on herbal blends that enhance healing. For example, classical Chinese herbal medicine uses formulas such as Xiao Feng San, which combines several herbs to address certain types of complex digestive or bodily imbalances. These historical blends are the result of hundreds of years of

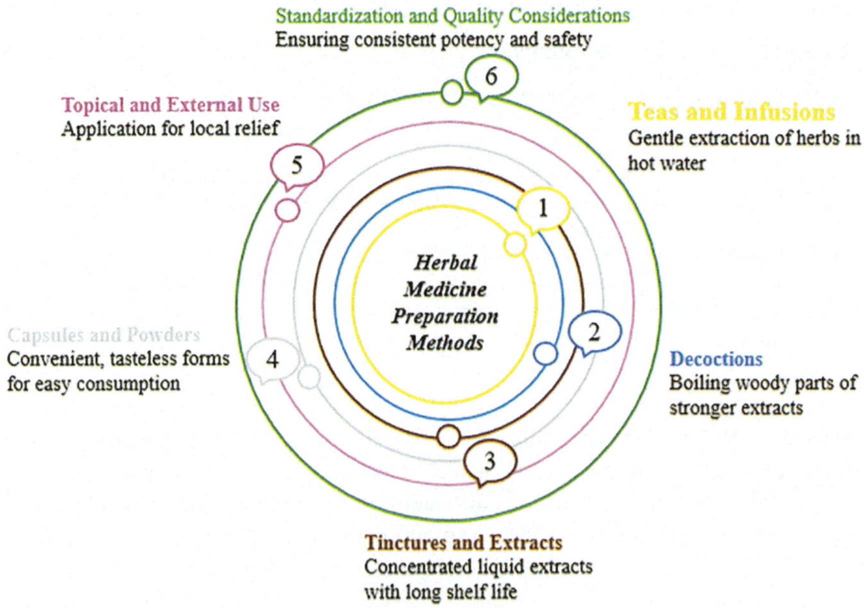

FIGURE 5.3 Herbal medicine preparation methods.

Various preparation methods of herbal drugs.

observational knowledge. They are meant to be blended in a way that allows each herb to work in synchrony, optimizing the medicinal properties of the others (Zhang et al., 2010).

5.8.2 Evidence-based combinations

Research within the modern sciences increasingly indicates that the combined use of herbal therapies is the most effective approach. Research suggests that combinations of herbs, such as ginger and peppermint, are more effective at alleviating nausea, indigestion, and stomach discomfort than single herbs. The availability of evidence-based herb combinations recognizes the inclusion of traditional wisdom and clinical experience and further validates their role in contemporary healthcare (Jugran et al., 2021).

5.8.3 Herb–nutrient synergy

Just like many herbs have an additive effect with nutrients to support digestive health, prebiotic and probiotic herbs may be used together to support beneficial gut flora. Similarly, herbs that promote bile flow may promote the digestion and

Table 5.2 Connections among digestive herbs.

Concept	Summary
Historical blends	Traditional systems combine multiple herbs for synergistic effects; centuries-old formulations optimize collective healing actions.
Evidence-based combinations	Modern research supports herbal combinations (e.g., ginger + peppermint) as more effective than single herbs for digestive relief.
Herb–nutrient synergy	Herbs work alongside nutrients, prebiotics, and probiotics to enhance gut flora, bile flow, and nutrient absorption.

absorption of fats and fat-soluble vitamins. The practical applications of herbs underscore the importance of a holistic, integrated approach to digestive health (Slavin, 2013). Table 5.2 illustrates the major mechanisms through which digestive herbs exert their therapeutic effects, providing a clear overview of their functional roles.

5.9 Safety, contraindications, and drug interactions

5.9.1 General guidelines for safety

Many individuals consider herbal treatments safe because they are often referred to as "natural." It is important to use herbal therapeutics responsibly. Patients should be encouraged to contact a qualified healthcare and/or alternative health professional before using an herb as treatment, especially when there is an underlying medical condition or concomitant prescriptions. Take recommended dosages, record potential side effects, and remember the therapeutic limits of the herb to help ensure safe and effective use (Posadzki et al., 2013).

5.9.2 Herb–drug interactions

Some herbs may increase the metabolic rate of certain medications, leading to higher or lower therapeutic effects. For example, St. John's Wort induces an increased metabolism of drugs like antidepressants, anticoagulants, and several immunosuppressants, thereby reducing the efficacy of these drugs. Herb–drug interactions are important factors in patient care, helping prevent complications and achieve the best treatment outcomes. Careful evaluation and good rapport between patients and healthcare providers are essential to avoid adverse reactions (Zhou et al., 2005).

5.9.3 Special populations (children, pregnancy, elderly)

Some groups may be more sensitive to herbal interventions. Children, older adults, and pregnant or breastfeeding women may need different dosages or may need to avoid certain herbs altogether. For example, some botanicals can stimulate uterine contractions or cross the placenta, making them potentially dangerous during pregnancy. Healthcare providers should be consulted before herbal recommendations are given to these populations to ensure safety and appropriateness (Ahmed et al., 2018).

5.9.4 Toxicity concerns and misuse

Herbs can be dangerous if they are used improperly or taken in excess. For instance, comfrey contains pyrrolizidine alkaloids, which are considered toxic to the liver when taken internally. Using stimulant laxative herbs too much can lead to electrolyte imbalances or dependency. Proper education for consumers regarding appropriate use, quality products, and adherence to evidence-based dosing could help minimize the risks of toxicity and misuse (Stickel and Seitz, 2000).

5.9.5 Regulatory aspects of herbal medicines

Regulators worldwide have broadly similar objectives about herbal medicine, ensuring the safety of the public, guaranteeing the quality of products and proper labeling, and basing the requirement of safety and/or efficacy evidence on the extent of the risk involved. Still, they differ in the breadth of issues covered and the methods employed. The World Health Organization (WHO) has issued documents offering guidance, urging member states to establish regulatory frameworks, set quality standards, implement pharmacovigilance systems, and implement good manufacturing practice (GMP) for herbal products. The WHO scenarios for herbal medicines stress the following: unambiguous terminology; monographs or pharmacopeial standards for plant identity and markers; setting maximum allowable levels for different types of contamination (heavy metals, pesticides, microbial load); providing stability and shelf-life information; and possessing systems for adverse event reporting.

At a national level, there are differences in regulatory approaches: some countries classify herbal products as foods/dietary supplements, thereby subjecting them to minimal premarket controls, whereas others require registration as medicines with dossier reviews, clinical data, and stricter GMP. The AYUSH system of India (Ayurveda, Yoga and Naturopathy, Unani, Siddha, and Homeopathy) is a formal, regulatory, and educational framework for traditional medicines that, among other things, provides manufacturing standards, pharmacopoeias, and a distinction between classical and proprietary formulations. The purpose of AYUSH is to bring together traditional knowledge, quality control, and safety monitoring. However, it is challenging in terms of standardizing across the vast herbal diversity, as well as ensuring that modern clinical evidence is available where appropriate.

Regulation revolves largely around quality control issues. The safety and efficacy of the products are compromised by the variability of raw materials (species substitution, different chemotypes), inconsistency of phytochemical content, adulteration with undeclared pharmaceuticals, and contamination (heavy metals, microbial, pesticide residues). The regulatory framework hence focuses on botanical authentication (botanical taxonomy, macroscopic/microscopic and DNA barcoding where available), determination of marker compounds, batch standardization, extraction and manufacturing processes validation, and product monitoring after launch. The implementation of supply chain traceability and Good agricultural and collection practices will help reduce variability at the source. Generally speaking, regulators weigh historical data on use against current scientific evidence to determine appropriate premarket applications and postmarket supervision.

5.9.6 Herb–herb interactions

Herb–herb interactions refer to situations where components of two or more botanicals affect each other's pharmacokinetics (how the body absorbs, distributes, metabolizes, and excretes a substance) or pharmacodynamics (whether the combined effect is additive, synergistic, or antagonistic). Mechanisms include: drug-metabolizing enzyme induction or inhibition (mainly CYP450 isoforms) and transporter (P-gp) modulation; receptor competition; exacerbation of toxicities, such as liver toxicity; and changes to the gut microbiota that alter phytochemical biotransformation.

Examples and clinical significance: Taking more than one stimulant laxative herb (senna + cascara) at the same time is likely to increase the risk of cramping, electrolyte loss, and tolerance; the use of several anticoagulant modulating herbs (high-dose garlic, ginkgo, ginseng) at the same time can intensify the risk of hemorrhage through additive antiplatelet or anticoagulant effects; using herbs with compounded hepatotoxic potential (kava, high-dose senna anthraquinones in vulnerable patients) can lead to liver damage. Traditionally, polyherbal formulations have been concocted so that their constituents balance each other and mitigate adverse effects, but such combinations pose significant challenges for modern safety assessment.

Risk mitigation strategies: get complete product ingredient lists, refrain from unnecessary polypharmacy of botanicals that have similar adverse profiles, implement combinations supported by scientific evidence, restrict doses/duration of potentially toxic herbs, check relevant labs (electrolytes, liver enzymes) when necessary, and advise patients to disclose their herb use to the healthcare provider. Both manufacturers and practitioners should emphasize standardized extracts with quantified marker compounds, and regulators should demand transparent labeling to promote safe combined use (Calixto, 2000).

5.10 Clinical applications and case examples

5.10.1 Herbal remedies for indigestion

Herbs such as ginger and peppermint are internationally recognized for their therapeutic potential, particularly for relieving indigestion symptoms. An abundance of published clinical experience and case reports documents their application in reducing bloating, gas, and general upper abdominal discomfort. Herbal remedies are an attractive adjunct and/or alternative to conventional antacids and prokinetics (Baek et al., 2025).

5.10.2 Constipation protocol

The therapy of constipation with herbal remedies typically centers around stimulant laxatives such as senna and bulk-forming fibers such as psyllium. Evidence from relevant clinical practice guidelines supports the recommendation to manage constipation and maintain regular bowel habits. This therapy, when appropriately prescribed, can help with stool consistency and bowel frequency, making it useful for both acute and chronic constipation (Icks et al., 2005).

5.10.3 Herbs for symptoms or signs of irritable bowel syndrome

Peppermint oil, chamomile, and other relaxing herbs have significant potential for treating several IBS-related symptoms. Clinical evidence has shown improvement in abdominal pain, bloating, and overall GI comfort. Improvement in quality of life can be achieved through herbal remedies within a treatment paradigm tailored to the individual (Cash et al., 2016).

5.10.4 Digestive problems associated with stress

Stress influences digestive function in many ways, and adaptogenic herbs like ashwagandha and valerian root counter this influence through their mechanisms of action, which help the body adapt to stress. The incorporation of adaptogens with digestive therapies may lead to more complete relief, as both emotional stressors and physical toxins can become triggers for symptomatic GI complaints (Salve et al., 2019).

5.10.5 Integrative strategies in clinical practice

More clinicians are integrating herbal medicine into their practice alongside traditional treatment protocols, and the literature includes case studies showing that geriatric or integrative models of care increase patient satisfaction and clinical outcomes. Combining traditional wisdom and modern-day clinical practices may prompt patients to seek individualized, comprehensive treatment strategies (Choudhary et al., 2017).

5.11 Current research and future directions

5.11.1 Recent clinical trials

Recent clinical studies explored the safety and efficacy of various herbs for GI conditions. These current trials expand the evidence base, shedding light on how individual botanicals may relieve disorders such as IBS, functional dyspepsia, and acid reflux. The current literature supports continued integration of herbal medicine into the care of digestive health (Lacy et al., 2022).

5.11.2 Advances in phytochemistry

Modern advances in phytochemistry, with the clear elucidation of the molecular composition of herbs, enable the isolation of active compounds that may explain their therapeutic properties. The research eventually leads to the development of standardized extracts, and then ultimately to more specific herbal products (Akhgarjand et al., 2022).

5.11.3 New botanical therapies

Several new plant formulations and herbal composites are being investigated for their effects on digestive health (i.e., plant-based health products). Researchers are investigating new combinations, plant extracts, and delivery systems to improve effects and bioavailability. These new therapies are expected to offer additional choices for patients and clinicians alike in the future (Wang et al., 2023).

5.11.4 Research gaps and challenges

Scientific interest in herbal medicine is growing, but there are still some substantial gaps in research at this point. The limitations include a lack of large-scale clinical trials, inconsistent product quality, and limited data on long-term safety and herb–drug interactions. Overcoming such limitations is extremely necessary for the advancement of this field and its complete acceptance in modern healthcare (Zhou et al., 2019).

5.12 Conclusion

Traditional wisdom, integrated with modern scientific research, offers a sound foundation for enhancing digestive health, combining thousands of years of empirical herbal knowledge with today's standards for safety and efficacy to develop richer treatment strategies and improve patient care. The need for clear guidelines on the safe and effective use of herbal remedies will encourage patients to consult professionals, adhere to proper dosages, and understand potential interactions. At the same time, healthcare providers give evidence-based advice to ensure responsible use. In general, when used appropriately and evidence-based, herbal modalities can serve as meaningful facilitators of digestive well-being and be complementary to traditional medicine. As knowledge continues to expand, herbal approaches will likely become an increasingly important aspect of integrative healthcare's future, enabling holistic contributions to digestive health with proper public education, adherence to safe practices, and the continued pursuit of scientific inquiry.

References

Ahmed M, Hwang JH, Hasan MA, Han D. Herbal medicine use by pregnant women in Bangladesh: a cross-sectional study. BMC Complementary and Alternative Medicine 2018;18(1). Available from https://doi.org/10.1186/s12906-018-2399-y, http://www.biomedcentral.com/bmccomplementalternmed/.

Ahmed S, Anuntiyo J, Malemud CJ, Haqqi TM. Biological basis for the use of botanicals in osteoarthritis and rheumatoid arthritis: a review. Evidence-based Complementary and Alternative Medicine 2005;2(3):301–8. Available from https://doi.org/10.1093/ecam/neh117.

Aimad A, Mssillou I, Siddique F, Mohmmed B, Abdelkrim A, Kara M, et al. Phytochemical analysis and pharmacological activities of essential oils extracted from *Zingiber officinale* (*Roscoe*) used in mediterranean diet: in vitro and in silico studies. International Journal of Food Properties 2024;27(1):1180–99. Available from https://doi.org/10.1080/10942912.2024.2387937.

Akhgarjand C, Asoudeh F, Bagheri A, Kalantar Z, Vahabi Z, Shab-bidar S, et al. Does Ashwagandha supplementation have a beneficial effect on the management of anxiety and stress? A systematic review and meta-analysis of randomized controlled trials. Phytotherapy Research 2022;36(11):4115–24. Available from https://doi.org/10.1002/ptr.7598.

Ananthakrishnan AN, Cheng S–C, Cai T, Cagan A, Gainer VS, Szolovits P, et al. Serum inflammatory markers and risk of colorectal cancer in patients with inflammatory bowel diseases. Clinical Gastroenterology and Hepatology 2014;12(8):1342. Available from https://doi.org/10.1016/j.cgh.2013.12.030.

Asiri AS, Algarni SS, Althubaiti AQ, Alzubaidi MA, Alghamdi JA, Almalki GA. Fecal calprotectin and organic gastrointestinal disease: a systematic review. Cureus 2023;15(9). Available from https://doi.org/10.7759/cureus.45019.

Badgujar SB, Patel VV, Bandivdekar AH. Foeniculum vulgare mill: a review of its botany, phytochemistry, pharmacology, contemporary application, and toxicology. BioMed Research International 2014;2014. Available from https://doi.org/10.1155/2014/842674, http://www.hindawi.com/journals/biomed/.

Baek HI, Ha NR, Kim C, Im TJ, Kim YY, Hwang SH, et al. Efficacy and safety of steamed ginger extract for gastric health: a randomized, double-blind, placebo-controlled multicenter clinical trial. Food and Function 2025;16(18):7316–29. Available from https://doi.org/10.1039/d5fo01172h, http://pubs.rsc.org/en/journals/journal/fo.

Bent S. Herbal medicine in the united states: review of efficacy, safety, and regulation. Journal of General Internal Medicine 2008;23(6):854–9. Available from https://doi.org/10.1007/s11606-008-0632-y.

Biswas TK, Pandit S, Chakrabarti S, Banerjee S, Poyra N, Seal T. Evaluation of *Cynodon dactylon* for wound healing activity. Journal of Ethnopharmacology 2017;197:128–37. Available from https://doi.org/10.1016/j.jep.2016.07.065.

Black CJ, Ford AC. Global burden of irritable bowel syndrome: trends, predictions and risk factors. Nature Reviews Gastroenterology and Hepatology 2020;17(8):473–86. Available from https://doi.org/10.1038/s41575-020-0286-8, http://www.nature.com/nrgastro/index.html.

Bone K, Mills S, editors. Principles and practice of phytotherapy. Elsevier eBooks; 2013. Available from https://doi.org/10.1016/c2009-0-48725-7.

Calixto JB. Efficacy, safety, quality control, marketing and regulatory guidelines for herbal medicines (*Phytotherapeutic agents).* Brazilian Journal of Medical and Biological Research 2000;33(2):179–89. Available from https://doi.org/10.1590/s0100-879x2000000200004.

Camilleri M. Gastrointestinal motility disorders in neurologic disease. Journal of Clinical Investigation 2021;131(4). Available from https://doi.org/10.1172/JCI143771, https://www.jci.org/articles/view/143771/pdf.

Candan F, Unlu M, Tepe B, Daferera D, Polissiou M, Sökmen A, et al. Antioxidant and antimicrobial activity of the essential oil and methanol extracts of *Achillea millefolium* subsp. millefolium Afan. (Asteraceae). Journal of Ethnopharmacology 2003;87(2-3):215–20. Available from https://doi.org/10.1016/s0378-8741(03)00149-1.

Cash BD, Epstein MS, Shah SM. A novel delivery system of peppermint oil is an effective therapy for irritable bowel syndrome symptoms. Digestive Diseases and Sciences 2016;61(2):560–71. Available from https://doi.org/10.1007/s10620-015-3858-7, www.wkap.nl/journalhome.htm/0163-2116.

Chaachouay N, Zidane L. Plant-derived natural products: a source for drug discovery and development. Drugs and Drug Candidates 2024;3(1):184–207. Available from https://doi.org/10.3390/ddc3010011.

Chao A, Vazquez JA. Fungal infections of the gastrointestinal tract. Gastroenterology Clinics of North America 2021;50(2):243–60. Available from https://doi.org/10.1016/j.gtc.2021.02.009.

Choudhary D, Bhattacharyya S, Joshi K. Body weight management in adults under chronic stress through treatment with ashwagandha root extract. Journal of Evidence-Based Complementary & Alternative Medicine 2017;22(1):96–106. Available from https://doi.org/10.1177/2156587216641830.

Curtin BF, Abichandani S, von Rosenvinge EC. Toasted: the perils of home remedy. Gastroenterology 2017;153(2):355–6. Available from https://doi.org/10.1053/j.gastro.2017.01.038, http://www.journals.elsevier.com/gastroenterology/.

Dai YL, Li Y, Wang Q, Niu FJ, Li KW, Wang YY, et al. Chamomile: a review of its traditional uses, chemical constituents, pharmacological activities and quality control studies. Molecules (Basel, Switzerland) 2023;28(1). Available from https://doi.org/10.3390/molecules28010133, http://www.mdpi.com/journal/molecules.

Dekker E, Chiu HM, Lansdorp-Vogelaar I, Caro LE, Dominitz JA, Halloran S, et al. Colorectal cancer screening in the novel coronavirus disease-2019 Era. Gastroenterology 2020;159(6):1998–2003. Available from https://doi.org/10.1053/j.gastro.2020.09.018, http://www.journals.elsevier.com/gastroenterology/.

Delfino BM, Pereira TM, Martins AC, Guimarães AS, Araújo TS. Author's response to reviews: Nutritional status of children under 5 years of age in the Brazilian Western Amazon before and after the Interoceanic Highway paving: a population-based study. 2013.

Drossman DA, Hasler WL. Rome IV - functional GI disorders: disorders of gut-brain interaction. Gastroenterology 2016;150(6):1257–61. Available from https://doi.org/10.1053/j.gastro.2016.03.035, http://www.journals.elsevier.com/gastroenterology/.

El Mihyaoui A, Esteves da Silva JCG, Charfi S, Candela Castillo ME, Lamarti A, Arnao MB. Chamomile (*Matricaria chamomilla* L.): a review of ethnomedicinal use, phytochemistry and pharmacological uses. Life (Chicago, Ill.: 1978) 2022;12(4):479. Available from https://doi.org/10.3390/life12040479.

Gupta V, Gupta I, Park J, Bram Y, Schwartz RE. Hedgehog signaling demarcates a niche of fibrogenic peribiliary mesenchymal cells. Gastroenterology 2020;159(2):624. Available from https://doi.org/10.1053/j.gastro.2020.03.075, http://www.journals.elsevier.com/gastroenterology/.

Hart J. CAM and medical education. Alternative and Complementary Therapies 2009;15(6):287–91. Available from https://doi.org/10.1089/act.2009.15602.

Hop NQ, Son NT. The medicinal plant agrimonia pilosa ledeb.: botanical description, traditional use, phytochemistry and pharmacology. Combinatorial Chemistry & High Throughput Screening 2023;26(9):1660–88. Available from https://doi.org/10.2174/1386207325666220928163930.

Hopkins SR. Ventilation/perfusion relationships and gas exchange: measurement approaches. Comprehensive Physiology 2020;10(3):1155–205. Available from https://doi.org/10.1002/cphy.c180042, https://onlinelibrary.wiley.com/doi/epdf/10.1002/cphy.c180042.

Hossain CM, Gera M, Ali KA. Current status and challenges of herbal drug development and regulatory aspect: a global perspective. Asian Journal of Pharmaceutical and Clinical Research 2022:31–41. Available from https://doi.org/10.22159/ajpcr.2022.v15i12.46134.

Hu ML. Effect of ginger on gastric motility and functional dyspepsia symptoms. World Journal of Gastroenterology 2010;17(1):105. Available from https://doi.org/10.3748/wjg.v17.i1.105.

Hughes RL, Alvarado DA, Swanson KS, Holscher HD. The prebiotic potential of inulin-type fructans: a systematic review. Advances in Nutrition 2022;13(2):492–529. Available from https://doi.org/10.1093/advances/nmab119, http://advances.nutrition.org/.

Hwang BK, Yoon Y-J, Han SY, Lee Han Chae SJ. Validity of yin-yang temperament in sasang personality questionnaire. Integrative Medicine Research 2018;7(1):77–84. Available from https://doi.org/10.1016/j.imr.2018.01.001.

Icks A, Rathmann W, Haastert B, John J, Löwel H, Holle R, et al. Cost-effectiveness of type 2 diabetes screening: results from recently published studies. Das Gesundheitswesen 2005;67(S 01):167–71. Available from https://doi.org/10.1055/s-2005-858232.

In the literature. Clinical Infectious Diseases 2016;62(10):iii–. Available from https://doi.org/10.1093/cid/ciw071.

Isbrucker RA, Burdock GA. Risk and safety assessment on the consumption of licorice root (*Glycyrrhiza* sp.), its extract and powder as a food ingredient, with emphasis on the pharmacology and toxicology of glycyrrhizin. Regulatory Toxicology and Pharmacology 2006;46(3):167–92. Available from https://doi.org/10.1016/j.yrtph.2006.06.002.

Iwuji SC, Ogbonna CV, Iwu CI, Okafor WC, Chibuike EC. Comparative effects of solvents on the herbal extraction of antidiabetic phytochemicals. Journal of Pharmaceutical

Research International 2021:149–59. Available from https://doi.org/10.9734/jpri/2021/v33i28b31549.

Jackson EA. Resources for information on herbal medicinals and dietary supplements. Journal of Herbal Pharmacotherapy 2002;2(1):63–70. Available from https://doi.org/10.1080/J157v02n01_09.

Jacobs J, Richter JE. Opening the bird's beak: tips and tricks for effective pneumatic dilation for Achalasia. American Journal of Gastroenterology 2016;111(2):157–8. Available from https://doi.org/10.1038/ajg.2015.435, http://www.nature.com/ajg/archive/index.html.

Jugran AK, Rawat S, Devkota HP, Bhatt ID, Rawal RS. Diabetes and plant-derived natural products: from ethnopharmacological approaches to their potential for modern drug discovery and development. Phytotherapy Research 2021;35(1):223–45. Available from https://doi.org/10.1002/ptr.6821, http://onlinelibrary.wiley.com/journal/10.1002/(ISSN)1099-1573.

Kelber O, Bauer R, Kubelka W. Phytotherapy in functional gastrointestinal disorders. Digestive Diseases 2018;35(Suppl. 1):36–42. Available from https://doi.org/10.1159/000485489.

Lacy BE, Chey WD, Epstein MS, Shah SM, Corsino P, Zeitzoff LR, et al. A novel duodenal-release formulation of caraway oil and L-menthol is a safe, effective and well tolerated therapy for functional dyspepsia. BMC Gastroenterology 2022;22(1). Available from https://doi.org/10.1186/s12876-022-02181-5, http://www.biomedcentral.com/bmcgastroenterol/.

Langley-Evans S. Nutrition: A Lifespan Approach. New York, NY: John Wiley & Sons; 2013.

Lete I, Allué J. The effectiveness of ginger in the prevention of nausea and vomiting during pregnancy and chemotherapy. Integrative Medicine Insights 2016;11. Available from https://doi.org/10.4137/imi.s36273.

Litke-Wager C, Delaney H, Mu T, Sawyer T. Impact of task-oriented role assignment on neonatal resuscitation performance: a simulation-based randomized controlled trial. American Journal of Perinatology 2021;38(09):914–21. Available from https://doi.org/10.1055/s-0039-3402751.

Mahendran G, Rahman LU. Ethnomedicinal, phytochemical and pharmacological updates on peppermint (*Mentha* × *piperitas* L.)—a review. Phytotherapy Research 2020;34(9):2088–139. Available from https://doi.org/10.1002/ptr.6664, http://onlinelibrary.wiley.com/journal/10.1002/(ISSN)1099-1573.

Matara DN, Nguta JM, Musila FM, Mapenay IO, Ali HM, Omambia VM. Botanical description, ethnomedicinal uses, phytochemistry and pharmacological effects of Croton dichogamus Pax (*Euphorbiaceae*). The Journal of Phytopharmacology 2021;10(1):42–7. Available from https://doi.org/10.31254/phyto.2021.10109.

Menghini L, Ferrante C, Leporini L, Recinella L, Chiavaroli A, Leone S, et al. An hydroalcoholic chamomile extract modulates inflammatory and immune response in HT29 cells and isolated rat colon. Phytotherapy Research 2016;30(9):1513–8. Available from https://doi.org/10.1002/ptr.5655.

Moayyedi PM, Lacy BE, Andrews CN, Enns, Howden CW, Vakil N. ACG and CAG clinical guideline: management of dyspepsia. American Journal of Gastroenterology 2017;112(7):988–1013. Available from https://doi.org/10.1038/ajg.2017.154, http://www.nature.com/ajg/archive/index.html.

Pai RK, Hartman DJ, Rivers CR, Regueiro M, Schwartz M, Binion DG, et al. Complete resolution of mucosal neutrophils associates with improved long-term clinical outcomes of patients with ulcerative colitis. Clinical Gastroenterology and Hepatology

2020;18(11):2510. Available from https://doi.org/10.1016/j.cgh.2019.12.011, http://www.elsevier.com/inca/publications/store/6/7/2/7/4/3/index.htt.

Pastorino G, Cornara L, Soares S, Rodrigues F, Oliveira MBPP. Liquorice (*Glycyrrhiza glabra*): a phytochemical and pharmacological review. Phytotherapy Research 2018; 32(12):2323–39. Available from https://doi.org/10.1002/ptr.6178, http://onlinelibrary.wiley.com/journal/10.1002/(ISSN)1099-1573.

Peyrin-Biroulet L. Will the quality of research remain the same during the COVID-19 pandemic? Clinical Gastroenterology and Hepatology 2020;18(9):2142. Available from https://doi.org/10.1016/j.cgh.2020.03.054.

Posadzki P, Watson L, Ernst E. Herb–drug interactions: an overview of systematic reviews. British Journal of Clinical Pharmacology 2013;75(3):603–18. Available from https://doi.org/10.1111/j.1365-2125.2012.04350.x.

Salve J, Pate S, Debnath K, Langade D. Adaptogenic and anxiolytic effects of ashwagandha root extract in healthy adults: a double-blind, randomized, placebo-controlled clinical study. Cureus 2019;11(12). Available from https://doi.org/10.7759/cureus.6466.

Samarh SN, Khalaf NA, Hajhamad MM. Evidence based medical use of aloe vera extracts, short review of literature. International Journal of Research in Medical Sciences 2017;5(10):4198. Available from https://doi.org/10.18203/2320-6012.ijrms20174547.

Sayiner M, Stepanova M, Pham H, Noor B, Walters M, Younossi ZM. Assessment of health utilities and quality of life in patients with non-alcoholic fatty liver disease. BMJ Open Gastroenterology 2016;3(1):e000106. Available from https://doi.org/10.1136/bmjgast-2016-000106.

Shi G, Jiang H, Feng J, Zheng X, Zhang D, Jiang C, et al. Aloe vera mitigates dextran sulfate sodium-induced ulcerative colitis in rats by potentiating the colon mucus barrier. Journal of Ethnopharmacology 2021;279:114108. Available from https://doi.org/10.1016/j.jep.2021.114108. Apr 8.

Simões DM, Malheiros J, Antunes PE, Figueirinha A, Cotrim MD, Fonseca DA. Vascular activity of infusion and fractions of *Cymbopogon citratus* (DC) Stapf. in human arteries. Journal of Ethnopharmacology 2020;258. Available from https://doi.org/10.1016/j.jep.2020.112947, www.elsevier.com/locate/jethpharm.

Slavin J. Fiber and prebiotics: mechanisms and health benefits. Nutrients 2013;5(4):1417–35. Available from https://doi.org/10.3390/nu5041417, http://www.mdpi.com/2072-6643/5/4/1417/pdf.

Spechler SJ, Souza RF. Barrett's esophagus. New England Journal of Medicine 2014;371(9): 836–845. Available from https://doi.org/10.1056/NEJMra1314704, http://www.nejm.org/doi/pdf/10.1056/NEJMra1314704.

Stickel F, Seitz HK. The efficacy and safety of comfrey. Public Health Nutrition 2000;3(4a):501–8. Available from https://doi.org/10.1017/s1368980000000586.

Strate LL, Gralnek IM. ACG clinical guideline: management of patients with acute lower gastrointestinal bleeding. American Journal of Gastroenterology 2016;111(5):755. Available from https://doi.org/10.1038/ajg.2016.155.

Straumann A, Schoepfer A. Positioning of dilation in eosinophilic oesophagitis. Nature Reviews Gastroenterology & Hepatology 2016;13(4):192–4. Available from https://doi.org/10.1038/nrgastro.2016.38.

Suares NC, Ford AC. Prevalence of, and risk factors for, chronic idiopathic constipation in the community: systematic review and meta-analysis. American Journal of Gastroenterology 2011;106(9):1582–91. Available from https://doi.org/10.1038/ajg.2011.164.

Thielman NM, Guerrant RL. Acute infectious diarrhea. New England Journal of Medicine 2004;350(1):38–47. Available from https://doi.org/10.1056/NEJMcp031534.

Tilg H, Grander C. Microbiota and diabetes: an increasingly relevant association. Polish Archives of Internal Medicine 2018;128(6):333–5. Available from https://doi.org/10.20452/pamw.4286.

Uygun A, Kadayifci A, Isik AT, Ozgurtas T, Deveci S, Tuzun A, et al. Metformin in the treatment of patients with non-alcoholic steatohepatitis. Alimentary Pharmacology and Therapeutics 2004;19(5):537–44. Available from https://doi.org/10.1111/j.1365-2036.2004.01888.x.

Vardakou M, Mercuri A, Barker SA, Craig DQM, Faulks RM, Wickham MSJ. Achieving antral grinding forces in biorelevant In Vitro models: comparing the USP dissolution apparatus II and the dynamic gastric model with human In Vivo data. AAPS PharmSciTech 2011;12(2):620–6. Available from https://doi.org/10.1208/s12249-011-9616-z, http://www.springerlink.com/content/1530-9932/.

Walker GJ, Chanchlani N, Thomas A, Lin S, Moore L, Heerasing NM, et al. Primary care faecal calprotectin testing in children with suspected inflammatory bowel disease: a diagnostic accuracy study. Archives of Disease in Childhood 2020;105(10):957–63. Available from https://doi.org/10.1136/archdischild-2019-317823, http://adc.bmjjournals.com.

Wang H, Chen Y, Wang L, Liu Q, Yang S, Wang C. Advancing herbal medicine: enhancing product quality and safety through robust quality control practices. Frontiers in Pharmacology 2023;14. Available from https://doi.org/10.3389/fphar.2023.1265178.

Wang X, Niu C, Zhang X, Dong M. Emodin suppresses activation of hepatic stellate cells through p38 mitogen-activated protein kinase and Smad signaling pathways in vitro. Phytotherapy Research 2018;32(12):2436–46. Available from https://doi.org/10.1002/ptr.6182.

Weerts ZZRM, Masclee AAM, Witteman BJM, Clemens CHM, Winkens B, Brouwers JRBJ, et al. Efficacy and safety of peppermint oil in a randomized, double-blind trial of patients with irritable bowel syndrome. Gastroenterology 2020;158(1):123–36. Available from https://doi.org/10.1053/j.gastro.2019.08.026, http://www.journals.elsevier.com/gastroenterology/.

Wynn SG, Fougère BJ. Veterinary herbal medicine: a systems-based approach. Elsevier BV; 2007. p. 291–409. Available from http://doi.org/10.1016/b978-0-323-02998-8.50024-x.

Yoon H, Kim N. Diagnosis and management of high risk group for gastric cancer. Gut and Liver 2015;9(1):5–17. Available from https://doi.org/10.5009/gnl14118.

Zhang A, Sun H, Wang Z, Sun W, Wang P, Wang X. Metabolomics: towards understanding traditional Chinese medicine. Planta Medica 2010;76(17):2026–35. Available from https://doi.org/10.1055/s-0030-1250542.

Zhang X, Wu J, Ye B, Wang Q, Xie X, Shen H. Protective effect of curcumin on TNBS-induced intestinal inflammation is mediated through the JAK/STAT pathway. BMC Complementary and Alternative Medicine 2016;16(1). Available from https://doi.org/10.1186/s12906-016-1273-z, http://www.biomedcentral.com/bmccomplementalternmed/.

Zhou S, Chan SY, Goh BC, Chan E, Duan W, Huang M, et al. Mechanism-based inhibition of cytochrome P450 3A4 by therapeutic drugs. Clinical Pharmacokinetics 2005;44(3):279–304. Available from https://doi.org/10.2165/00003088-200544030-00005.

Zhou X, Afzal S, Wohlmuth H, Münch G, Leach D, Low M, et al. Synergistic anti-inflammatory activity of ginger and turmeric extracts in inhibiting lipopolysaccharide and interferon-γ-induced proinflammatory mediators. Molecules (Basel, Switzerland) 2022;27(12):3877. Available from https://doi.org/10.3390/molecules27123877.

Zhou X, Li C-G, Chang D, Bensoussan A. Current status and major challenges to the safety and efficacy presented by chinese herbal medicine. Medicines 2019;6(1):14. Available from https://doi.org/10.3390/medicines6010014.

Zhuang Y, Wu H, Wang X, He J, He S, Yin Y. Resveratrol attenuates oxidative stress-induced intestinal barrier injury through PI3K/Akt-mediated Nrf2 signaling pathway. Oxidative Medicine and Cellular Longevity 2019;2019:1–14. Available from https://doi.org/10.1155/2019/7591840.

PART

3

Botanicals for chronic conditions

CHAPTER

6 Botanicals for pain management

Praveen Kumar Vemuri[1], Pradeepthi Katragadda[2], Mahima Swaroopa Mandava[3], Kanaka Durga Devi Nelluri[2] and Sai Prakash Donka[4]

[1]*Department of Biotechnology, Koneru Lakshmaiah Education Foundation, Guntur, Andhra Pradesh, India,*

[2]*Department of Pharmaceutics and Biotechnology, KVSR Siddhartha College of Pharmaceutical Sciences, Vijayawada, Andhra Pradesh, India,*

[3]*Department of Pharmacy Practice, KL College of Pharmacy, KL Deemed to be University, Vijayawada, Andhra Pradesh, India,*

[4]*University of Wisconsin, Milwaukee, WI, United States*

6.1 Introduction

Pain is a very personal, complicated, and multifaceted thing; it is one of the most common reasons people go to the doctor. There is the physical part of pain, but there are also emotional and mental aspects. Pain helps us identify the injuries to our bodies. The International Association for the Study of Pain has defined pain as "an unpleasant experience that involves both sensory and emotional components, and it indicates that there is (or may be) a tissue injury", meaning that pain is a way to alert us to an injury. Acutely, one experiences rapid pain from a tissue injury, which eventually goes away; however, chronic pain is longer-lasting and can be considered a disease in itself. Chronic pain may occur because of conditions like arthritis, cancer, diabetes, neuropathy, musculoskeletal conditions, or inflammatory diseases. Chronic pain can significantly impact an individual's ability to function, cause considerable psychological distress, and decrease their quality of life (Cohen et al., 2021).

Numerous medications such as opioids, nonsteroidal antiInflammatory drugs (NSAIDs), OPIs, anticonvulsants, and antidepressants are currently being used for the treatment of pain across the medical field. The majority of the pain medications available today are either NSAIDs or prescription opiates, which healthcare professionals frequently prescribe. Antidepressants and anticonvulsants are also alternative treatments to manage immediate pain relief. Still, both classes typically have long-term negative side effects (exacerbating current problems), limiting their use by patients after the acute phase of their pain has subsided. Although NSAIDs have potential for several different gastrointestinal complications, including ulcers, as well as renal and lung toxicity, opioids have significant potential for addiction, tolerance, and severe respiratory depression due to their effects on the central nervous system (CNS). They are generally

Healing Botanicals. DOI: https://doi.org/10.1016/B978-0-443-44295-7.00009-4

seldom effective for chronic pain. If you suffer from chronic pain and desire a different kind of pain management, please consult with your physician about other alternatives may be available to help manage your chronic pain condition. Pain relief through the use of plants has been recognized for many years by numerous people around the world. Throughout history, people have used plant-based remedies to manage pain and develop medicines for a wide range of health conditions, including the Ayurvedic tradition (India), traditional Chinese herbal medicine, Unani medicine (Middle Eastern), and countless other indigenous healing modalities developed to help people cope with their particular environments. Most of the healing systems discussed above take a holistic and comprehensive approach to pain relief, providing patients with avenues not only to reduce pain but also to address the source of their pain and restore balance to their bodies. The majority of the plant materials used to alleviate pain contain a variety of additional molecules that have therapeutic effects through their mechanisms of action (e.g., alkaloids, flavonoids, terpenoids, polyphenols, glycosides, and so forth). Many plant-derived compounds have a range of therapeutic effects, including pain relief, antiinflammatory action, antioxidant effects, and/or neuromodulatory effects, through various mechanisms (Olalekan et al., 2024; Hu et al., 2022).

With the dissemination of advanced technologies and the advancements in pharmacology, phytochemistry, and molecular biology, there has been a growing trend toward incorporating scientific research into the evidence-based validation of traditional medicine. The results of scientific research indicate that some natural substances promote the synthesis of several important biochemicals, including, but not limited to, prostaglandins, cytokines, and nitric oxide (NO). These are believed to play an important role in regulating the body's inflammatory response. In addition, by inhibiting the cyclooxygenase (COX)-1 and COX-2 enzyme pathway (i.e., the enzymes that produce these mediators) and blocking the lipoxygenase (LOX) pathway, natural substances have been found to stop the production of the reactive oxygen species (ROS). Inhibition of COX enzymes reduces the production of additional prostaglandins, cytokines, and NO through the use of natural medicinal compounds. The majority of conventional pain-relief medicines target nociceptors; however, results show that many botanicals are effective not only for treating nociceptive but also neuropathic pain. Advancements in quality control, standardization, and information on bioavailability and herbal-drug interactions will continue to impede the broad use of botanical analgesics in clinical practice. Thus research must provide the necessary evidence that botanical analgesics can be a reasonable contributor to modern-day pain management through a rigorous, scientifically based study with conclusions supported by evidence regarding their effectiveness. The goal of this chapter is to provide a comprehensive review of current scientific information regarding the effectiveness of botanical analgesics and traditional knowledge regarding their use in treating pain, the mechanism(s) of action of botanicals used as analgesics, and the existing preclinical and clinical evidence regarding the effectiveness of botanicals as a treatment for pain, along with the potential future development of scientifically validated, clinically safe, and effective treatments for the management of pain (Forouzanfar and Hosseinzadeh, 2018; Blake, 2018).

6.2 Traditional use of botanicals in pain management

6.2.1 Ayurveda

In the Ayurvedic tradition, pain (shoola) is understood as rooted in an individual's unique constitution and qualities. Its classification is based on four major subcategories: (1) the cause, (2) where pain is located anatomically, (3) how long someone has had pain, and (4) the associated dosha (vata, pitta, kapha) imbalance of the three doshas, vata contributes most to a person's ability to feel pain because it affects nerve conduction, sensory perception, and movement throughout the body. Therefore the consequences of disturbed vata, especially when associated with increased pitta-induced inflammation, generally result in diseases such as osteoarthritis, sciatica, migraines, neuropathic pain, and degenerative musculoskeletal disorders. A person's treatment should include herbs/medicinal plants that possess vatahara (vata pacifying), shothahara (antiinflammatories), and vedanasthapana (painkillers). An example of an herb that possesses these qualities is *Boswellia serrata*, which has antiinflammatory properties by stopping leukotriene synthesis. Another example of an herb that possesses some of these same qualities is *Curcuma longa* (turmeric), which reduces inflammatory cytokines and oxidative stress in the body. The herb *Withania somnifera* is classified as a rasayana (herbal remedy) that may help protect the nervous system, reduce stress, and improve pain tolerance. It is important to note that in Ayurveda, when using multiple herbs together to achieve synergistic effects and/or enhanced bioavailability, while reducing the risk of adverse reactions, Ayurvedic practitioners often formulate polyherbal remedies. Additionally, external treatment options include the application of medicated oils, poultices, and heat therapy, which further complement the internal use of these herbs and together, extend Ayurveda's multifaceted approach to relieving pain (Thakur et al., 2020; Agarwal, 2024).

6.2.2 Traditional Chinese medicine

Traditional Chinese medicine (TCM) holds that pain results from a disruption in the flow of Qi (energy) and blood through the body's meridian system. When Qi and blood are flowing properly, tissues are supplied with nutrients, and there is no pain; however, if there is an obstruction or insufficient flow, pain will occur. These disruptions can be caused by external pathogenic factors such as wind, cold, dampness, and heat, which invade the body and create blockages in the meridian channels, or by Internal or Qi/blood imbalances. For example, a person may have a Qi or blood deficiency or stagnation due to chronic illness, emotional distress, or old age. The treatment goal of TCM for pain management is to eliminate obstructions, improve Qi/blood circulation, and achieve a balance of Yin and Yang energies within the body. The treatment is based on syndrome differentiation; therefore therapies can be modified depending on the type of pain a person has and its source. The use of plant medicines in TCM is an important factor in both providing symptomatic relief of pain and treating the disharmony that causes it, due to the way they work together and complement each other.

Pain relief in TCM is classified into several categories based on energetics (e.g., hot, warm, neutral, cool, cold), taste (e.g., sweet, sour, bitter, salty, pungent), and meridian connections. Pain-relieving herbs are used in the treatment of different types of pain, such as visceral pain, musculoskeletal pain, and traumatic pain. Among all the pain-relieving herbs in TCM, *Corydalis yanhusuo* is known as a highly potent pain reliever and is often used to treat any pain related to Qi stagnation and/or blood stagnation. *Angelica sinensis* (also known as Dang Gui) is also frequently used in TCM for pain relief and is especially valued for its ability to nourish and invigorate the blood. *A. sinensis* is frequently prescribed for gynaecological pain (eg, dysmenorrhea) as well as ischemic types of pain due to reduced blood flow. TCM is structured using the "Chief – Deputy – Assistant – Envoy" method to combine several herbs, resulting in greater potency, greater efficacy, and reduced toxicity. This type of herbal formulation parallels current combination drug therapy in modern Western medicine (Jiang et al., 2022; Matos et al., 2021).

6.2.3 Traditional and folk medicine systems of indigenous people

Indigenous communities have amassed extensive ethnobotanical knowledge from nature, which has led to the establishment of their traditional and folk medicine systems. The knowledge base concerning medicines developed over many centuries includes many pharmaceutical and chemical plant sources that are still used by indigenous peoples today. Medicinal plants have been identified through empirical observation based on how well they alleviate symptoms, such as pain and inflammation, and facilitate natural healing. Furthermore, medicinal plant use was established as safe and effective long before modern pharmacology existed.

Traditional and folk medicine systems are heavily reliant on botanical remedies for providing relief from pain via topical, oral, or inhalation routes. The method of administration is based on the type and severity of pain, the culture and region available, and the availability of plant material for making the remedy. Generally, topical applications (e.g., poultices, ointments, and compresses) are preferred for musculoskeletal discomforts, as well as for treating wounds and inflammatory conditions, because topically applied products act only at the site of application and have little to no systemic effect (i.e., through the bloodstream). Oral remedies (e.g., decoctions, infusions, and powdered forms) are primarily used to treat chronic and abdominal pain and systemic inflammatory disorders. Certain cultures also use inhalation of aromatic vapours or smoke to relieve headaches, neuropathic (nerve) pain, and stress-related pain. A large number of contemporary analgesics have their origins in tradition and folk medicine; traditional knowledge enabled the identification of plants with analgesic properties, which were later developed into modern drugs. An example is *Salix* species (willow bark), which has been used in European and Asian folk medicine to treat pain and fever. These uses have been scientifically tested and found to be associated with salicylates, the active ingredient related to the development of aspirin, a widely used analgesic and antiinflammatory drug.

In the same vein, *Capsicum* species have also been used by various cultures for the topical treatment of muscle and joint pain, leading to pain relief through excitation and subsequent desensitization of nociceptive sensory nerves, namely via TRPV1 channels. This has led to the development of modern capsaicin-based topical analgesics, which are widely used to treat osteoarthritis and neuropathic pain. One of the most interesting substances with historical importance for the development of current interpretations of pain relief is the herbal remedy *Harpagophytum procumbens* (commonly known as devil's claw), which is found in Southern Africa and used by many Indigenous peoples for inflammation. More recently, devil's claw has gained global recognition for its effectiveness in reducing arthritis and lower back pain due to its antiinflammatory and analgesic properties. Clinical and pharmacological trials have validated the efficacy of devil's claw for pain relief and improved mobility. Overall, traditional and folk medicine exemplify how indigenous knowledge represents the foundation for many modern pain management methodologies. Today, traditional methods continue to influence modern therapies; they provide an expanding body of potential bioactive compounds and holistic treatment approaches. The importance of traditional practice in conjunction with modern scientific validation for the continuing relevance of ethnobotanical knowledge is documented in the emerging evidence supporting the development and approval of safe and effective pain therapies (Kleinman, 2022; Jahromi et al., 2021).

6.3 Current use of botanicals in modern pain management

The limitations and side effects of conventional pain medications, such as the possibility of developing tolerance and/or addiction to opioid painkillers, have created new opportunities to study other methods for controlling pain, including the use of botanicals.

6.3.1 Integration into modern and western medicine

The contemporary integration of botanicals into practice (nutritional supplements, standardized herbal extracts, topical preparations, and adjunctive therapies) is increasing in Western medicine. Unlike botanicals used in historical practice, those used in modern healthcare are produced in a regulated manner, with quality controls, standardized production methods, and scientific evaluation to assess their uniformity, safety, and reproducibility of therapeutic benefits. The combination of these regulatory types has allowed botanicals to gain greater acceptance among healthcare practitioners and the regulatory bodies that govern their use. Some of the most commonly used analgesics in Western medicine are derived directly from natural sources or have been synthesized from naturally occurring compounds. Some examples include salicylic acid (the active ingredient in aspirin), which is produced from the bark of willows (Salix species), and capsaicin (which is used in topical applications for both neuropathic and orthopedic pain), derived from pepper plants (Capsicum species). Additionally, morphine (one of the most widely used medications for severe pain) is derived from the opium poppy (*Papaver somniferum*). Each of these naturally derived compounds has demonstrated strong clinical efficacy through known mechanisms of action, including inhibition of prostaglandin synthesis, modulation of nociceptive ion channels, and stimulation of opioid receptors.

There has been a growing trend towards the increased use of botanicals, also referred to as "Phytochemicals", in addition to standalone phytochemicals. These products have been used anecdotally for pain control, but there is now evidence of their effectiveness in chronic inflammatory or musculoskeletal disorders. For example, curcumin (from *C. longa*), boswellic acid (from *B. serrata*), and gingerol (from *Zingiber officinale*) have demonstrated both antiinflammatory and analgesic activity through several mechanisms, including suppression of cytokine production, inhibition of COX and LOX pathways, and modulation of oxidative stress. In practice, all botanical products are indicated for a variety of clinical conditions, including osteoarthritis, rheumatoid arthritis, and lower back pain. They may be prescribed alone for mild cases or as adjunctive treatments in conjunction with NSAIDs for more severe cases. Regulation of herbal products is evolving across Europe, Asia, and North America. Therefore to successfully incorporate herbal products into existing healthcare systems, regulatory agencies have established guidelines for the development of herbal products that meet established safety, quality, and clinical efficacy criteria. These regulatory guidelines have led to the development of evidence-based herbal products and increased practitioners' confidence in recommending and prescribing them. It is now common to see integrative medicine evolving through the combination of botanical therapies, with other conventional medical practices such as physiotherapy, acupuncture, and prescription analgesics. This multimodal approach will help maximize therapeutic outcomes and subsequently improve patients' quality of life while reducing the incidence of adverse reactions that may arise from the continued use of synthetic medications (Nema, 2025; Dalamagka, 2024).

6.3.2 Commonly used medicinal plants for pain relief

Medicinal plants have been used to manage pain in many traditional medicine systems for hundreds of years. As pharmacological research advanced and many of these traditional medicines were validated through clinical research, they have also received scientific validation, enabling their increased use in modern medicine. As research on botanical medicines continues to grow, the number of medicinal plants available for therapeutic use in modern clinical settings has increased, providing new alternatives or complementary methods for managing pain. The wide range of bioactive compounds in *C. longa* (turmeric) and their powerful antioxidant and antiinflammatory actions have made this medicinal plant one of the most widely researched for pain relief. Curcumin, the active component of turmeric, can modulate various inflammatory pathways by inhibiting proinflammatory cytokines, COX, and NF-kB. Numerous clinical trials have confirmed that turmeric extract has successfully reduced joint pain, swelling, and stiffness in patients with osteoarthritis and rheumatoid arthritis compared with conventional NSAIDs, with significantly fewer gastrointestinal side effects. The use of *B. serrata* (Indian frankincense) for its analgesic and antiinflammatory properties has been well documented. Boswellic acids are the active components in Boswellia resin. These compounds have been shown to inhibit 5-LOX, which is necessary for the synthesis of proinflammatory leukotrienes and chemokines. Clinical studies have shown that Boswellia reduces the severity of pain-related discomfort and movement limitations, and improves functional status in patients with arthritic-type joint disorders (as well as other musculoskeletal disorders), making this herb widely recommended for chronic pain conditions.

As history shows, ginger (*Z. officinale*) has long been used for medicinal purposes before becoming an accepted dietary supplement for pain relief. The active ingredients in ginger, which are thought to exert this effect, are named gingerol and shogaol. These substances inhibit prostaglandin synthesis by blocking COX and LOX pathways. There is considerable evidence from many clinical studies that demonstrates the efficacy of ginger for the treatment of postoperative pain, menstrual pain (dysmenorrhea), muscular/joint discomfort, and inflammation caused by competition or advancing degenerative joint diseases. The use of botanicals topically has become increasingly popular, as they offer localized pain relief with very few, if any, systemic side effects. One of the most popular topical agents is capsaicin, which comes from certain Capsicum species. Topical preparations containing capsaicin are used to manage neuropathy and other forms of neuropathic pain, such as diabetic and postherpetic neuralgia. Capsaicin activates TRPV1 receptors on nociceptive neurons; with repeated application, these receptors become desensitized, reducing pain signal transmission. *H. procumbens* (devil's claw), *Arnica montana* (arnica), and *Salix alba* (white willow) are also widely used in topical applications for the treatment of back pain, soft tissue injuries (muscle strains and sprains), and inflammatory challenges. Devil's claw works by inhibiting mediators that cause inflammation; arnica reduces bruising and muscle soreness; and willow bark has analgesic properties due to the salicylates it contains, which are natural precursors to aspirin. The increasing use of these medicinal plants for clinical pain management indicates that botanical treatments are being more widely integrated into traditional healthcare. Research studies and the growing number of patients who prefer herbal remedies or other complementary healing methods have strengthened support for their use in clinical settings (Yousofvand and Moloodi, 2023; Silva-Correa et al., 2021).

6.4 Phytochemical classes with analgesic potential

Medicinal plants exert analgesic and antiinflammatory effects largely because to the wide variety of phytochemicals they contain, whereas synthetic analgesics typically only interact with one receptor or pathway. Plant-based compounds can generally interact with multiple pathways and receptors, which may provide greater therapeutic options and possibly fewer side effects than synthetic analgesics, as mentioned in Table 6.1

6.4.1 Alkaloids

Alkaloids are compounds containing nitrogen that have a high level of pharmacological activity but are especially effective on the CNS and peripheral nervous system (PNS). In addition, they provide analgesic, antiinflammatory, and neuro-modulatory effects through their interactions with opioid receptors, voltage-gated ion channels, and neurotransmitter systems (Jin et al., 2022; Salehi et al., 2023).

6.4.1.1 Mechanism of action

Alkaloids exert their analgesic effects on the CNS via interactions with various types of opioid receptors; however, the predominant mechanism involves interaction with the μ-opioid receptor, stimulating that receptor similarly to endogenous opioid agonizts (endorphins and enkephalins). When activated, these receptors inhibit pain signal transmission by decreasing neuronal excitability at both spinal and supraspinal levels, thereby reducing neurotransmitter release. Alkaloids may also modulate sodium (Na+) and calcium (Ca2+) voltage-gated channels, thereby decreasing nociceptive signals. Alkaloids influence numerous neurotransmitter systems, such as the dopaminergic, adrenergic, and serotonergic systems, which are intimately related to the mesolimbic pain pathways and the higher-order emotional or reactive components of pain. In addition to inhibiting receptors involved in pain regulation, many alkaloids are known to have antiinflammatory actions by blocking the production and release of inflammatory cytokines and other mediators, such as tumour necrosis factor-alpha (TNF-α), interleukins, and prostaglandins. These actions produce both peripheral and central analgesia.

6.4.1.2 Examples

Numerous alkaloids have been studied and used for the treatment of pain, including morphine and codeine, both derived from the plant *P. somniferum* (the opium poppy). These alkaloids are traditional opioids and are still considered two of the most potent drugs to treat moderate to severe pain, especially in situations where a patient is recovering from surgery or suffering from a terminal illness or palliative care. Both drugs work by activating the μ-opioid receptors. Berberine is an isoquinoline alkaloid found in plants of the genus *Berberis*. It has remarkable antiinflammatory and analgesic properties. Its ability to reduce pain results from its

Table 6.1 Phytochemical classes with analgesic potential.

Phytochemical class	Key mechanisms of action	Representative examples	Clinical relevance
Alkaloids	μ-opioid receptor activation; modulation of central nervous system (CNS)/peripheral nervous system (PNS) neurotransmitters; inhibition of proinflammatory cytokines	Morphine; codeine (*Papaver somniferum*); berberine; tetrahydropalmatine	Potent analgesics; effective in severe pain; risk of tolerance and dependence
Flavonoids	Cyclooxygenase (COX)/lipoxygenases (LOX) inhibition; cytokine suppression; antioxidant activity	Quercetin; luteolin; kaempferol; rutin; hesperidin	Adjunct therapy in inflammatory and neuropathic pain; good safety profile
Terpenoids	Transient receptor potential channel modulation; CB2 receptor activation; GABAergic effects	Menthol; camphor; β-caryophyllene; ginsenosides	Common in topical and systemic pain formulations
Polyphenols	Reactive oxygen species scavenging; nuclear factor kappa B, COX, LOX inhibition; neuroprotection	Curcumin; resveratrol; ellagic acid	Useful in chronic inflammatory and neuropathic pain; bioavailability-enhanced formulations
Glycosides	COX/LOX modulation; CNS/PNS effects; prodrug activation	Salicin; harpagoside; ginseng glycosides	Widely used in standardized herbal products for musculoskeletal pain

modulation of signalling pathways involved in inflammation and oxidative stress; therefore it has been studied and tested for the treatment of inflammatory and neuropathic pain. A separate alkaloid, tetrahydropalmatine, has also been isolated from the tuber of the *C. yanhusuo* plant. A growing body of evidence suggests that this compound has potent analgesic activity based on experimental studies to date. The mechanism by which tetrahydropalmatine acts is thought to involve modulation of dopaminergic receptors and inhibition of neurotransmitters associated with the sensation of pain, consistent with its traditional use as a remedy for visceral and musculoskeletal pain.

6.4.1.3 ***Clinical significance***

Alkaloids are highly potent (especially in their role in the clinical management of pain), requiring a high degree of monitoring and careful dosing to lessen their probability of adverse effects. Alkaloids, by definition, can produce toxicity, tolerance, physical dependence, and addiction. To combat this, current pharmaceutical efforts focus on controlled-release and targeted delivery systems, which maintain consistent and stable blood concentrations of therapeutic agents while reducing peak-related side effects. Alkaloids have also been investigated for their analgesic properties, with reduced potential for abuse, as part of the ongoing development of multimodal pain management approaches. Overall, alkaloids remain an important group of bioactive substances, reflecting both the historical use of herbal medicine and the current use of pharmacotherapy for pain management.

6.4.2 **Flavonoids**

Fruits, vegetables, and medicinal plants are the primary sources of flavonoids and polyphenolic compounds. Their activity is attributed to their well-established antioxidant and antiinflammatory properties. These two activities play a significant role in the development of analgesic properties in the body (Ferraz et al., 2020; Uddin et al., 2020).

6.4.2.1 ***Mechanism of action***

Flavonoids inhibit the formation of inflammatory substances as part of their mechanism of action for pain relief (analgesic). By decreasing prostaglandin levels (a well-known class of inflammatory substances), flavonoids effectively inhibit both human- and animal-induced pain and inflammation. Flavonoids reduce both the expression and activity of COX and LOX, which produce prostaglandins and leukotrienes, by inhibiting their ability to generate these products. Additionally, flavonoids downregulate proinflammatory cytokines such as TNF-α, IL-1β, and IL-6, which could limit inflammation. Flavonoids also have strong antioxidant properties by neutralizing free radicals and enhancing the body's endogenous antioxidant defenses. Reducing oxidative stress is especially important when treating neuropathic and chronic pain conditions, as ROS contribute to neuronal damage and sensitization.

Additionally, flavonoids have been shown to modulate both the peripheral and CNSs by altering the activity of ion channels (such as voltage-gated sodium and calcium) and neurotransmitter receptors (including GABAergic and serotonergic), thereby affecting the transmission/perception of pain.

6.4.2.2 Examples

Flavonoids provide excellent sources of medicinal materials for use as analgesics in a variety of ways. Examples of the analgesic actions of flavonoids include quercetin, luteolin, and kaempferol. All three flavonoids can decrease both nociceptive and neuropathic pain in experimental animal models. The analgesic effects of these flavonoids are primarily due to their antioxidant properties and their ability to modulate inflammation and other pathophysiological processes. Most flavonoids can be found in fruits, vegetables, and medicinal plants. Several flavonoids are also associated with analgesic activity and the ability to reduce the average time animals spend displaying pain-related behaviors. Several flavonoids, including hesperidin and rutin, have also been shown to have promising antiinflammatory and analgesic capabilities. They work by decreasing the release of inflammatory mediators, improving blood flow to the painful area, and reducing oxidative stress in nerve tissue. Thus the traditional and new uses of these flavonoids in the treatment of pain are consistent with the modern-day use of analgesics.

6.4.2.3 Clinical significance

Currently, there is growing interest in using phytochemical extracts containing high levels of flavonoids as adjunctive treatment options in chronic pain management. *Ginkgo biloba*, green tea (*Camellia sinensis*), and *Passiflora* spp., for example, are among many natural plants whose extracts are being investigated for their potential therapeutic effects on neuropathic and inflammatory types of chronic pain. Because flavonoids can be safely utilized and because of their multiple mechanisms of action, they may provide long-term chronic pain relief to individuals who otherwise would be intolerant of standard medication therapies. To summarize, flavonoids are a promising new category within natural therapies that offer a nutritionally based, pharmacologically active approach to addressing pain and support multimodal, collaborative approaches to managing chronic pain.

6.4.3 Terpenoids

Terpenoids (also known as isoprenoids) are a class of structurally diverse compounds that produce the aroma, flavour, and bioactivity of many medicinal plants. They exhibit analgesic, antiinflammatory, and neuromodulatory properties (Devi et al., 2024; Nesterkina et al., 2020).

6.4.3.1 Mechanism of action

Terpenoids have a multimodal mechanism of action (MOA) that mediates several molecular routes and neuronal pathways for their analgesic effects. The MOA of terpenoids includes modulation of transient receptor potential (TRP) channels. TRP channels are a significant component of the sensory system that regulates sensory perception, including both pain and temperature. Many terpenoids can activate or desensitize TRP channels, thereby reducing the transmission of nociceptive signals. For instance, menthol can activate the TRPM8 channel, providing a cooling effect that counteracts pain by modulating sensory perception and inhibiting central nociceptive processing. The effects of terpenoids on the endocannabinoid system (ECS) are very strong. The two cannabinoid receptors in the human body, CB1 and CB2, play a significant role in the ECS for modulating pain and inflammation; β-caryophyllene, a sesquiterpene found in clove oil and many other plants, selectively binds to the CB2 receptor, which results in a decrease in the production of inflammatory mediators and the mitigation of neuropathic pain. Importantly, activation of the CB2 receptor does not induce psychoactive effects, which makes β-caryophyllene particularly appealing for use in a clinical setting. Terpenoids will enhance neuronal inhibition in the CNS via GABAergic pathways, reducing neuronal excitability during pain perception. The majority of terpenoids also exhibit antiinflammatory properties by decreasing the production of inflammatory mediators, including Prostaglandins, NO, and cytokines. Also, terpenoids exhibit antioxidant properties, thereby reducing oxidative stress in individuals with chronic or neuropathic pain.

6.4.3.2 Examples

Numerous terpenoid compounds are found in many different types of nonprescription and prescription pain relievers. Menthol, a product of the peppermint plant (*Mentha piperita*), is often found in topical preparations used to treat muscle and joint pain because it has a cooling effect and produces sensations of counter-irritation. Camphor, derived from the camphor tree (*Cinnamomum camphora*), is a natural product that provides a warming sensation, increases blood flow and circulation to the area of use and thereby relieving pain and stiffness. β-Caryophyllene, which is present in clove oil (*Syzygium aromaticum*), black pepper, and cannabis, has been shown to have analgesic and antiinflammatory effects in animal models of inflammatory and neuropathic pain. Other terpenoids with topical applications include menthol and camphor; additionally, ginsenosides (a type of triterpenoid saponin obtained from *Panax ginseng*), which have both analgesic and antiinflammatory effects when used topically, are another type of terpenoid with systemic analgesic and antiinflammatory effects. Ginsenosides may also help stimulate the immune response to inflammation, decrease oxidative stress, and regulate neuronal activity. Therefore ginsenosides may provide effective relief for chronic inflammatory and degenerative pain syndromes.

6.4.3.3 *Clinical significance*

Terpenoids find utility in the clinical setting because they have an excellent safety profile, are easy to formulate, and exhibit a wide range of therapeutic effects. Topical formulations of terpenoids, such as ointments, oils, creams, balms, and patches, can be safely applied to treat musculoskeletal pain associated with arthritis, sports injuries, and neuropathy. Their local activity limits, to some extent, systemic drug exposure and results in fewer potential side effects, thus making prolonged and multiple uses feasible. Terpenoids offer the potential to add numerous pharmacological activities to a therapy regimen. These compounds can serve as a complementary or adjunctive approach to treating mild to moderate pain alongside traditional medications. Over time, terpenoids have the potential to provide an alternative source of medication for pain management using both traditional and modern methods.

6.4.4 Polyphenolic compounds

Polyphenols include phenolic acids, tannins, stilbenes, and many others. They are plant secondary metabolites known to exhibit potent antioxidant and antiinflammatory capabilities. Polyphenols can modulate multiple signalling pathways, making them a viable option for treating both acute and chronic pain (El-Mekkawy et al., 2020; Chlif et al., 2022).

6.4.4.1 *Mechanism of action*

The analgesic activity of polyphenolic compounds is related to their ability to modulate oxidative and inflammatory pathways involved in the development and maintenance of pain. The antioxidant effects of polyphenolic compounds can help protect against the development of chronic pain by preventing neuronal sensitization through scavenging oxidants that cause oxidative damage. In addition, the antioxidant properties of polyphenolic compounds have been shown to protect both the peripheral and CNSs against the damaging effects of oxidative stress, via their ability to scavenge free radicals and enhance the body's natural antioxidant defense systems. They also reduce the activity of major inflammatory signalling pathways, such as the nuclear factor kappa B (NF-κB) pathway, which regulates the transcription of many proinflammatory genes. Consequently, inhibition of NF-κB signalling reduces the production of inflammatory cytokines, chemokines, and adhesion molecules. Moreover, polyphenols inhibit COX and LOX activity, thus decreasing the endogenous production of prostaglandins and leukotrienes, which are the major mediators of pain and inflammation. By combining the mechanisms mentioned above, polyphenols can reduce both nociceptive and neuropathic pain, prevent inflammation, and support the health of the entire nervous system.

6.4.4.2 *Examples*

Polyphenols have several well-studied examples, with curcumin, isolated from *C. longa*, being a major focus of research, due to its broad array of antiinflammatory and analgesic effects. Curcumin affects several inflammatory pathways, including NF-κB, COX-2, and cytokines that promote inflammation. Curcumin has also been shown to help relieve pain and improve function in osteoarthritis, rheumatoid arthritis, and neuropathic pain. Resveratrol is a stilbene found in *Vitis vinifera* (grapes) and has potent antioxidant properties in addition to antiinflammatory properties as well as neuroprotective activity. Like curcumin, resveratrol can modulate pain-related signalling pathways and reduce inflammation in animal models of inflammatory and neuropathic pain. Ellagic acid from *Punica granatum* (pomegranate) is also a significant contributor to pain prevention by decreasing oxidative stress and the production of inflammatory mediators through its antiinflammatory and antioxidant effects.

6.4.4.3 *Clinical significance*

In recent years, there has been an increase in the use of polyphenolic compounds in integrative pain management programs (IPMP) as standardized herbal extract formulations. Historically, their clinical use was limited due to poor oral bioavailability and extensive metabolic processing of most polyphenols; however, polyphenolic compounds have been incorporated into IPMP. Advanced drug delivery systems (e.g., nanoparticles, phytosomes) and combinations of bioenhancers, such as piperine with polyphenolic compounds, have provided promising strategies for improving the bioavailability, stability, and therapeutic activity of polyphenolic compounds through innovative formulations. The continued development of polyphenolic compound formulations will enhance their clinical potential by allowing for greater predictability of therapeutic outcomes and will encourage the use as adjunct or alternative therapies for the management of chronic inflammatory and neuropathic pain due to their excellent safety profile and ability to act through multiple mechanisms of pain relief.

6.4.5 Glycosides

Glycosides contain a sugar molecule (called the "ring") that is connected to a chemical compound known as an aglycone (the "active" part of a glycoside). They provide multiple pathways through which glycosides produce pain relief, reduce inflammation, and support neuromodulation (Khan et al., 2020; Tian et al., 2022).

6.4.5.1 *Mechanism of action*

Analgesic and antiinflammatory properties of glycosides occur via multiple types of biological mechanisms. Glycoside interaction with COX and LOX pathways reduces peripheral inflammation and nociceptive pain by modulating cytokines, leukotrienes, and prostaglandins. Some glycosides also affect receptors (CNS and PNS) to decrease the transmission and perception of pain. The defining feature of all prostaglandins, for example, is that they can only be made when they are present in the body in a chemical form called a glycoside. For example, salicin is metabolically converted to salicylate and has well-established antiinflammatory and analgesic properties as a prodrug. By functioning as prodrugs, glycosides allow prolonged therapeutic effects while minimizing gastrointestinal upset and the systemic toxicity that can occur with their active metabolites.

6.4.5.2 Examples

Many glycosides have been investigated for their pain-relieving and treatment purposes. Salicin from the white willow tree (*S. alba*) is an excellent example. When metabolized, salicin converts to salicylic acid; this molecular transition served as the foundation for the development of aspirin, one of the most widely used pain medications and antiinflammatories available today. Harpagoside is a major constituent of *H. procumbens* (devil's claw) plant root structure. It has been extensively studied for its effectiveness as both an antiinflammatory agent and a pain reliever. It is often used to treat arthritis, lumbar (lower back) pain, and other musculoskeletal conditions by inhibiting the formation of inflammatory mediators. The *Digitalis* plant family also contains cardiac glycosides, which are used to treat cardiovascular disease, underscoring the wide range of therapeutic effects glycosides can have on the body.

6.4.5.3 Clinical significance

As compared with other types of plant compounds (e.g., components that exhibit antiinflammatory properties), glycoside-type phytochemicals (e.g., *Panax ginseng*) have received increased attention in the treatment of pain and discomfort related to the various forms of arthritis and lower back pain/chronic inflammatory diseases. They may be used alone or in combination with other analgesics and pain-relieving treatments. With the availability of several commercially produced standardized herbal products containing glycosides, combined with increasing evidence of their benefit and safety when used as directed, glycosides are emerging as a significant source of pain relief and antiinflammatory activity for practitioners of integrative or multimodal approaches to treating pain. In particular, glycosides possess prodrug properties, allowing them to deliver long-lasting analgesic and antiinflammatory effects and to enhance patient tolerance when taken as directed or in combination with conventional pain management approaches.

6.5 Mechanisms of action of botanical analgesics

Botanical analgesics relieve pain in multiple ways, which demonstrates that pain can be caused by more than one type of problem. In contrast to standard pain relievers, which usually affect only one part of the body, plant-based products act on multiple

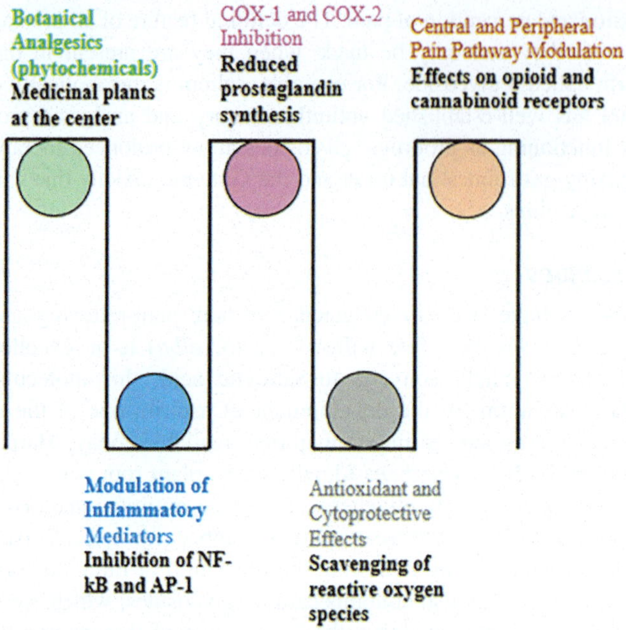

FIGURE 6.1 Multitarget mechanisms of botanical analgesics.

Various mechanisms for botanical analgesics.

systems involved in inflammation, oxidative stress, and nociceptive signalling, both centrally and peripherally. By acting on multiple targets, it is possible to treat both short and long-term pain with fewer side effects, as shown in Fig. 6.1.

6.5.1 Modulation of inflammatory mediators

The modulation of inflammatory mediators is one way in which inflammation induces, maintains, and prolongs pain by sensitizing peripheral pain receptors and activating the central pain pathway. Many inflammatory mediators, such as prostaglandins, leukotrienes, bradykinin, cytokines (TNF-α, IL-1β, IL-6), and NO, amplify pain signalling at the site of injury and within the CNSby enhancing and prolonging nociceptive transmission. Botanicals appear to inhibit the inflammatory response by downregulating inflammatory signalling pathways. Four phytochemicals, using curcumin, *Boswellia*, flavonoids, and terpenoids, were identified and found to inhibit two important transcription factors, NF-κB and AP-1, thus inhibiting the expression of proinflammatory genes. Finally, botanical agents ($n = 8$) were shown to decrease the activity of iNOS and COX-2, resulting in decreased levels of NO and prostaglandins, decreased levels of inflammatory mediators, reduced nociception, and prolonged improved quality of life, especially for patients with inflammatory disorders and musculoskeletal pain (Galvão et al., 2018; Kahkhaie et al., 2019).

6.5.2 Cyclooxygenase (COX-1 and COX-2) inhibition

The enzymes COX-1 and COX-2 play an important role in the production of pain by converting 2-arachidonoyl glycerides into prostaglandins that sensitize pain pathways (nociceptors) and promote inflammatory processes. COX-1 is always present in the body (constitutively expressed), has physiological roles, and serves a general function in tissue homeostasis. Conversely, COX-2 is a form of COX that is made only when there is a stimulus (induced). COX-2 produces prostaglandins that are responsible for inflammatory pain and other pathologies associated with inflammation. Many plant-based substances can inhibit COX enzymes, leading to decreased prostaglandin production and therefore less pain. Salicin from willow bark, gingerols and shogaols from *Zingiber officials,* and flavonoids found in several species of medicinal plants all have documented COX-inhibitory properties. Compared to synthetic NSAIDs, COX inhibitors from plants have either a partial or selective effect on the enzyme COX, which significantly reduces the risk of developing gastrointestinal and cardiovascular side effects. This balance between the different effects of COX inhibition enhances the appeal of using plant-derived COX inhibitors for long-term treatment of chronic pain, including osteoarthritis and rheumatoid arthritis (Pannunzio and Coluccia, 2018; Vishwakarma and Negi, 2020).

6.5.3 Antioxidant and cytoprotective effects

The recognition of the detrimental effects of oxidative stress in the development of chronic pain and neuropathic pain, as well as neuroinflammation, has attracted the attention of researchers. The harmful effects of ROS overproduction can cause neuronal injury and impair mitochondrial function, leading to heightened pain sensitivity in the afflicted individuals. Botanicals with a high antioxidant content can mitigate these effects by neutralizing free radicals, thereby boosting the body's own antioxidant defenses. The antioxidant-rich polyphenols, including resveratrol, quercetin, catechins, and curcumin, can both scavenge ROS and upregulate the production of antioxidant enzymes such as superoxide dismutase, catalase, and glutathione peroxidase. The cytoprotective effects of these compounds can help maintain neuronal structural integrity and reduce central sensitization, which is associated with chronic pain. Antioxidant-mediated modulation of pain is significant in many neuropathic conditions, including diabetic neuropathy, chemotherapy-induced neuropathy, and various forms of neurodegeneration (Valentová, 2020).

6.5.4 Central and peripheral pain pathway modulation

The pathways for processing pain (the sensation of pain) are complex, including peripheral (nervous systems outside the brain and spinal cord) and central (nervous systems in the brain and spinal cord). Botanical analgesics modulate these pathways by binding to and interacting with various neurotransmitters, ion channels, and receptors associated with pain perception (nociception). Both alkaloids and terpenoids act on opioid receptors. They enhance the body's own way of processing pain, known as endogenous analgesic mechanisms, without producing the adverse effects associated with using synthetic opioids. Some botanical remedies work by affecting TRP channels throughout the peripheral and CNSs, especially TRPV1 and TRPM8, which are involved in transmitting pain signals from the periphery to the CNS. Capsaicin is an example of a compound that desensitizes TRPV1 receptors and relieves neuropathic pain. Another compound, β-caryophyllene, activates the cannabinoid CB2 receptor, which will inhibit inflammation and pain without causing a high. Botanical analgesics also affect neurotransmitters such as serotonin, GABA, and glutamate, therefore producing central inhibition of pain and increased pain tolerance through reduced CNS pain signal transmission (Yao et al., 2023; Lee and Neumeister, 2020).

6.6 Preclinical evidence supporting botanical analgesics

Through preclinical research, verifiable scientific evidence of the pain-relieving efficacy of botanicals historically used to alleviate pain has been established. Preclinical studies utilize in vitro and in vivo models to identify and determine mechanisms of action, safety profiles, and the ability of specific plant compounds to exert their therapeutic activity through multiple receptor pathways in both nociceptive and neuropathic pain models, as mentioned in Table 6.2.

6.6.1 Experimental models of nociceptive pain

Nociceptive pain can result from tissue injury, followed by acute inflammation, and is mediated by the stimulation of peripheral nociceptors. To evaluate the effects of botanical analgesics in preclinical studies, acute and subacute pain models such as the acetic acid-induced writhing test, formalin test, hot plate test, tail-flick test, and carrageenan-induced paw edema model are used to differentiate between peripheral and central effects of analgesics and antiinflammatory compounds. Numerous botanical extracts and isolated phytochemicals demonstrate significant analgesic effects in these models of acute and subacute injury. For instance, extracts of *C. longa* (turmeric) reduce pain-associated behaviors (e.g., licking, biting, or guarding) in formalin-induced and carrageenan-induced inflammation models, most likely through inhibition of prostaglandin synthesis, cytokine release, and COX-2 expression. Extracts from *B. serrata* (Indian frankincense) containing large amounts of boswellic acids have powerful antiinflammatory and analgesic effect. This effect is achieved by inhibiting leukotriene biosynthesis, thereby reducing edema and decreasing pain sensitivity.

Various ginger extracts, which contain the compounds gingerol and shogaol, exhibit dose-dependent analgesic effects in the writhing and hot-plate tests. This indicates that both peripheral antiinflammatory and central analgesic mechanisms

Table 6.2 Preclinical evidence for botanical analgesics.

Pain type/model	Experimental models	Botanical examples	Key analgesic effects	Major mechanisms
Nociceptive/ inflammatory pain	Writhing, formalin, hot plate, tail-flick, carrageenan paw edema	*Curcuma longa*, *Boswellia serrata*, *Zingiber officinale*	Reduced pain behaviors, edema, and thermal sensitivity	Cyclooxygenase (COX)-2 and prostaglandin inhibition; cytokine suppression
Central analgesia	Hot plate, tail-flick tests	*Corydalis yanhusuo*, *Papaver* spp.	Increased pain threshold	Opioid receptor activation
Neuropathic pain	Chronic constriction injury, spinal nerve ligation, partial nerve injury	Curcumin, resveratrol, quercetin	Reduced allodynia and hyperalgesia	Nuclear factor kappa B inhibition; antioxidant activity
Diabetic/ chemotherapy neuropathy	DPN and CIPN models	Polyphenols, curcumin	Neuroprotection and pain reduction	Reduced oxidative stress; mitochondrial protection
Receptor-targeted models	Transient receptor potential and CB receptor assays	Capsaicin, β-caryophyllene	Sustained analgesia	TRPV1 desensitization; CB2 activation
Translational relevance	Safety and dose-response studies	Multiple botanicals	Favorable safety profile	Multitarget modulation; standardization challenges

contribute to the analgesia. Various alkaloid-containing extracts from *C. yanhusuo* and the various *Papaver* species display powerful central analgesic activity, as assessed in several thermal pain models, suggesting that these botanicals modulate analgesic action by activating opioid receptors as the pain-relieving mechanism. Collectively, the current evidence indicates that botanical preparations may have broad potential to modulate both acute and inflammatory pain responses (Thiam et al., 2019; Middleton et al., 2021).

6.6.2 Experimental models of neuropathic pain

The nervous system can be a target of injury or dysfunction, which gives rise to the condition known as neuropathic pain, typified by symptoms such as allodynia (Increased sensitivity) and hyperalgesia (excessive sensitivity). and about its treatment. The current standard of care is not usually effective for treating neuropathic pain; therefore new approaches need to be developed. Preclinical models of neuropathic pain can be established using various experimental procedures, including the following: the chronic constriction injury (CCI) model of the sciatic nerve, the spinal nerve ligation model, the partial nerve injury model, the diabetes peripheral neuropathy model, and the chemotherapy peripheral neuropathy model. etc. Multiple studies have documented the efficacy of various botanical compounds in various neuropathic pain models by targeting three primary events in the CNS: neuroinflammatory processes, oxidative stress, and neuronal sensitization. For example, curcumin has demonstrated good efficacy in both CCI models and diabetic peripheral neuropathy models for reducing mechanical allodynia and thermal hyperalgesia. These effects may be attributed to NF-κB inhibition, reduced proinflammatory cytokine levels, and stimulation of antioxidant enzyme activity.

Studies have shown that polyphenols (resveratrol and quercetin) have beneficial effects on the nervous system and reduce pain in animal models of neuropathic pain by decreasing oxidative stress and protecting mitochondria, as well as by decreasing glial cell activation in the spinal cord. Capsaicin has been extensively studied for its ability to diminish the sensitivity of TRPV1 receptors on sensory neurons. This reduction in sensitivity to neuropathic pain continues for a longer period compared with the initial reduction from capsaicin treatment. Terpenoids, including β-caryophyllene, can also provide analgesic effects by selectively binding to cannabinoid CB2 receptors and decreasing neuroinflammation, without causing psychoactive effects (Kankowski et al., 2021; Yeh et al., 2020).

6.6.3 Translational relevance of animal studies

Based on numerous animal models, many studies have confirmed the analgesic effects of botanical medicine, which offer the practical benefit of targeting multiple pathways involved in pain generation and persistence, especially in chronic and neuropathic discomfort. Additionally, safety evaluations conducted during preclinical testing have indicated that most botanical medicines exhibit very low toxicity when used at appropriate therapeutic doses (e.g., under a physician's prescription). There are still barriers to translating the results of animal studies into successful human clinical trials. These include variations among the plants used in studies, the lack of standardized protocols, disparities in dosing regimens, and the limited bioavailability of botanical medicines. Therefore standardizing extract preparation methods, developing more advanced delivery systems, and creating precise protocols for use in animal studies will be critical to improving the translation of findings from animal studies to human clinical trials (Huerta et al., 2025).

6.7 Clinical evidence and human studies

Preclinical research provides mechanistic understanding and proof of concept, while clinical studies are required to demonstrate the safety, efficacy, and clinical relevance of medicinal plants for treating pain in humans. Based on this evidence, botanicals are now considered an important component of current evidence-based and integrative approaches to pain treatment.

6.7.1 Evidence from controlled clinical trials

There have been numerous clinical trials assessing herbal products for pain treatment over the past 20 years. Curcumin (*C. longa*) has received significant research attention as an antiinflammatory as well as a way to treat osteoarthritis and inflammatory joint conditions, with promising results showing reduced levels of pain intensity, morning stiffness, and inflammation markers when compared to NSAIDs, along with better gastrointestinal safety. *B. serrata* (boswellic acid) extracts consistently provide benefits for treating pain and increasing functional activity by decreasing leukotriene synthesis. *Z. officinale* (Ginger) has been studied for the treatment of several types of pain, including postoperative recovery, menstrual cycle-related discomfort, and other arthritic and muscular disorders, with documented improvements in pain levels and demand for pain relief medications. Capsaicin, a topical product, has been approved for patients suffering from postherpetic neuralgia and diabetic neuropathy, providing long-lasting analgesia following regular usage. Devil's claw and willow bark extracts have also been found to be effective for treating low back pain and osteoarthritis (Malone and Tsai, 2018; Jahromi et al., 2021).

6.7.2 Safety, tolerability, and clinical limitations

Botanical analgesics may have a favorable safety record when used appropriately, according to most clinical trials. Most adverse effects associated with botanical analgesics have been mild (e.g., gastrointestinal discomfort, headaches, and localized skin irritation when used topically). Compared with the opioid painkillers and NSAIDs, botanical analgesics have lower incidences of dependency, gastrointestinal ulcers, and cardiovascular complications. There are still significant limitations to the use of botanical analgesics, e.g., differences in the composition of different extracts, lack of standardized dosing protocols, and insufficient long-term safety data. Potential herb-drug interactions may also occur in patients taking anticoagulants, antihypertensives, and antidiabetic medications; therefore caution should be exercised when using botanical analgesics in these patients. Rigorous trial designs and standardized manufacturing methods will be necessary to address these limitations and enable wider adoption of botanical analgesics in clinical practice (Kumar and Ashaq, 2021; Al-Shuhaib and Al-Shuhaib, 2025).

6.8 Challenges and limitations

Botanicals have great potential as viable alternatives or adjuncts to traditional pain medications; however, numerous barriers to their successful clinical translation exist, including scientific, technical, and regulatory challenges. Understanding these barriers is important for a rational and evidence-based approach to the use of botanicals for pain treatment, as shown in Fig. 6.2

6.8.1 Standardization and quality control

The most challenging aspect of developing and using botanical-based medicines is creating standardized products. It is through the standardization of botanical-based

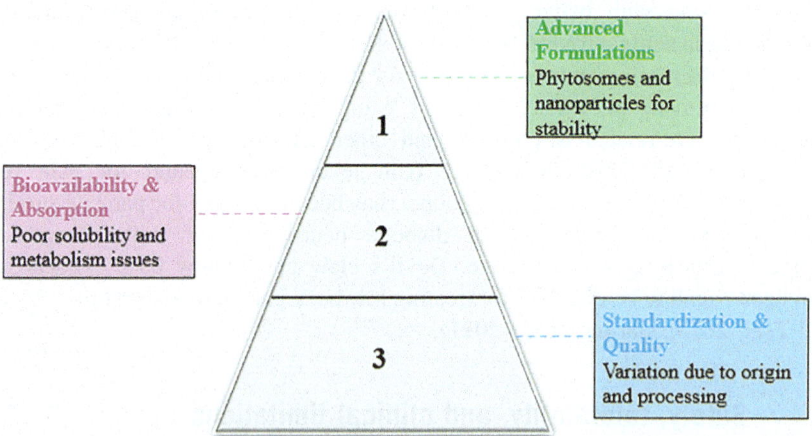

FIGURE 6.2 Botanical analgesic development challenges.

Botanicals for treating pain.

Akili (medicinal) plants that their therapeutic efficacy can be determined. The therapeutic efficacy of any botanical-based Akili (medicinal) plant relies on the amount, as well as the type, of active phytochemicals found within that plant. Multiple factors affect the concentration and composition of phytochemicals across various botanical species and subspecies. These include: geographic origin, climate, soil, harvest timing, and postharvest processing. Other factors contributing to phytochemical content include the extraction method employed and the Solvent Selection (i.e., water, alcohol, or oil). As a result of formulation variations between batches of products, there will be between-batch variation, leading to inconsistent amounts of therapeutic benefit derived from each product. In addition, there are concerns about contamination of products with heavy metals, pesticides, microbial toxins, or synthetic drug adulteration if quality control is inadequate. Advanced analytical technologies, including HPLC, GC-MS, and DNA barcoding, are increasingly used to ensure that the intended botanical identity and purity, as well as the product's strength, are accurate. The establishment of pharmacopeial standards and GMP are necessary to enhance the reliability and credibility of botanical-based analgesics (Mukherjee, 2019; Sarwa et al., 2021).

6.8.2 Bioavailability and absorption issues

As mentioned above, the bioavailability of these compounds is affected negatively by: (1) water solubility, (2) intestinal barrier permeability, (3) inappropriate formulations, and (4) high levels of "first-pass" metabolism. Three polyphenols with antiinflammatory and analgesic effects have been described. Curcumin, resveratrol, and quercetin exhibit antiinflammatory and analgesic activities in vitro but are poorly absorbed when taken orally. Botanical formulas contain many phytochemicals, which create an added challenge for their preparation, as they have adrenal actions when metabolized, digested, and converted into their bioactive forms in each person's intestine. Because all humans have unique and diverse gut microbiomes, the types and levels of phytochemicals a person metabolizes in their liver before they enter the body also differ significantly. This has led many scientists to continue developing new and better methods to deliver these compounds in the highest possible doses, with the greatest stability, and to the appropriate tissues. Some of the alternative delivery systems currently under development include phytosomes, nanoparticle systems, liposome systems, and solid lipid carrier systems(Hu et al., 2025; Oladimeji et al., 2018).

6.9 Future perspectives and research directions

To advance the field of botanical medicine in pain management, we must continue to conduct rigorous scientific investigations, whilst also maintaining the "Holistic" approach of botanical medicine and/or its "Multi-Targeted" effect of plant-based therapies as shown in Fig. 6.3

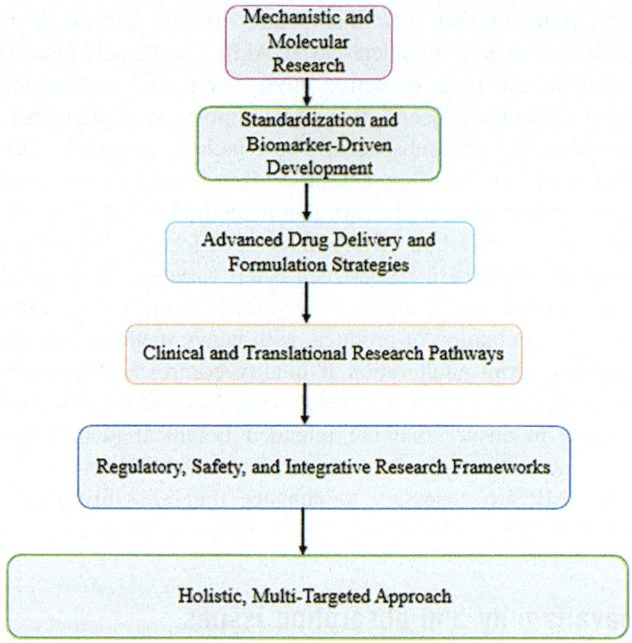

FIGURE 6.3 Future perspectives and research directions.

Perspectives and directives of research.

6.9.1 Mechanistic and molecular research priorities

Understanding the different ways plants relieve pain is crucial for developing plant-derived medicines. The vast majority of plant compounds first discovered for their antioxidant, antiinflammatory, or related properties are used to relieve pain. All forms of symptoms associated with pain and discomfort result from an extremely complex biological process that can occur through many channels in both the peripheral and CNSs. Most recently, research has shown that many types of plant extracts work together to relieve pain by affecting multiple pathways simultaneously (including enzymes, ion channels, and neurotransmitters); however, as studies progress, researchers will continue to learn more about these pathways. To advance botanical studies and the subsequent development of new pain management approaches, the next steps in this line of research should include determining botanical extract interactions with COX-1 and COX-2, as well as LOX, and determining how these enzymes regulate the production of prostaglandins and leukotrienes, also known as inflammation mediators. The next phase of the research should focus on TRPV1, TRPA1, and TRPM8 channels, which generate electrical impulses that transmit signals from peripheral nerves (nociceptors) to the brain, producing pain and the sensation of inflammatory pain. Another area of further research will be interactions between voltage-gated sodium and voltage-gated calcium channels, as these channels affect the transmission of nociceptive signals in the nervous system, that is, they play a role in how the body perceives and responds to pain through an inflammatory response. Integrating these methods will enable researchers to identify the most significant molecular targets, key regulatory pathways, and master switches underlying the analgesic effects of botanicals.

6.9.2 Standardization and biomarker-driven development

The current primary issue in botanical analgesic research is the lack of a reproducible, reliable method to produce clinical-grade products from a botanical analgesic formulation. The next major step in research will be the establishment of standardized methods for validating botanical analgesics using both chemical and biological markers. The process of using a chemical standardization to create a botanical product would likely involve finding and measuring the presence of several specific marker compounds or "fingerprints" within a botanical product; however, using a single marker compound as an indicator of activity may not be sufficient because multiple marker compounds will most likely work together synergistically. Using a biomarker-based developmental approach for botanical analgesics would involve relating phytochemical profiles to pharmacological effects. The current goal of research is to identify associations between unique metabolomic patterns and biological responses, such as the inhibition of inflammatory mediators, modulation of pain-related receptors, and altered pain behaviors, in preclinical models. Developing strategies that prioritize biomarker use will help optimize dosing, produce consistent batches, and improve clinical outcomes with botanical analgesics. In addition to using chemical and biological markers to validate botanical analgesic products, the identification of patient-specific biomarkers, including inflammatory cytokines (interleukin-1, interleukin-6), biomarkers of oxidative stress, or genetic polymorphisms that influence pain perception, will help create a schema for the stratification of patients and the development of personalized botanical analgesic therapy for patient care. Identifying chemical, biological, and clinical biomarkers for botanical analgesic products will be critical for obtaining regulatory support and increasing confidence within the healthcare community regarding the use of botanical analgesics.

6.9.3 Advanced drug delivery and formulation strategies

Numerous botanical analgesics are now being tested for their ability to treat pain; however, there are numerous limitations on their use in practice based on how they interact with the body's systems. Many properties of botanical analgesics compromise their effectiveness, including low water solubility, limited absorption in the digestive system, rapid clearance from the body, and diminished concentration in the bloodstream.

A comprehensive evaluation of novel approaches will help future researchers identify more effective methods to alleviate pain using botanical-derived materials. Advances in technology, including but not limited to nanotechnology, have led to the increasing use of nanoparticles, nanoemulsions, phytosomes, liposomes, and solid lipid dispersions, providing opportunities to improve botanical-derived material technology through increased solubility, stability, and bioavailability. In addition, by utilizing targeted delivery mechanisms to specifically target inflamed tissues and/or CNS nerve cells, these technologies can increase their therapeutic potential while decreasing the potential for systemic reactions to the botanical-derived material. Additionally, controlled release of these compounds can decrease the frequency with which patients must take their medications, while also providing prolonged therapeutic concentrations to enhance compliance for individuals suffering from chronic pain. Furthermore, there is significant promise in combining these technologies, as opportunities exist to create unique synergistic effects among different botanical compounds while achieving optimal pharmacokinetics through combination delivery systems. As formulation science continues to develop and advance, it will serve as the bridge between the powerful experimental capabilities of botanical-based compounds and clinically significant outcomes.

6.9.4 Clinical research and translational pathways

Clinical validation of botanical products for their use in evidence-based pain management will require considerable effort. Future clinical studies must consist of large-scale, multicenter randomized controlled trials using standardized botanical preparations and validated pain assessment instruments. Furthermore, future studies must measure not only the intensity of pain but also associated functional outcomes, patients' quality of life, and long-term safety outcomes, especially those concerning patients suffering from chronic pain and neuropathic pain conditions. Using translational research, which combines preclinical (animal) and clinical (human) data, creates predictive value to help lower the risk of product failure when developing new botanical pain therapeutics. The use of adaptive trial designs, real-world evidence, and digital tools that monitor patients' pain levels will also improve the quality and real-world clinical relevance of the resulting data. It is also necessary to conduct long-term observational studies to determine the long-term efficacy of botanical therapies and the safety and compliance of patients using these products in everyday life.

6.9.5 Regulatory, safety, and integrative research directions

The lack of uniformity and agreement among regions regarding regulatory requirements has hampered the global acceptance of botanical analgesics. Future studies must establish harmonized legal frameworks to address the unique characteristics of botanical products and establish acceptable standards for quality, safety, and efficacy. In particular, establishing stronger pharmacovigilance and adverse event reporting systems would help ensure the long-term safety of botanical analgesics. The methodology for systematically evaluating herb-drug interactions is also an important research goal, given the widespread use of polypharmacy in pain management. Combining traditional medicine with today's advances in pharmaceutical science to develop integrative research models will promote the rational utilization of botanical drug products with conventional analgesics. Collaboration among traditional healers, researchers, clinical practitioners, and regulatory agencies will enable the establishment of a well-founded basis for developing safe, effective, and evidence-based integrative pain management protocols.

6.10 Conclusion

Botanicals have many mechanisms that can support and treat individuals with pain through analgesia, antiinflammation, antioxidant effects, and modulation of neurotransmission at various biological levels. There is documented evidence of medicinal plant use throughout history, including many plants that have been studied for their efficacy in treating pain. Medicinal plants are frequently used as adjunctive or alternative therapies for patients who are suffering from chronic or neuropathic pain, as conventional medications tend to be less effective for these types of pain than other methods may be. To further illustrate the impact of isolated compounds from plants in the treatment of pain, extensive preclinical investigations have shown that the compounds present in plants can influence the primary pathways associated with pain, including the production of inflammatory mediators, the formation of free radicals, the activity of multiple ionic channels, and both the peripheral and central mechanisms associated with pain perception. In the meantime, clinical research is ongoing, and there is considerable evidence from numerous published studies supporting the benefits and safety of multiple botanicals for treating various types of musculoskeletal, inflammatory, and neuropathic pain.

Furthermore, the multitargeted activity of the botanicals and the presence of numerous active constituents may offer opportunities to enhance clinical efficacy while reducing side effects. Although these advantages exist, several significant obstacles must be overcome before botanical extracts can be used clinically on a broader scale (e.g., regulatory approvals, safety assessments, bioavailability, and standardization). These barriers must be addressed by conducting large-scale mechanistic studies, developing better products, performing well-controlled clinical trials, and developing a uniform set of regulatory standards. The merging of traditional botanical knowledge with modern science, through the application of botanical neurological evidence, has strong potential to produce safer, more effectively integrated treatments for pain management across many patient categories.

References

Agarwal V. Ayurvedic protocols of chronic pain management: spatiotemporality as present moment awareness and embodied time. Frontiers in Pain Research 2024;5. Available from https://doi.org/10.3389/fpain.2024.1327393, https://www.frontiersin.org/journals/pain-research.

Al-Shuhaib MBS, Al-Shuhaib JMB. Assessing therapeutic value and side effects of key botanical compounds for optimized medical treatments. Chemistry & Biodiversity 2025;22(1). Available from https://doi.org/10.1002/cbdv.202401754.

Blake HM. Botanical treatments in cancer pain management essentials of interventional cancer pain management. United States: Springer International Publishing; 2018. p. 503–6. Available from https://doi.org/10.1007/978-3-319-99684-4_54, http://www.springer.com/in/book/9783319996820.

Chlif N, Bouymajane A, Oulad El Majdoub Y, Diouri M, Rhazi Filali F, Bentayeb A, et al. Phenolic compounds, in vivo anti-inflammatory, analgesic and antipyretic activities of the aqueous extracts from fresh and dry aerial parts of Brocchia cinerea (Vis. Journal of Pharmaceutical and Biomedical Analysis 2022;213 Available from https://doi.org/10.1016/j.jpba.2022.114695, www.elsevier.com/locate/jpba.

Cohen SP, Vase L, Hooten WM. Chronic pain: an update on burden, best practices, and new advances. The Lancet 2021;397(10289):2082–97. Available from https://doi.org/10.1016/S0140-6736(21)00393-7, http://www.journals.elsevier.com/the-lancet/.

Maria I. Dalamagka. Integrating traditional medicine into a modern health care system. International Journal of Science and Research Archive 2024;12(1):2372–5. Available from https://doi.org/10.30574/ijsra.2024.12.1.1046.

Devi M, Bamrah PK, Goyal R, Choudhary M, Chopra H. Insights on the emerging therapeutic potential of terpenoids as anti-inflammatory agents: a scoping review. Journal of Bio-X Research 2024;7(1). Available from https://doi.org/10.34133/jbioxresearch.0006, https://spj.science.org/journal/jbioxresearch.

El-Mekkawy S, Shahat AA, Alqahtani AS, Alsaid MS, Abdelfattah MA, Ullah R, et al. A polyphenols-rich extract from moricandiasinaicaboiss. exhibits analgesic, anti-inflammatory and antipyretic activities in vivo. Molecules (Basel, Switzerland) 2020;25:5049.

Ferraz CR, Carvalho TT, Manchope MF, Artero NA, Rasquel-Oliveira FS, Fattori V, et al. Therapeutic potential of flavonoids in pain and inflammation: mechanisms of action, pre-clinical and clinical data, and pharmaceutical development. Molecules (Basel, Switzerland) 2020;25(3). Available from https://doi.org/10.3390/molecules25030762, https://www.mdpi.com/1420-3049/25/3/762.

Forouzanfar F, Hosseinzadeh H. Medicinal herbs in the treatment of neuropathic pain: a review. Iranian Journal of Basic Medical Sciences 2018;21(4):347–58. Available from https://doi.org/10.22038/ijbms.2018.24026.6021, http://ijbms.mums.ac.ir/article_10461_e65a82e3e47a88a9546a1a30d4660c1e.pdf.

Galvão I, Sugimoto MA, Vago JP, Machado MG, Sousa LP. Mediators of inflammation immunopharmacology and inflammation. Brazil: Springer International Publishing; 2018. p. 3–32. Available from https://doi.org/10.1007/978-3-319-77658-3_1, http://www.springer.com/in/book/9783319776576.

Hu L, Luo Y, Yang J, Cheng C. Botanical flavonoids: efficacy, absorption, metabolism and advanced pharmaceutical technology for improving bioavailability. Molecules (Basel, Switzerland) 2025;30(5). Available from https://doi.org/10.3390/molecules30051184, http://www.mdpi.com/journal/molecules.

Hu WL, Hung YC, Tsai MY, Yen HR, Delfino DV. Herbal medicines in pain management. Frontiers in Pharmacology 2022;13:955743.

Huerta MÁ, Roza P, Cisneros E, Alique M, Roza C. Enhancing clinical translation of analgesics for neuropathic pain: a systematic review and meta-analysis on the role of non-evoked pain assessment in preclinical trials. Pain 2025;166(11). Available from https://doi.org/10.1097/j.pain.0000000000003697.

Jahromi B, Pirvulescu I, Candido KD, Knezevic NN. Herbal medicine for pain management: efficacy and drug interactions. Pharmaceutics 2021;13(2):1–43. Available from https://doi.org/10.3390/pharmaceutics13020251, https://www.mdpi.com/1999-4923/13/2/251/pdf.

Jiang W, Tang M, Yang L, Zhao X, Gao J, Jiao Y, et al. Analgesic alkaloids derived from traditional chinese medicine in pain management. Frontiers in Pharmacology 2022;13. Available from https://doi.org/10.3389/fphar.2022.851508, http://www.frontiersin.org/Pharmacology.

Jin Q, Zhao YL, Liu YP, Zhang RS, Zhu PF, Zhao LQ, et al. Anti-inflammatory and analgesic monoterpenoid indole alkaloids of Kopsia officinalis. Journal of Ethnopharmacology 2022;285 Available from https://doi.org/10.1016/j.jep.2021.114848, www.elsevier.com/locate/jethpharm.

Kahkhaie KR, Mirhosseini A, Aliabadi A, Mohammadi A, Mousavi MJ, Haftcheshmeh SM, et al. Curcumin: a modulator of inflammatory signaling pathways in the immune system. Inflammopharmacology 2019;27(5):885–900. Available from https://doi.org/10.1007/s10787-019-00607-3, https://rd.springer.com/journal/10787.

Kankowski S, Grothe C, Haastert-Talini K. Neuropathic pain: spotlighting anatomy, experimental models, mechanisms, and therapeutic aspects. European Journal of Neuroscience 2021;54(2):4475–96. Available from https://doi.org/10.1111/ejn.15266, http://onlinelibrary.wiley.com/journal/10.1111/(ISSN)1460-9568.

Khan H, Pervaiz A, Intagliata S, Das N, Nagulapalli Venkata KC, Atanasov AG, et al. The analgesic potential of glycosides derived from medicinal plants. Journal of Pharmaceutical Sciences 2020;28(1):387–401. Available from https://doi.org/10.1007/s40199-019-00319-7, https://rd.springer.com/journal/40199.

Kleinman A. Indigenous systems of healing: questions for professional popular, and folk care alternative medicines: popular and policy perspectives. United States: Taylor and Francis; 2022. p. 138–64. Available from https://www.routledge.com/Alternative-Medicines-Popular-and-Policy-Perspectives/Salmon/p/book/9781032279879.

Kumar N, Ashaq M. Safety and toxicity of botanical medicines: a critical appraisal. International Journal of All Research Education, Science and Methods 2021;9:2455–6211.

Lee GI, Neumeister MW. Pain: pathways and physiology. Clinics in Plastic Surgery 2020;47(2):173–80. Available from https://doi.org/10.1016/j.cps.2019.11.001, http://www.elsevierhealth.com.

Malone M, Tsai G. The evidence for herbal and botanical remedies, Part 1. Journal of Family Practice 2018;67(1):10–6. Available from: https://www.mdedge.com/sites/default/files/Document/December-2017/JFP06701010.PDF.

Matos LC, Machado JP, Monteiro FJ, Greten HJ. Understanding traditional chinese medicine therapeutics: an overview of the basics and clinical applications. Healthcare 2021;9(3):257. Available from https://doi.org/10.3390/healthcare9030257.

Middleton SJ, Barry AM, Comini M, Li Y, Ray PR, Shiers S, et al. Studying human nociceptors: from fundamentals to clinic. Brain 2021;144(5):1312–35. Available from https://doi.org/10.1093/brain/awab048, http://brain.oxfordjournals.org/.

Mukherjee PK. undefined Quality control and evaluation of herbal drugs: evaluating natural products and traditional medicine quality control and evaluation of herbal drugs: evaluating natural products and traditional medicine. Elsevier; 2019. p. 1–784. Available from https://www.sciencedirect.com/book/9780128133743, https://doi.org/10.1016/C2016-0-04232-8.

Nema N. Integrating traditional and modern medicine: advancements in complementary therapies. Holistic and Integrative Medicine Science Journal 2025;1:7–12.

Nesterkina M, Ognichenko L, Shyrykalova A, Kravchenko I, Kuz'min V. QSAR models for the prediction of analgesic activity of terpenes and their derivatives. Structural Chemistry 2020;31:947–54.

Oladimeji FA, Adegbola AJ, Onyeji CO. Appraisal of bioenhancers in improving oral bioavailability: applications to herbal medicinal products. Journal of Pharmaceutical Research International 2018;24(4):1–23. Available from https://doi.org/10.9734/jpri/2018/45330.

Olalekan SO, Okwute PG, Osonuga IO, Adeyanju MM. A review of botanical interventions for neuropathy and neuropathic pain. Clinical Phytoscience 2024;10(1). Available from https://doi.org/10.1186/s40816-024-00385-8.

Pannunzio A, Coluccia M. Cyclooxygenase-1 (COX-1) and COX-1 inhibitors in cancer: a review of oncology and medicinal chemistry literature. Pharmaceuticals 2018;11(4):101. Available from https://doi.org/10.3390/ph11040101.

Salehi A, Ghanadian M, Zolfaghari B, Jassbi AR, Fattahian M, Reisi P, et al. Neuropharmacological potential of diterpenoid alkaloids. Pharmaceuticals 2023;16(5):747. Available from https://doi.org/10.3390/ph16050747.

Sarwa KK, Patel D, Rudrapal M, Bhattacharya S, Saraf S, Jain V, et al. Standardization and quality evaluation of botanicals with special reference to marker components evidence based validation of traditional medicines: a comprehensive approach. India: Springer Singapore; 2021. p. 405–26. Available from: https://link.springer.com/book/10.1007/978-981-15-8127-4 https://doi.org/10.1007/978-981-15-8127-4_20.

Silva-Correa CR, Campos-Reyna JL, Villarreal-La Torre VE, Calderón-Peña AA, González Blas MV, Aspajo-Villalaz CL, et al. Potential activity of medicinal plants as pain modulators: a review. Pharmacognosy Journal 2021;13(1):248–63. Available from https://doi.org/10.5530/pj.2021.13.35, http://www.phcogj.com/article/1358.

Thakur R, Nanda GC, Gupta A, Bharali BK. Pain management in ayurveda with special reference to angamarda prashamana and vedana sthapana mahakashaya of charaka: a review. International Journal of Research in Ayurveda and Pharmacy 2020;11(2):103–8. Available from https://doi.org/10.7897/2277-4343.110240.

Thiam P, Bellmann P, Kestler HA, Schwenker F. Exploring deep physiological models for nociceptive pain recognition. Sensors 2019;19(20):4503. Available from https://doi.org/10.3390/s19204503.

Tian MM, Li YX, Liu S, Zhu CH, Lan XB, Du J, et al. Glycosides for peripheral neuropathic pain: a potential medicinal components. Molecules (Basel, Switzerland) 2022;27(1). Available from https://doi.org/10.3390/molecules27010255, https://www.mdpi.com/1420-3049/27/1/255/pdf.

Uddin MS, Mamun AA, Rahman MA, Kabir MT, Alkahtani S, Alanazi IS, et al. Exploring the promise of flavonoids to combat neuropathic pain: from molecular mechanisms to therapeutic implications. Frontiers in Neuroscience 2020;14:1–18. Available from: https://doi.org/10.3389/fnins.2020.00478, https://www.frontiersin.org/journals/neuroscience#.

Valentová K. Cytoprotective activity of natural and synthetic antioxidants. Antioxidants 2020;9(8):1–4. Available from https://doi.org/10.3390/antiox9080713, https://www.mdpi.com/2076-3921/9/8/713/pdf.

Vishwakarma RK, Negi DS. The development of COX-1 and COX-2 inhibitors: a review. IInternational Journal of Pharmaceutical Sciences and Research 2020;11:3544.

Yao D, Chen Y, Chen G. The role of pain modulation pathway and related brain regions in pain. Reviews in the Neurosciences 2023;34(8):899–914. Available from https://doi.org/10.1515/revneuro-2023-0037.

Yeh TY, Luo IW, Hsieh YL, Tseng TJ, Chiang H, Hsieh ST. Peripheral neuropathic pain: from experimental models to potential therapeutic targets in dorsal root ganglion neurons. Cells 2020;9(12). Available from https://doi.org/10.3390/cells9122725, https://www.mdpi.com/2073-4409/9/12/2725/pdf.

Yousofvand N, Moloodi B. An overview of the effect of medicinal herbs on pain. Phytotherapy Research 2023;37(3):1057–81. Available from https://doi.org/10.1002/ptr.7697, http://onlinelibrary.wiley.com/journal/10.1002/(ISSN)1099-1573.

CHAPTER

7 Herbal approaches to mental health: anxiety, depression, and mood disorders

Alekhya Dandamudi[1], Sai Vishnu Mopidevi[1], Kanaka Durga Devi Nelluri[2], Monica Singuluri[2] and Sasi Priya Dodda[3]

[1]*College of Pharmacy, Krishna University, Machilipatnam, Andhra Pradesh, India,*
[2]*Department of Pharmaceutics and Biotechnology, KVSR Siddhartha College of Pharmaceutical Sciences, Vijayawada, Andhra Pradesh, India,*
[3]*Department of Biotechnology, Florida Institute of Technology, Melbourne, FL, United States*

7.1 Introduction

The COVID-19 pandemic created additional mental health challenges for individuals, with a major effect on elderly citizens. Prolonged periods of social distancing and confinement of citizens to their homes through quarantine and periods of solitary heightened levels of loneliness and mental illness. These are two major mental health challenges faced by elderly citizens are major risk factors for anxiety and depression. Elderly citizens also face challenges in accessing mental health-care and treatment for mental health illnesses. The barriers are high and include cultural and religious beliefs about mental illness, stigmatization of mental illness, poor mental health literacy among citizens, and stigmatization of mental illness and its treatment (Hudson et al., 2007). These have created a critical concern and require a supportive mental health-care system owing to the increasing growth rates of the elderly citizen demographics. Several other social factors contributing to the increased mental health disorder problem include poverty, unemployment, financial security, and future insecurity. Mental health disorders and issues clearly underscore a significant global public health problem that not only affects individuals but also extends across families, the health-care industry, and beyond.

What is even more alarming and warrants much more consideration and thoughtfulness in this area, has been documented in recent findings. It was highlighted that depressive, anxious, and stressful feelings were rampant in the elderly population, to the tune of 19.2%, 16.5%, and 13.9%, respectively. On this note, greater interest in Complementary and Alternative Medicine (CAM) and, more specifically, herbal medicine has increased by leaps and bounds within the population. Herbal medicine has a long history in ancient healing practices in Asia, Africa, and other parts of the world; this system of herbal healing dates back

Healing Botanicals. DOI: https://doi.org/10.1016/B978-0-443-44295-7.00005-7

centuries and has been used for the management of emotional health problems, sleep disturbances, and affective states of mind. Another reason for its popularity in contemporary health-care practices is its relationship with cultural beliefs; this system of therapy is also viewed as safe because it uses natural substances for the management of psychiatric problems, with relative accessibility to patients in comparison to conventional psychiatric care for mental health problems. Available scientific evidence points to the fact that various natural substances, including *Hypericum perforatum* (St. John's Wort), *Withania somnifera* (ashwagandha), and *Valeriana officinalis* (valerian root), might provide relief by reducing stress in patients with anxiety or depressive states of mind by affecting neurotransmitter levels, exerting stress-reducing activity, and improving sleep disturbances (Garg, 2025; Yadav et al., 2024). It must also be pointed out that adverse outcomes are less common with this system of therapy; it might also be more acceptable for elderly patients with a higher predisposition to adverse outcomes associated with medication. In the face of the international trend toward an increasing number of individuals suffering from mental health disorders, herbal medicine provides an approach that can fit within culturally appropriate service delivery on an international stage in terms of cost-effectiveness and evidence-based delivery, thereby helping to improve the accessibility and comprehensiveness of services for people suffering from mental health disorders.

7.2 Understanding mental health disorders

As depicted in Fig. 7.1, generalized anxiety disorder, major depressive disorder (MDD), bipolar disorder, and social anxiety disorder are some of the significant

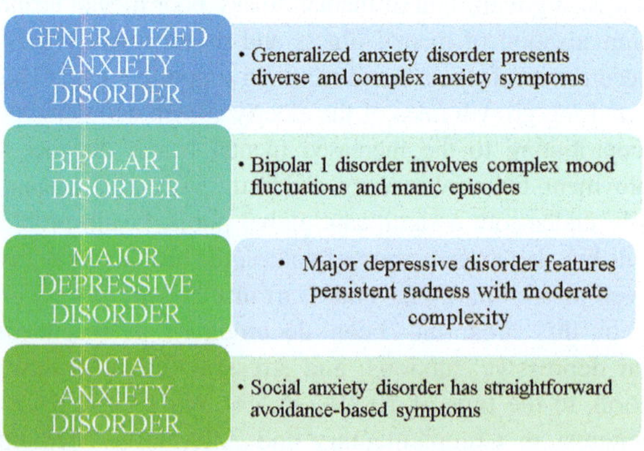

FIGURE 7.1 Classification of common mental health disorders.

Mental health disorders classification.

psychiatric illnesses that have their own distinct yet overlapping phenotypes and may be subject to potential phytotherapeutic interventions. Among these phenotypes of psychiatric illnesses, anxiety disorders comprise the most prevalent type of psychiatric illnesses across the world, characterized by excessive fear or worry that is persistent in nature. The clinical symptoms of anxiety disorders may vary, as they may be characterized by excessive sweating, tachycardia, tremors, fatigue, palpitations, and cognitive disturbances, such as excessive reflection, reduced concentration, fear of anticipation, as well as behavioral manifestations such as avoidance of situations that may induce anxious behaviors in an individual and social withdrawal, leading to impaired functioning in individuals whose vulnerability to coexisting psychiatric illnesses may be high. Nevertheless, the pathophysiology of anxiety disorders is complex and involves intricate interrelations among several important neurotransmitters. The three principal neurotransmitters implicated in anxiety disorders are serotonin, norepinephrine, and gamma-aminobutyric acid (GABA). A decrease in the inhibitory function of GABA contributes to neuronal hyperexcitability. In contrast, the intricate interrelation between serotonin and norepinephrine influences neurotransmitters activity associated with anxiety disorders. An inhibitory imbalance in the latter's function may lead to neuronal hyperexcitability in the brain. On the other hand, the intricate interrelation between serotonin and norepinephrine influences anxiety-related neural signaling. These complex alterations are further exacerbated by hyperfunction of the amygdala, a primary nervous center for fear responses. In effect, the abnormal functioning of the latter diminishes the ability to filter out irrational fear and to dampen perceptions of threats. Anxiety affects 264 million people worldwide and has nearly reached a staggering prevalence of up to 28.8% of the population in the United States. The next major mental disorder is MDD. This disorder is another major mental health problem characterized by persistent low mood and low self-esteem, and affected individuals are often not interested in activities or things they previously enjoyed. The DSM-5 describes MDD as a mental health disorder in which individuals exhibit certain symptoms daily for at least two weeks. The symptoms of this disorder include low moods, low energy, and significant weight and/or appetite change. Other symptoms include sleep difficulties and either retardation or agitation in individuals with this disorder. The symptoms also include fatigue or fatigue-related difficulties as well as thoughts of death and/or suicide. The nature of this disorder causes affected individuals to experience severe impairment in social and occupational functioning, as well as cognition. From a neurobiological point of view, major monoaminergic neurotransmitters, including serotonin, norepinephrine, and dopamine, are associated with depression. Serotonin helps maintain a stable mood and emotions; and its regulation is targeted by drugs known as selective serotonin reuptake inhibitors (SSRIs). Norepinephrine affects wakefulness, motivation, or stress responses. Dopamine controls the experience of reward or pleasure in life. Lapses in dopamine receptors or neurotransmitter systems are closely associated with anhedonia, or the inability to experience pleasure; this is a key feature of depression (Krishnan and Nestler, 2008). Current data also suggest neuroinflammation, oxidative stress, hypothalamo-pituitary-adrenocortical disturbances, and reduced neuroplasticity as causes of depression. This shift in knowledge provides an editorial rationale for identifying drugs or phytochemical agents that act on multiple biological systems rather than targeting a single neurotransmitter system. Bipolar disorders, falling under the spectrum of mood disorders, are characterized by significant mood, energy, and activity changes. Bipolar I disorder is characterized by the presence of mania or manic episodes that last seven or more days or require hospitalization, often with depressive episodes. Bipolar II disorder is characterized by hypomanic episodes that recur alongside depressive episodes, without any occurrences of mania. The key feature that sets bipolar disorders apart is that they involve emotional dysregulation. Emotional dysregulation refers to the inability to control emotions in social situations adequately. During hypomanic episodes, an individual may display characteristics such as increased energy, reduced sleep requirements, grandiosity, impulsive behaviors, and reckless behaviors; however, they may also exhibit characteristics of depressive episodes similar to those observed in MDD during the depressive episode phase. The etiology of bipolar disorder is complex and is proposed to be an interplay between genetic factors, neurobiological abnormalities, and environmental triggers. Brain network abnormalities related to the regulation of the body's circadian rhythms, central nervous system neurotransmitter systems, and intracellular signal transduction are among the implicated factors. In addition to these factors, the limbic system and the prefrontal cortex regions of the brain also show structural and functional abnormalities. Bipolar depression, being a chronic and episodic illness, requires long- and short-term management to tackle the symptoms of the illness. While pharmacotherapy involving the use of antipsychotics and mood stabilizers appears to be the main approach for tackling the symptoms of the illness, the use of adjunctive therapies such as phytotherapy to tackle the residual symptoms of the illness also holds potential for the management of residual symptoms. As a whole, these psychiatric illnesses emphasize the complexity of mental illnesses and underscore the importance of individualized treatment approaches. In fact, phytotherapeutic treatments offer complementary options with a broader range of neurobiological targets, including neurotransmitter modulation, stress responses, antiinflammatory effects, and neuroprotective properties (McIntyre and Soczynska, 2010). As demonstrated in Fig. 7.1, the identification of distinct clinical phenotypes of all three classes of mental illnesses—anxiety disorder, major depression, and bipolar disorder—is critical for the rational use of herbal medicines in modern psychiatric treatment.

7.3 Herbal medicine in psychiatry: an overview

7.3.1 Historical use of herbs in mental health and advantages and limitations of herbal therapies

Herbal medicine has been a major contributor to health-care delivery systems across cultures over time, especially in the treatment of mental and emotional illnesses. Ancient civilizations, such as those of China, Greece, and India, have used medicinal plants in their systems of medicine to treat disturbances of emotion, such as mood and behavioral disturbances. Agents such as *Panax ginseng*, *Ginkgo biloba*, and *H. perforatum*, commonly known as St. John's Wort, have been used for the traditional treatment of anxiety and depressive symptoms, enhancement of cognitive function, and restoration of emotional balance. Health care in medieval monastic communities was also instrumental in preserving traditional knowledge of herbs. The medicinal plants in the monks' gardens, along with the herbals that documented their medical properties, often blended botanical information with spiritual or religious remedies (Fuloria et al., 2022). It was a crucial period for the preservation of traditional medical knowledge, especially at a time when scientific understanding and research had not yet fully developed.

The Renaissance marked a renewed interest in medicinal plants, leading to extensive observations, classifications, and documentation of their therapeutic properties. Scientific advances gradually integrated traditional herbal knowledge with emerging medical concepts, leading to the development of the modern disciplines of pharmacognosy and phytotherapy (Fuloria et al., 2022). In recent years, the use of herbal medicine has received renewed interest in contemporary medical practice. Of interest to current medical practice is the domain of mental health care, as patients increasingly turn to herbal medicine options rather than the more dominant psychotropic medications of conventional medicine. Herbal medicine also offers the advantage of being more holistic, relieving symptoms, and achieving an optimal balance between physiological and psychological aspects. In addition to this is the advantages of availability and affordable, combined with a more desirable "side effect" profile compared to dominant synthetic psychotropics. Another desirable feature of the herbal medicine is its use for the long-term management of the chronic conditions.

Despite these advantages, various issues prevent the widespread use of herbal medicine for psychiatric care. For example, the standardization of herbal medicinal products remains a problem, since variability within plant species can have serious implications for the potency of the medicine. There is also a lack of conclusive clinical trials on the efficacy and potency of various herbal products, which prevents the large-scale evaluation of their efficacy and potency for psychiatric conditions. In addition to this, there exists a risk of herb-drug interaction that remains potentially dangerous for the individuals using traditional psychotropic drugs for condition management. There is significant variability in the regulatory jurisdictional of herbal products across countries worldwide. Nonetheless, current theory has established a significant base of knowledge that demonstrates the potential additional use of herbs in the prevention of mental illnesses. The fact that they have remained relevant shows that ancient knowledge continues to contribute to modern approaches to mental health (Ernst, 2002).

7.3.2 Pharmacognosy of psychoactive herbs

Many theories address the impact of herbs derived from natural sources on mental health. On the other hand, the basic foundation for the study of the impact of herbs on the mind was established through the field of pharmacognosy. Generally, pharmacognosy involves the use of herbs obtained from the environment to demonstrate their efficacy in affecting mental conditions at the physiological level. In general terms, psychoactive herbs function by regulating the physiological mechanisms that control the mind. Examples of herbs that are used to address depression include *H. perforatum*, commonly known as St. John's Wort. This type of herb has been used to treatdepression at the physiological level through the inhibition of important physiological mechanisms such as serotonin and the hormones norepinephrine and dopamine. Another significant herb is *Piper methysticum* or Kava, traditionally used by the people living in the Pacific Islands. It exhibits good anxiolytic and relaxing activity. This activity is due to the herb's kavalactone content. Kavalactones increase GABA activity in the brain. This reduces the excitability of brain cells while maintaining sufficient wakefulness. Similarly, *V. officinalis*, or valerian root, has traditionally been used for the treatment of insomnia or anxiety. The primary bioactive ingredients in the herb, such as valerenic acid, interact with GABA receptors in the brain and prevent the breakdown of GABA. These actions produce the desired therapeutic effects without causing a significant hangover effect. Unlike many conventional sedative medications, the hangover effect is less pronounced in patients. Unlike other related medicines on the market, the hangover effect in patients is less pronounced. *W. somnifera*, or ashwagandha roots, finds significance in Ayurvedic medicine. It belongs to the family of adaptogens; that is, the drug or herb increases the body's resistance to stress. According to studies in the domain of pharmacognosy, the contents of ashwagandha, namely withanolides, modulate activity in the hypothalamic-pituitary-adrenal (HPA) axis. These components reduce the cortisol level in the brain. Collectively, these herbs illustrate various mechanisms by which plant active principles can affect neurochemical pathways, alleviate oxidative stress, and modulate inflammatory processes associated with mental disorders. Currently, pharmacognosy has scientifically justified traditional herbal remedies for their rational use in modern psychiatric care. It enables the integration safer, complementary therapeutic strategies for the management of anxiety, depression, and stress-related disorders while preserving the holistic principles of natural medicine.

7.4 Herbal remedies for anxiety disorders

As indicated in Table 7.1, herbal compounds such as valerian (*V. officinalis*) and ashwagandha (*W. somnifera*) exhibit substantial anxiolytic effects, mainly through their influence on the GABA system or through adaptogenic effects that enable them to counteract stress factors. Anxiety disorders encompass one of the most prominent mental health challenges globally and transcend all ages and socioeconomic classes of human populations in various countries of the world. While pharmacologic management of anxiety disorders utilizing therapies in the class of

Table 7.1 Common herbal remedies for anxiety and their pharmacological actions.

Herb name	Botanical name	Key active constituents	Mechanism of action
Valerian	*Valeriana officinalis*	Valerenic acid, isovaleric acid	GABA modulation, sedative effects
Passionflower	*Passiflora incarnata*	Flavonoids, alkaloids	GABAergic activity, anxiolytic action
Chamomile	*Matricaria chamomilla*	Apigenin	Benzodiazepine receptor binding, antiinflammatory
Ashwagandha	*Withania somnifera*	Withanolides	Adaptogenic, cortisol-reducing

benzodiazepine compounds or SSRIs has dominated management strategies and has shown substantial efficacy in large numbers of patients worldwide, their use in managing anxiety in most patients in a long-term fashion has shown substantial adverse effects such as sedation, dependency, withdrawal symptoms, and interference with sexual capabilities among other unwanted effects of their prolonged use. Such effects have prompted substantial interest in herbal compounds as management strategies in anxiety disorders of different kinds in many defining periods of recent world history. Three herbal compounds known to have the most marked anxiolytic effects among those currently studied as treatments for anxiety states are valerian root, passionflower, and chamomile (Al-Attraqchi et al., 2020; Zaharie et al., 2022). Although the specific chemical compositions vary, each compound has distinct impact sites in the brain. Valerian root is native to Asia and parts of Europe and it has traditionally been used to treat nervous states, such as restlessness. Essentially, the observed anxiolytic effect is due to the action of the valerenic acid constituents in the plant. They delay the inactivation of the neurotransmitter GABA. Furthermore, GABA binds to the GABA-A receptors in the brain. As a result, the likelihood of brain excitation decreases dramatically. Unlike benzodiazepines, these plant compounds accomplish the same purpose while maintaining the cognitive capabilities in a fully operational state. Although the impact on the brain is similar to that of other compounds, such as benzodiazepines, the compound seems to be more appropriate for the management of anxiety states in older people. Research in the area has indicated that the compound has the potential to help reduce anxiety levels in patients. These states directly affect sleep quality. Passionflower (*Passiflora incarnata*) is another well-recognized herbal anxiolytic, traditionally used for nervous agitation, insomnia, and stress-related disorders. Modern pharmacological research supports its use, since several clinical studies have confirmed the efficacy of extracts of the plant in reducing anxiety, including situational anxiety like the one occurring before dental procedures or surgical interventions. These anxiolytic properties have been attributed to flavonoids and alkaloids, which facilitate GABA activity by reducing GABA reuptake and modulating GABA receptor functions (Velasquez et al., 2024; Appel et al., 2011). In addition, passionflower may act on other neurotransmitter systems, including serotonin and dopamine, thereby contributing to mood stabilization and regulation of emotional response. Administration compared to classical anxiolytics presents with fewer side effects; it causes fewer sedative effects and also entails less risk of dependence, making this plant a promising adjuvant in anxiety therapy. One such herbal supplement is chamomile (*Matricaria chamomilla*). It is frequently consumed as a herbal tea and is widely recognized for its mild anxiolytic activity. The anxiolytic activity of chamomile is believed to be due to the flavonoid compound apigenin which binds to benzodiazepine receptors in the brain. In addition to its effects on neurotransmitter system, chamomile also exhibits significant antioxidant and antiinflammatory activity. Thus such activity may be another significant contributor to anxiety relief. It has been seen in recent times that patients suffering from chronic anxiety may be display oxidantive stress as well as inflammation. These may be key areas where the antianxiety herbal drug chamomile may manifest its activity. Such activity may be significant in alleviating the patient's condition. Results based on clinical studies have indicated that patients with generalized anxiety disorder showed reduced severity of anxiety through the administration of chamomile (Mao et al., 2016; Amsterdam et al., 2020). In addition to GABAergic activity, these herbs exert their anxiolytic effects through several other mechanisms. The herbs produce their effects by altering and modulating various neurotransmitters in the monoamine system. For example, neurotransmitters serotonin and dopamine are very crucial in mood changes and emotional well-being. Alterations in these neurotransmitters have often been evidenced in patients with anxiety disorders. The herbal compounds' role in restoring balance to these neurotransmitters has therapeutic potential formanaging and reducing anxiety disorders. Additionally, some herbal compounds have adaptogenic activities that help in managing and coping with various forms of mental and physical stress. The adaptogenic activity of herbal compounds leads to the reduction of activation of the HPA axis and reduction in cortisol levels to prevent a cascade of anxiety disorders. On other fronts, there has been increased attention to reactive oxygen stress and neuroinflammation as causes of anxiety disorders. Polyphenolic and flavonoid compounds in herbal preparations are known to scavenge free radicals and have antiinflammatory effects in the central nervous system through their antioxidant activity. The neuroprotective activity of herbal preparations may have profound effects on managing anxiety disorders through their overall effects on neuronal functions. Indeed, as shown in Fig. 7.2, valerian root extract, passionflower extract, and St. John's Wort exert their effects through antianxiety activity that includes GABA modulation, neurotransmitter interactions, antiinflammatory activity, and antioxidant activity, among others. It can thus be stated conclusively that, in consideration of the general efficacy of herbal anxiolytics as a form of medication for various forms of anxieties or anxieties in general, this approach is promising for

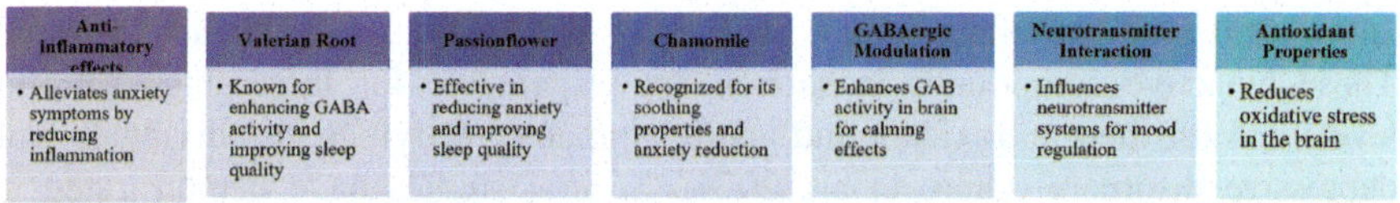

FIGURE 7.2 Herbal remedies and mechanisms in the management of anxiety.

Anxiety management herbal remedies.

individuals who experience adverse outcomes to the application of conventional forms of medication. As such, more studies would need to be conducted on a larger scale to confirm the general safety of these forms of medication for anxiety.

7.5 Herbal treatments for depression

Herbal preparations are being studied in various pharmacological studies as natural or auxiliary treatments for managing depressive states with noted interest. These treatments are likely to be of value for managing conditions of mild-to-moderate depressive states or those in which standard drugs for depression are poorly tolerated. *H. perforatum* (St. John's Wort), *Crocus sativus* (saffron), and *Rhodiola rosea* (golden root) are among the most extensively researched medicinal plants for managing depressive conditions, with apparently impressive pharmacological and clinical evidence supporting their antioxidant activities. John's Wort is one of the most extensively studied herbal antidepressants and has been used in traditional medicine for centuries in Europe. Its antidepressant effect has been attributed mainly to the bioactive compounds hypericin and hyperforin, which inhibit the reuptake of monoamine neurotransmitters, including serotonin, dopamine, and norepinephrine (Butterweck, 2003). The mechanism of this action resembles the action of conventional antidepressants, including SSRIs and SNRIs. In addition to modulating monoamines, St. John's Wort has antiinflammatory effects. It also interacts with other neurotransmitter systems, including receptors for GABA and glutamate, all of which are implicated in mood regulation and emotional stability (Zanoli, 2004). Results from clinical trials and metaanalyzes indicate that St. John's Wort is more effective than a placebo and as effective as standard antidepressants for mild-to-moderate depression; it often presents fewer side effects, such as sexual dysfunction and sedation (Laakmann et al., 2002; Di et al., 2008). However, despite these advantages, it must be used with care, as it induces cytochrome P450 enzymes and P-gp, which in turn reduce the efficacy of concomitant medications such as oral contraceptives, anticoagulants, immunosuppressants, and antiretroviral agents. Saffron, prepared from the stigmas of *C. sativus* plants, is another promising herbal drug with mounting evidence supporting its use as an effective adjunct therapy from RCTs. The major bioactive compounds in saffron, responsible for its antioxidant and antiinflammatory properties, are crocin, crocetin, and safranal. These bioactive compounds in saffron are responsible for increasing serotonin levels and for their antioxidant and antiinflammatory activities in individuals with depressive disorders (Otero et al., 2024). Several studies have demonstrated its effectiveness in improving depressive symptoms in individuals with mild-to-moderate depressive disorders. The findings of several clinical studies indicate its potential as a promising adjunct therapy, as it not only improves depressive symptoms in individuals but also exhibits effectiveness comparable to several antidepressants like fluoxetine and imipramine. The best part about using this herbal drug as a potential adjunct therapy is its lower potential to produce adverse effects as opposed to antidepressants. Its safe profile and tolerability make it a promising herbal drug with potential antidepressant activities. *R. rosea*, commonly known as golden root, is a herbal antidepressant in the adaptogen category. Adaptogens are medicines that improve the ability to resist physical, emotional, and environmental stressors, which are major precipitating and perpetuating factors in depressive disorders. *Rhodiola* exerts its antidepressant actions via several mechanisms, involving modulation of serotonin and dopamine pathways, regulation of the HPA axis, and the reduction of cortisol levels. In addition, high cortisol levels have often been associated with depression, leading to common symptoms such as a lack of energy, reduced cognition, and disrupted emotional states. The normalization of stress hormone levels by *Rhodiola* thus restores emotional balance and enhances stress tolerance. Furthermore, *Rhodiola* exhibits antioxidant and antiinflammatory activities, thereby contributing to neuroprotection and improved neuronal function. Clinical studies have so far demonstrated that supplementation with *Rhodiola* decreases depressive symptoms, mental fatigue, and stress-related exhaustion, ensuring good tolerability and a minimal side-effect profile. Collectively, these herbal compounds have some common pharmacological properties. The pharmacological actions of these herbal compounds underlie their antidepressant activity. Such actions include the regulation of monoamine compounds, antineuroinflammatory effects, antioxidative stress effects, and stress-modulating activity. Unlike the single-virus theory employed by other antidepressants to address the issue, herbal compounds possess multiple pharmacological targets. Multiple targets have the potential to confer broader benefits. Depression is a condition with multiple interrelated biological, psychological, and environmental components. As stated in Table 7.2, prominent herbal drugs like St. John's Wort or saffron reveal impressive efficacy in managing depressive symptoms, with fewer and less severe side effects in compared with standard pharmacotherapy for depression. Although the results of the research in this field are promising in their current form, several critical considerations need to be taken into account regarding the application of natural or herbal antidepressants in regular psychiatric settings. First and foremost, these include the standardization of the drugs concerned, as well as safety levels regarding possible interactions with pharmacotherapy. Nonetheless, this type of therapy remains an important part of integrative depression care because it provides

Table 7.2 Herbal antidepressants and their clinical efficacy.

Herb	Botanical name	Clinical use	Evidence from trials	Notable interactions
St. John's Wort	*Hypericum perforatum*	Mild to moderate depression	Comparable to SSRIs in some trials	May interact with SSRIs, CYP3A4 inducers
Saffron	*Crocus sativus*	Mild depression	Comparable to fluoxetine	Minimal, but requires further study
Turmeric	*Curcuma longa*	Depression, cognitive decline	Antiinflammatory effects aid mood	May potentiate anticoagulants

patients with additional viable measures of therapy with respect to the holistic management of depression as a mental health concern.

7.6 Herbs in the management of mood disorders

Herbal medicine still plays an important role in the management of mood disorders in complementary and integrative medicine, especially in traditional systems like Ayurveda, which view mood disorders as the result of imbalances in physical, psychological, and environmental states. Among the most studied Ayurvedic medicinal plants for their potential to manage mood disorders are *W. somnifera* (ashwagandha), *Bacopa monnieri* (brahmi), and *Curcuma longa* (turmeric), characterized by their cognitive-enhancing properties and active components that exert protracted effects in conditions like depression, with HPA axis-targeting components. Ashwagandha is commonly regarded as an adaptogenic plant; that is, it significantly supports and upregulates the body's ability to cope effectively with stress and environmental demands while maintaining homeostasis. Its primary method of functioning consists of modulating HPA axis function, a key system in handling stress and emotions, commonly imbalanced in various mood disorders, for example, depression and anxiousness. Prolonged HPA hyperactivity is related to hypercortisolism, a situation in which excessive levels of this major stress hormone cortisol affect various psychiatric and neurological states, as well as neuronal injury, difficulties in neurogenesis, and emotional instability. Ashwagandha reduces circulating cortisol levels, thereby lowering the body's stress responses and returning the neuroendocrine system to a balanced state (Singh et al., 2019; Zakrzewska et al., 2024). Ashwagandha also has a neuroprotective role, promoting new neuronal growth and likely counterbalancing the neurodegenerative processes that commonly manifest in various chronic mood disorders. The plant, in fact, supports improved stress tolerance, reduces anxiety and depression, promotes better sleep, and, overall, fosters more emotionally positive states, while at the same time proving safe and well-tolerated, as shown in various studies. *B. monnieri*, also called brahmi, is another mainstay of Ayurvedic neurotherapeutics. *B. monnieri* has been traditionally employed to enhance intellect, memory, and emotional steadiness. Although *Bacopa* is best known for its cognition-enhancing properties, its positive influence on mood regulation is also being recognized. *Bacopa* contains several bioactive molecules called bacosides. Bacosides have been found to possess robust antioxidant capacity. Bacosides scavenge free radicals and cause a reduction in oxidative stress in the central nervous system. Oxidative stress has been seen to contribute to the development of depression and anxiety disorders. Therefore *Bacopa* has been seen to reduce oxidative stress, thus maintaining healthy neuronal function. *B. monnieri* has also been shown to significantly affect major neurotransmitters such as serotonin and dopamine (Kean et al., 2022), both of which play critical roles in the regulation of emotions. *Bacopa* has been seen to increase IQ and enhance cognitive flexibility, executive function, and emotional function, thus being particularly promising in treating depression and anxiety disorders. Although *Bacopa* has a tonic effect on regulating emotions, its capacity to positively affect cognition also makes it particularly promising. *C. longa*, commonly known as turmeric, has gained considerable attention in psychiatric pharmacology due to the pharmacological properties of its first bioactive metabolite, curcumin. It has marked antiinflammatory and antioxidant properties that target two crucial biological processes in mood disorders. Neuroinflammation, defined by increased cytokine production, has been recognized to play an increased role in the development and maintenance of depressive symptomatology (Fuloria et al., 2022; Anand Shukla, 2021). Turmeric extracts reduce cytokine production and downregulate elevated microglial activity, thereby mitigating adverse effects on neural function. In addition, turmeric extracts also increase antioxidant levels, giving turmeric significant protective properties in regulating and reducing the adverse effects of free oxygen radicals that may compromise neurotransmitter biosynthesis and neural communication. More importantly, curcumin regulates monoamine neurotransmission by increasing the availability of serotonin and dopamine. Curcumin biosupplements have a significant effect in reducing depressive symptomatology in individuals who suffer from drug-resistant depressive disorders. Collectively, ashwagandha, *B. monnieri*, and *C. longa* have a multidirectional effect that promotes the therapeutic efficacies of these three medicinal herbs. Their combined pharmacotherapy alleviates and addresses the insidious neurobiology of MI. The effectiveness of this pharmacotherapy lies in its multidirectional effect when applied in concert to combat neurotransmitter imbalance, oxidative stress, neuroinflammation, and stress reactions. These neurobiological mechanisms correspond to and harmonize with current models of depression and anxiety. As depicted in Fig. 7.3, this class of herbs has been shown to play a critical role in regulating mood through various pathways that may be interconnected but certainly function independently. Ashwagandha has been demonstrated to act on stress hormones; *Bacopa* has been shown to affect cognitive and emotional function through

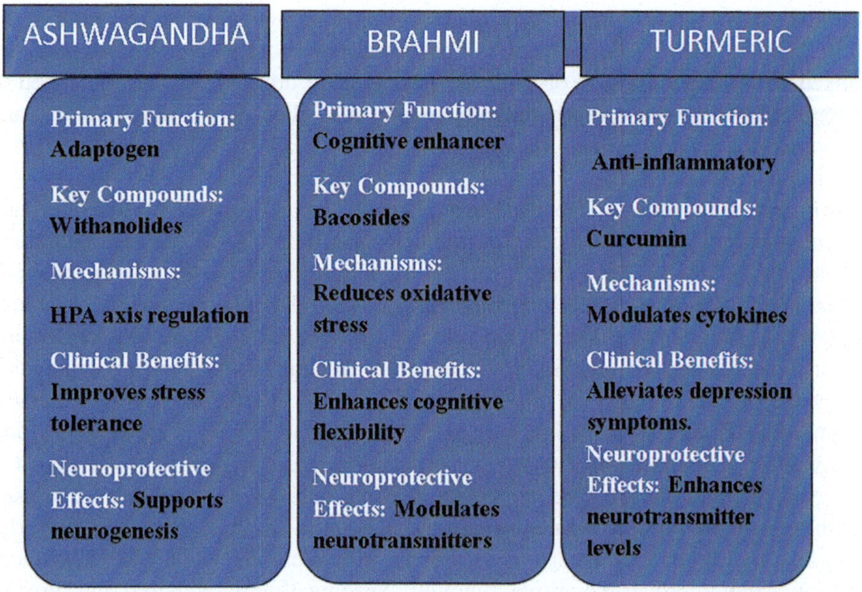

FIGURE 7.3 Herbal medicines for mood disorders—functions, compounds, and mechanisms.

Mood disorders herbal remedies.

antioxidant and neurotransmitter-modulating pathways; and finally, there is evidence that turmeric acts on inflammatory/oxidative pathways and regulates monoaminergic systems. This confluence of herb usage within integrative mental health care might provide a synergistic effect, mainly for patients who show partial response and/or side effects to standard pharmacologic care. In a word, a range of preclinical and emerging human studies provide evidence that supports the integration of AYUSH herbal medicines—in this case, ashwagandha, *B. monnieri*, and *C. longa*—into a holistic mental health management system. When properly standardized, formulated, and clinically monitored, each of these single herbs holds promise as a useful adjunct in managing mental health, increasing overall emotional well-being, and achieving better outcomes, ultimately culminating in better mental health outcomes for populations worldwide.

7.7 Synergistic and adjunctive use with conventional therapies

In the field of mental health, increasing scientific interest has been given to the overlap between herbal treatment and traditional pharmacological therapies for mood disorders.

An integrative strategy will likely increase the effectiveness of treatment and the tolerability for patients, as well as possibly alleviate some of the side effects commonly linked with the chronic use of synthetic psychoactive drugs. However, the skeptical approach is justified because herbal preparations may alter the pharmacodynamic and pharmacokinetic profiles of conventional agents. It has been reported that herbal components interact with a range of drug-metabolizing enzymes, thereby affecting drug absorption, distribution, metabolism, and excretion. As an example, *W. somnifera* and *B. monnieri* could have an impact on multiple cytochrome P450 enzymes and could change the plasma levels of simultaneously used antidepressants, anxiolytics, or mood stabilizers (Hooli et al., 2025; Mohamed Afsal, 2025). Therefore drug interactions may either enhance or decrease the therapeutic effect, which requires careful consideration during coadministration. In particular, curcumin, an antiinflammatory and antioxidant bioactive phenolic compound extracted from *C. longa* (turmeric), has demonstrated significant efficacy as an adjunct in mental health therapy by regulating neuroinflammation, a key factor in the etiology of mood-related disorders (Ali et al., 2024).

Nevertheless, the pro-bleeding effect of curcumin increases the risk when used with antiplatelet or anticoagulant treatment. Herbal medicine, when used properly, can be a useful complementary medicine that promotes emotional balance, reduces anxiety, and improves cognitive function. Specifically, *B. monnieri* has proven effective in treating anxiety and cognitive impairment and has become a supportive treatment option for the outpatient treatment of depression in the absence of cognitive disabilities (Bosco Don Bosco et al., 2024; Awuchi, 2023). To achieve safety and effectiveness in integrative therapeutic strategies, patient-centered care should be incorporated into practice. Drug interactions with other herbs should be clarified, strict observation guidelines should be established (primarily with drugs with narrow therapeutic indices or those that need careful adjustment of dosage), and strict monitoring programs should be introduced. Also, the establishment of elaborate clinical guidelines and strict quality control systems remains necessary to maintain the uniformity and reproducibility of herbal preparations. Moreover, as shown in Fig. 7.4, other agents with mood-supportive effects, including *W. somnifera* and *C. longa*, exert their effects by modulating the stress-response and inflammatory pathways, highlighting a potentially highly effective combination of herbal agents for a variety of affective disorders. Similarly, one can explain the preference for *W. somnifera* by citing its relatively favorable side-effect profile compared to traditional anxiolytics, as well as its adaptogenic effects.

7.8 Clinical evidence and trials

7.8.1 Clinical evidence from human and animal studies

Clinical and preclinical studies have shown that botanical interventions have the potential to improve anxiety- and depression-related disorders. The paucity of

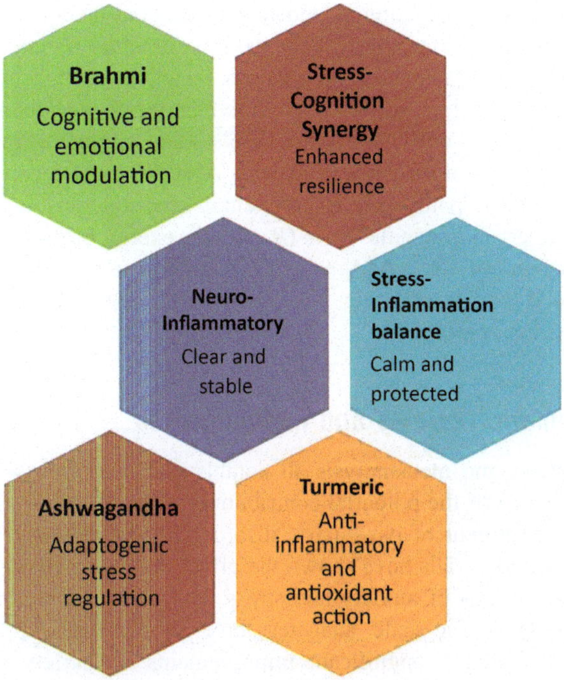

FIGURE 7.4 Overlapping mechanisms of herbal mood support.

Mood support herbal drugs mechanism of action.

randomized human trials, however, has limited their use across a variety of settings. *W. somnifera*, or ashwagandha, has proven to mitigate the effects of stress and anxiety strikingly in adulthood. Standardized instruments such as the Hamilton Anxiety Rating Scale and the Perceived Stress Scale were used in one recent study to assess reduced stress perception and changes in affective well-being. These outcome measures support the herb's efficacy in modulating stress and in potentially restoring physiological equilibrium. Another well-researched botanical, *H. perforatum*, also known as St. John's wort, is effective for treating mild-to-moderate depression. Clinical data suggest that its antidepressant effects are similar to those of SSRIs, presumably as a result of modulation of serotonergic neurotransmission.

Nevertheless, St. John's Wort is known to induce drug-metabolizing enzymes, altering the pharmacokinetics of concomitant medications and making it vitally important to monitor it carefully in clinical settings. Collectively, these data highlight the traditional use of botanical medicines in mental health practice. While mechanistic information from preclinical investigations is important, the lack of large-scale, rigorous clinical data currently makes it difficult to formulate definitive standards for integrating herbal agents into mental health assessment and management. Moreover, preclinical studies using animal models are set to hone in on the mechanisms of these clinical benefits. Experimental research in rodents has shown *W. somnifera* to have anxiolytic effects, which are demonstrated by reversal of anxiety-related behaviors, increased cognitive performance, and neuroprotection. These effects have been attributed to the herb's adaptogenic constituents and to its influence on gamma-aminobutyric acid type A (GABA-A) receptors and stress-related neurochemical pathways. Experimental investigations in rodents further reveal that another adaptogenic herb (*R. rosea*) exerts stress-mitigating actions (reduced biomarkers of stress and increased resilience to stress). These effects are mainly mediated by modulating the HPA axis, the principal regulator of physiological stress responses.

7.8.2 Systematic reviews and metaanalyses

Systematic reviews and metaanalysis all notably produce robust and increasingly compelling evidence of the potential contribution of herbal medicines as beneficial adjuncts in the treatment of psychiatric disorders. Comprehensive evaluations of randomized controlled trials have shown the effect of *W. somnifera* (ashwagandha) on decreased symptoms of anxiety and stress with a proven favorable safety and tolerability profile. Large-scale analyzes have shown that supplementation of ashwagandha is linked to significant improvements in anxiety scores and few adverse effects, suggesting that it is likely safe to incorporate ashwagandha into a conventional treatment protocol for anxiety-related disorders (Arumugam et al., 2024; Bachour et al., 2025).

Similarly, evidence from systematic reviews and controlled clinical studies shows the efficacy of *C. sativus* (saffron) in the management depressive symptoms. These analyzes demonstrate that the antidepressant effects of saffron are comparable to those of standard pharmacotherapies and are associated with fewer side effects. Consequently, saffron is a promising therapeutic choice for patients who are intolerant to conventional antidepressants or patients choosing to undertake complementary treatment. *H. perforatum* (St John's Wort) is consistently ranked amongst the most effective herbal interventions for mild to moderate depression in systematic evaluations, especially in European clinical practice, where prescription of St John's Wort is more widespread. Metaanalytic data indicate that its efficacy is similar to that of conventional antidepressants.

Nevertheless, St John's Wort is associated with a high potential for clinically relevant drug-drug interactions, particularly through strong induction of cytochrome P450 enzymes and drug transporters. Therefore its application must be done with great care and careful monitoring, particularly among those patients taking several concomitant drugs. Systematic reviews and metaanalyzes are contributing to the expanding understanding of herbal medicines as potentially effective complementary treatments for psychiatric disorders. When herbal therapies are chosen based on evidence, administered at appropriate doses, and supervised clinically, they may enhance treatment outcomes while reducing the likelihood of adverse effects in specific patient groups.

7.8.3 Efficacy and safety considerations

It has been observed that the therapeutic value of herbal drug constituents in mental health disorders varies depending upon the disease condition and the phytoconstituents present in medicinal herbs. It has been observed that a drug such as *W. somnifera* (ashwagandha) has produced significant effects on mental well-being by reducing stress-related biochemical markers and contributing to psychological well-being (Kaptchuk, 2000). Its adaptogenic properties may play a significant role in balancing neuroendocrine responses and alleviating stress. *C. sativus* (saffron) has significant antidepressant properties, which may be explained by its rheostatic function in modulating serotonin neurotransmission and alleviating mood. *R. rosea* has proven its efficacy in building tolerance to stress and emotional resistance. Therefore *R. rosea* holds promise for the management of stress-related and mood disorders. *Rhodiola's* efficacy is related to the regulation of HPA activity and the modulation of stress-induced alterations in neurochemistry. So, the therapeutic potential for *R. rosea* in the management of psychiatric illness needs to be endorsed. Most of the herbal medicines discussed are generally considered safe at the doses of evidence-based therapeutics. Ashwagandha is generally well tolerated; however, large doses may cause digestive problems, such as nausea or abdominal pain. Saffron is safe for consumption in low to moderate amounts, but high doses could be toxic and cause adverse neurologic effects or gastrointestinal effects. St John's Wort, though an established herbal antidepressant, needs to be given with particular caution because it possesses a strong potential for clinically significant interactions between drugs and herbs. It may interfere with commonly prescribed drugs such as SSRIs, anticoagulants, and oral contraceptives. These findings consequently underscore the importance of evidence-based dosing, professional supervision, and cautious monitoring of herb-drug interactions in psychiatric practice. For rational integration of herbal therapies, an individual approach should be adopted based on clinical evidence, including safety considerations and patient-specific considerations (Machín et al., 2023). As indicated in the summary contained in Table 7.3, some important randomized controlled studies concerning the evaluation of herbal preparations in the intervention of anxiety and depression have found statistically significant evidence to support the potential utilization of such drugs in the mental health care intervention.

7.9 Regulatory and quality control aspects

The safe and optimal use of herbs in handling mental health disorders, such as anxiety, depression, and mood variation, requires stringent supervision and quality

Table 7.3 Summary of randomized controlled trials on herbal therapies for mental health disorders.

Herb used	Condition studied	Sample size	Outcome
Valerian	Generalized anxiety	36	Reduced anxiety versus placebo
Saffron	Depression	230	Significant mood improvement
Ashwagandha	Stress and anxiety	64	Cortisol reduction, improved mood
Turmeric	Major depressive disorder	56	Enhanced antidepressant effect

management. It is believed that herbs, as an alternative medicine, are clinically safe because of their naturally derived constituents. The quality of herbs thus greatly determines their efficiency and potency, as well as ensuring safe handling of mental health-related disorders. One of the biggest demands in this regard is related to standardizing herbs. The standardization of herbs primarily involves determining, measuring, and maintaining optimal levels of their active constituents that trigger a therapeutic response to health issues. The proportion of active constituents in medicinal plants varies substantially depending upon aspects that range from variations in plant genetics, soil, climatic, and seasonal components to the methods of harvesting, processing, and so forth. A good example of a well-recognized medicinal herb that benefits from a wide application in the treatment of mild-to-moderate depression, as well as MDDs, through the influence of bioactive compounds like hypericin and hyperforin, which act through an influence on the monoamine neurotransmitter systems that stabilize mood, is *H. perforatum* or St. John's Wort. However, there is considerable variation in the content of these compounds from one formulation to another, and a lack of proper standardization will likely result in underdosing or overdosing, thus compromising not only the efficacy or effectiveness of a drug but also its safety (Adeniyi Grace et al., 2024). To address these challenges, various advanced analytical techniques are employed for evaluating the quality and composition of herbal products. Conventional laboratory methods, such as high-performance liquid chromatography (HPLC) and GC-MS, are commonly used to identify active constituents, detect contaminants, and verify purity and stability in herbal formulations. These approaches enable precise quantification of phytochemicals, thereby helping ensure batch-to-batch consistency while minimizing the risk of adulteration and contamination with pesticides, heavy metals, and microbial toxins. This kind of analytical validation is especially important in the case of herbal medicines designed for long-term application in psychiatric conditions, where sustained exposure requires very high safety limits. Also equally important is the observance of good manufacturing practices (GMP). This encompasses the entire process of producing the herbal product, from the procurement of raw materials to packaging for customer consumption. Such compliance with GMP helps ensure that herbal products are produced under controlled conditions that follow standardized procedures (Jain et al., 2024; Shinde et al., 2023). Additionally, the entire process is regulated by quality assurance standards that provide guidelines for packaging and storing products for customer consumption. In the case of the mentally ill population who require prolonged periods of medication use, the need for observance by these regulations becomes even more pronounced for the purpose of building confidence in the efficacy of the medication within the pharmaceutical community and, generally, for the purpose of protecting the health of the mentally ill population who require such periods for consumption of the medication at hand for the purpose of seeking cures for the various forms of illnesses that afflict the mentally. Yet another important factor influencing the effectiveness of herbal medicine is its formulation and route of administration. The herbal medicine is made available to patients through various routes of administration, including capsules/tablets, tea infusions, and tinctures. Liquid routes of administration are typically preferred over tablets and capsules, as they ensure rapid and maximum drug bioavailability. This is attributed to the fact that these routes bypass some of the digestion periods typically followed by tablets and capsules (Pannu et al., 2025). This is critical in mental health, as symptom relief is desired as quickly as possible. The concept of rapid relief is mainly sought in anxiety and sleep disturbances, but may not be desired during depressive disorder treatment. Another crucial factor in the safe administration of herbal medicines is determining doses. Herbal medicines, unlike conventional medicines, do not entirely resolve this issue, as the administration of standard dosages for some of these drugs remains subjective in practice. A correct dosing regimen is a combination of factors, including a patient's age, general well-being, health status, the route of administration, and the active ingredient. When considering a dosing regimen, elderly persons and those with comorbid health conditions may require a smaller amount of a medicine to avoid unwanted side effects. The dosages of a medicine should ideally be determined through a series of clinical trials, as this would help identify the most effective and safest levels. Along with dosing, careful monitoring is essential if herbal medicines are taken in addition to conventional psychotropic drugs. Certain herbs, especially St. John's Wort, are known to alter the activity of drug-metabolizing enzymes and transporters, potentially altering the pharmacokinetics of coadministered medications (Sarris and Karamacoska, 2013). The importance of health professional supervision at the point of integrating herbal therapies into mental health treatment plans cannot be understated. It is equally important to educate patients on the proper use of herbal medicines, possible interactions, and the issue of not exceeding recommended doses for safe outcomes. While herbal medicines offer considerable promise as complementary and alternative therapies for mental disorders, their safety and effectiveness depend on rigorous quality control measures, standardized active constituents, validated analytical testing, GMP compliance, appropriate formulation selection, and evidence-based dosing strategies. Strengthening regulatory frameworks and expanding clinical research will further enhance the credibility and integration of herbal medicines into modern psychiatric care, supporting safer, more reliable, and patient-centered options in mental health treatment.

7.10 Future prospects and research gaps

Herbal medicine has been established as an innovative practice that has gained popularity in recent times in the management of mental health conditions. The interest in the use of complementary drugs in managing depression, anxiety, and other mood-related conditions has been on the increase, justifying the growing interest in this medical discipline. Over the years, significant progress has been made in phytopharmacology and ethnobotany. Various constituents of medicinal plants have been identified as potentially active in the management of mental health conditions. These plants have been used traditionally over time to manage mental health-related conditions. *W. somnifera* (ashwagandha) has been demonstrated in clinical studies to be active in stress reduction, improving sleep, and enhancing mood stability through regulation of the HPA axis, reduction in cortisol levels, and modulation of GABAergic and serotonergic neurotransmission. *R. rosea* has been demonstrated as an adaptogen, enhancing resistance to stress and fatigue and improving mood regulation through catecholaminergic system regulation while stabilizing neuronal excitability.

C. sativus (saffron) has also shown antidepressant effects, which might be mediated through serotonergic, antioxidant, and antiinflammatory effects, once again highlighting the multifaceted nature through which phytochemicals can affect the underlying neuropsychopharmacology of psychiatric disorders. These findings not only validate traditional medicine but also pave the way for the integration of phytomedical interventions into an evidence-based healthcare framework. This optimal translation into clinical practice depends significantly on processes such as quality control and standardization. Also, recent advances in technology have enabled accurate determination, quantification, and confirmation of the bioactive constituents in an herbal formulation using methods such as HPLC, mass spectrometry (MS), and nuclear magnetic resonance spectroscopy. The standardization of herbal drugs is important, as phytoconstituents can vary significantly across various parameters, such as the type of drug, growing conditions, harvest time, etc. Besides the accurate preparation of an herbal drug in the form of a capsule/tablet or a liquid/tincture, it significantly affects the rate of bioavailability and absorption patterns, which accordingly necessitates careful consideration in the context of therapeutic planning. While liquid extracts are more active in terms of absorption, other tablet or capsule forms may be more convenient for administering an herbal drug appropriately. Another exciting and promising facet of investigation is the use of synergistic preparations. This type of therapy consists of using multiple herbs simultaneously or using herbal supplements along with psychopharmacological agents. The goal of this type of therapy is to maximize benefits and minimize adverse effects as far as possible. This allows for a very effective and tailored approach towards drug therapy and fulfills the aim of personalized therapy. Thus using herbal supplements alongside conventional psychopharmacological agents may be highly effective and can fill this niche well. However, there are many barriers and bottlenecks to implementing this type of therapy at a global level. The major barriers are the lack of scientific evidence-based studies to support its use at a large clinical level and limited practitioner awareness of herbal therapy. A major obstacle to its use is also the lack of acceptance of unconventional approaches within mainstream clinical treatment protocols. The best way to overcome these hurdles is to implement well-informed awareness campaigns at a global or, at least, national level. Looking ahead, the integration of herbal medicine into mental health care is expected to support the trend toward more holistic, patient-centered models of care, in which healthcare providers address not only the biochemical aspects of mental health disorders but also their emotional, cognitive, and social dimensions. This model of care will support more effective conventional mental health care with fewer adverse side effects, as well as more effective patient-centered care with fewer adverse side effects for clients. Furthermore, research on the biochemical relations of phytochemicals, their neuroprotective effects on mental health care services generally, and their interactions with pharmaceutical agents in mental health care services also promises to provide opportunities for new models of integrated mental health care services in the future (Balkrishna et al., 2024).

7.11 Conclusion

Herbal medicine has emerged as an alternative and complementary practice for the management and regulation of mental illnesses such as anxiety, depression, and mood swings. Years of practice and use of herbal medicine are being supported by modern scientific discoveries that reveal the potential ability of certain herbs to effectively cure anxiety, depression, and mood swings more naturally and causally compared to the use of psychotropic pharmacotherapies commonly associated with the management of these conditions. Such herbs include *W. somnifera* (ashwagandha) and *H. perforatum* (St. John's Wort), used to alleviate and prevent anxiety and depression by regulating the mind to achieve psychological homeostasis. Such regulation and prevention are achieved through the mechanisms by which these herbs interact with the principal chemical transmitters in the body, known as neurotransmitters, involved in the management of mood swings and depression. Such mechanisms also involve reducing neuroinflammation and increasing antioxidant levels to support neuronal well-being.

Furthermore, evidence is now available on the role of the gut-brain axis in regulating mood. Thus the beneficial effects of these herbal compounds may be due to modulation of the gut microbiota, antiinflammatory effects on the gastrointestinal system, and the regulation of various metabolic and endocrine pathways, thereby exerting an overall effect on the subject rather than simply on the central nervous system. However, despite the good safety profiles, the use of herbal medicine also requires careful management, especially regarding the administered dose and the quality control of medicinal herbs. It has also been reported that the quality of medicinal herbs can affect their potency, with the type of plants, their mode of growth, time of harvesting, and the sequence of preparation influencing potency. The useful integration of an herbal approach into mainstream psychiatric health care approach will also provide several benefits, including an individual, culturally sensitive, and holistic approach to treatment, where an herbal approach, when combined with other treatment modalities like allopathic or pharmacological approaches, will provide a beneficial approach to helping patients meet their needs, thereby ensuring reproducibility, efficacy, and safety. Looking ahead, herbal medicine may assume an increasingly central place in integrative, patient-centered mental health care. Building on evidence-based herbal interventions, alone or in combination with conventional pharmacological and psychological therapies, clinicians may adopt holistic approaches that target the biological, emotional, cognitive, and social dimensions of mental health. Integration can enable culturally responsive care, offering the patient therapeutic choices aligned with their personal beliefs and preferences. As research endeavors continue to expand, herbal medicine could transition from a complementary adjunct to an intrinsic part of comprehensive mental health strategies aimed at effective, accessible, and low-risk interventions by modulating emotional resilience, cognitive enhancement, and overall psychological well-being, while tending towards personalized and preventive mental health approaches., they offer the opportunity to address factors related to lifestyle, dietary patterns, and psychosocial factors important for mental well-being, minimizing the disadvantages to maximize the benefits and hence matching the recent wellness movement for prevention, well-built resilience strategies for dealing with life challenges and day-to-day self-care.

Furthermore, for herbal medicine to gain current-day popularity at the clinical level, it necessitates ongoing research to standardize herbs, along with the necessary education for patients and clinical staff. Proper clinical trials need to be conducted to evaluate the dosages that should be used by patients at the cellular level, along with potential interactions between the herbs and drugs patients use. Similarly, the mechanisms by which the herbs function should be well elucidated at the molecular level. However, for herbs to benefit patients, they should also be educated on their use, including possible interactions with the prescription drugs they take.

References

Adeniyi Grace P, Igbokwe Christopher O, Mahady Gail B, Adeniyi Bolanle A. The role of plants having antimicrobial properties in the management of mental health disorders: a review. Journal of Complementary and Alternative Medical Research 2024;25(11):125–46. Available from https://doi.org/10.9734/jocamr/2024/v25i11588.

Al-Attraqchi OHA, Deb PK, Al-Attraqchi NHA. Review of the phytochemistry and pharmacological properties of valeriana officinalis. Current Traditional Medicine 2020;6(4):260–77. Available from https://doi.org/10.2174/2215083805666190314112755, https://benthamscience.com/journals/current-traditional-medicine/volume/6/issue/4/page/260/.

Ali D, Verma S, Malviya R, Mishra S, Sundram S. Implications of herbal components in the treatment of neurological disorders. Current Nutrition & Food Science 2024;20(6):677–86. Available from https://doi.org/10.2174/1573401319666230821102546.

Amsterdam JD, Li QS, Xie SX, Mao JJ. Putative antidepressant effect of chamomile (*Matricaria chamomilla* L.) oral extract in subjects with comorbid generalized anxiety disorder and depression. The Journal of Alternative and Complementary Medicine 2020;26(9):815–21. Available from https://doi.org/10.1089/acm.2019.0252.

Anand Shukla D. Turmeric: the golden spice of India. International Journal of Scientific Engineering and Research 2021;9(2):12–4. Available from https://doi.org/10.70729/se21217135616.

Appel K, Rose T, Fiebich B, Kammler T, Hoffmann C, Weiss G. Modulation of the γ-aminobutyric acid (GABA) system by *Passiflora incarnata* L. Phytotherapy Research 2011;25(6):838–43. Available from https://doi.org/10.1002/ptr.3352.

Arumugam V, Vijayakumar V, Balakrishnan A, B Bhandari R, Boopalan D, Ponnurangam R, et al. Effects of Ashwagandha (*Withania Somnifera*) on stress and anxiety: a systematic review and meta-analysis. EXPLORE 2024;20(6):103062. Available from https://doi.org/10.1016/j.explore.2024.103062.

Awuchi CG. Plants, phytochemicals, and natural practices in complementary and alternative system of medicine for treatment of central nervous system disorders. International Journal of Food Properties 2023;26(1):1190–213. Available from https://doi.org/10.1080/10942912.2023.2205039, https://www.tandfonline.com/loi/ljfp20.

Bachour G, Samir A, Haddad S, Houssaini MA, El Radad M. Effects of ashwagandha supplements on cortisol, stress, and anxiety levels in adults: a systematic review and meta-analysis. BJPsych Open 2025;11(S1):S39. Available from https://doi.org/10.1192/bjo.2025.10136.

Balkrishna A, Sharma N, Srivastava D, Kukreti A, Srivastava S, Arya V. Exploring the safety, efficacy, and bioactivity of herbal medicines: bridging traditional wisdom and modern science in healthcare. Future Integrative Medicine 2024;3(1):35–49. Available from https://doi.org/10.14218/fim.2023.00086.

Bosco Don Bosco R, Selvan Christyraj JRS, Yesudhason BV. Synergistic activity of nootropic herbs as potent therapeutics for Alzheimer's disease: a cheminformatics, pharmacokinetics, and system pharmacology approach. Journal of Alzheimer's Disease Reports 2024;8(1):1745–62. Available from https://doi.org/10.1177/25424823241307019.

Butterweck V. Mechanism of action of St John???s wort in depression. CNS Drugs 2003;17(8):539–62. Available from https://doi.org/10.2165/00023210-200317080-00001.

Di YM, Li CG, Xue CC, Zhou SF. Clinical drugs that interact with St. John's wort and implication in drug development. Current Pharmaceutical Design 2008;14(17):1723–42. Available from https://doi.org/10.2174/138161208784746798Australia, http://www.ingentaconnect.com/content/ben/cpd/2008/00000014/00000017/art00010.

Ernst E. The role of complementary and alternative medicine in psychiatry. Psychology Medicine 2002;32:849–55. Available from https://doi.org/10.1017/S0033291702005664 Jul.

Fuloria S, Mehta J, Chandel A, Sekar M, Rani NNIM, Begum MY, et al. A comprehensive review on the therapeutic potential of *Curcuma longa* Linn. in relation to its major active constituent curcumin. Frontiers in Pharmacology 2022;13 Available from https://doi.org/10.3389/fphar.2022.820806, http://www.frontiersin.org/Pharmacology.

Garg A. Nature's mental health toolkit: herbal remedies for anxiety and depression. Research Journal of Pharmacognosy and Phytochemistry 2025:45–8. Available from https://doi.org/10.52711/0975-4385.2025.00008.

Hooli S, Sutar KP, Sammasagi SS. Revisiting Parkinson's disease management: a comprehensive review of herbal approaches and modern delivery strategies. International Journal on Science and Technology 2025;16(2). Available from https://doi.org/10.71097/ijsat.v16.i2.5077.

Hudson JI, Hiripi E, Pope HG, Kessler RC. The prevalence and correlates of eating disorders in the national comorbidity survey replication. Biological Psychiatry 2007;61(3):348–58. Available from https://doi.org/10.1016/j.biopsych.2006.03.040.

Jain SD, Shrivastava SK, Agrawal A, Gupta AK. WHO guidelines for quality control of herbal medicines: from cultivation to consumption. International Journal of Pharmaceutical Chemistry and Analysis 2024;11(3):212–25. Available from https://doi.org/10.18231/j.ijpca.2024.031.

Kaptchuk TJ. Effectiveness of placebo treatments: a systematic review. JAMA 2000;283:1971–6. Available from https://doi.org/10.1001/jama.283.15.1971.

Kean JD, Downey LA, Sarris J, Kaufman J, Zangara A, Stough C. Effects of *Bacopa monnieri* (CDRI 08®) in a population of males exhibiting inattention and hyperactivity aged 6 to 14 years: a randomized, double-blind, placebo-controlled trial. Phytotherapy Research 2022;36(2):996–1012. Available from https://doi.org/10.1002/ptr.7372, http://onlinelibrary.wiley.com/journal/10.1002/(ISSN)1099-1573.

Krishnan V, Nestler EJ. The molecular mechanisms of depression. Nature 2008;455:894–7. Available from https://doi.org/10.1038/nature07455. Oct 9.

Laakmann G, Jahn G, Schüle C. *Hypericum-perforatum*-Extrakt bei der Behandlung leichter bis mittelschwerer depressionen Klinische und pharmakologische Aspekte. Der Nervenarzt 2002;73(7):600–12. Available from https://doi.org/10.1007/s00115-002-1300-9.

Machín RP, Florido M, Chirino-Godoy R, López-Rios L. Adaptogenic botanicals with emphasis on *Rhodiola rosea* and *Withania somnifera*. European Journal of Medicinal Plants 2023;34(11):20–39. Available from https://doi.org/10.9734/ejmp/2023/v34i111168.

Mao JJ, Xie SX, Keefe JR, Soeller I, Li QS, Amsterdam JD. Long-term chamomile (*Matricaria chamomilla* L.) treatment for generalized anxiety disorder: a randomized clinical trial. Phytomedicine 2016;23(14):1735–42. Available from https://doi.org/10.1016/j.phymed.2016.10.012, www.urbanfischer.de/journals/phytomed.

McIntyre RS, Soczynska JK. The role of emotional dysregulation in the pathophysiology of bipolar disorder. *Journal of Affective Disorders* 2010;126:1–10. Available from https://doi.org/10.1016/j.jad.2010.01.041.

Mohamed Afsal A. Herbs for cognitive enhancement: pharmacological insights into memory and neuroprotection. International Journal for Research in Applied Science and Engineering Technology 2025;13(5):5159–65. Available from https://doi.org/10.22214/ijraset.2025.71383.

Otero MC, Ceric F, Miranda-Rojas S, Carreño C, Escares R, Escobar MJ, et al. Documentary analysis of *Hypericum perforatum* (St. John's Wort) and its effect on depressive disorders. Pharmaceuticals 2024;17(12). Available from https://doi.org/10.3390/ph17121625, https://www.mdpi.com/journal/pharmaceuticals.

Pannu A, Goyal RK, Sharma K. The potential role of herbal plants in the management of depression: pre-clinical and clinical evidence. Current Nutraceuticals 2025;6. Available from https://doi.org/10.2174/0126659786306271240905054531.

Sarris J, Karamacoska D. Herbal medicine in psychiatry: a review of the evidence. Journal of Psychopharmacology 2013;27:469–79. Available from https://doi.org/10.1177/0269881112473794. May [cited 2025 Jul 27].

Shinde M, Bais SK, Devmare D. Review on standardization of herbal drug and formulation. International Journal of Pharmacognosy and Life Science 2023;4(1):20–4. Available from https://doi.org/10.33545/27072827.2023.v4.i1a.69.

Singh H, Sharma P, Kumar D. In sighting role of ashwagandha: pharmacological review. Indo Global Journal of Pharmaceutical Sciences 2019;09(02):145. Available from https://doi.org/10.35652/igjps.2019.92s43.

Velasquez ACA, Tsuji M, dos Santos Cordeiro L, Petinati MFP, Rebellato NLB, Sebastiani AM, et al. Effects of *Passiflora incarnata* and *Valeriana officinalis* in the control of anxiety due to tooth extraction: a randomized controlled clinical trial. Oral and Maxillofacial Surgery 2024;28(3):1313–20. Available from https://doi.org/10.1007/s10006-024-01259-6, https://www.springer.com/journal/10006.

Yadav K, Rani A, Dua K. Potential herbal remedies for treatment of depression: a mini review. The Natural Products Journal 2024;14(6). Available from https://doi.org/10.2174/0122103155271537231006074854.

Zaharie D, Achim L, Muntean L. Therapeutic effects of valerian (*Valeriana officinalis* L). Hop and Medicinal Plants 2022;30(1-2):216–23. Available from https://doi.org/10.15835/hamp14454.

Zakrzewska N, Bychowski M, Kwaśna J, Załęska A, Kaźmierczyk I, Lenart K, et al. The multifaceted effects of ashwagandha (*Withania somnifera*) root extract on physical performance, cognitive function, mental health, and safety in adults. Quality in Sport 2024;36:56521. Available from https://doi.org/10.12775/qs.2024.36.56521.

Zanoli P. Role of hyperforin in the pharmacological activities of St. John's wort. CNS Drug Reviews 2004;10(3):203–18. Available from https://doi.org/10.1111/j.1527-3458.2004.tb00022.x.

CHAPTER

8 Botanicals for cardiovascular health: lowering cholesterol and blood pressure

Bhaskar Rao Bollineni[1], Lakshmi Lalitha Danturty[1], Bharghava Bhushan Rao Pathange[2], Vineela Nirmala Nalam[3] and Sunjana Nelluri[1]

[1]Great Eastern Medical School and Hospital, Srikakulam, Andhra Pradesh, India,
[2]A.M.Reddy Memorial College of Pharmacy, Narasaraopeta, Andhra Pradesh, India,
[3]Department of Pharmaceutics and Biotechnology, KVSR Siddhartha College of Pharmaceutical Sciences, Vijayawada, Andhra Pradesh, India

8.1 Introduction

Cardiovascular diseases (CVDs) are the world's largest cause of illness and mortality, accounting for 17.9 million deaths yearly, or almost one-third of all fatalities worldwide (Chong et al., 2025). Low- and middle-income nations bear a disproportionate amount of the burden and mortality from CVD; more than three-quarters of all deaths worldwide take place in these countries, mostly as a result of delayed diagnosis, restricted treatment options, and an inability to receive preventive care. CVDs include coronary artery disease, cerebrovascular disease, heart failure, and peripheral arterial disease, all of which are major and costly health, social, and economic problems. Urbanization, sedentary living, poor diet, and an increase in life expectancy have only hastened the CVD epidemic. The common global problem of CVD is significant, and despite improving diagnostic modalities, continued growth of interventional cardiology, and an ever-growing array of pharmacological therapies, the problem continues due to residual cardiovascular risk, adverse drug effects, and poor long-term adherence. Some of the several risk factors for CVD that can be changed include high blood pressure and elevated cholesterol, especially high-density lipoprotein cholesterol (HDL-C). Hypercholesterolemia contributes to the accumulation of lipids within the arterial wall, fostering inflammation and endothelial dysfunction, and atherosclerotic plaque formation. Hypertension, often asymptomatic until significant complications develop, counteracts the normal response to vascular injury, leading to increased vascular inflammation, vascular damage, left ventricular hypertrophy, and arterial stiffness. All the above combine to result in the faster development of vascular injury, myocardial infarction, stroke, CVD, and heart failure in conjunction. Conventional pharmacotherapies, such as statins or antihypertensive agents (including diuretics and beta-blockers), or a combination of both,

Healing Botanicals. DOI: https://doi.org/10.1016/B978-0-443-44295-7.00002-1

remain the standard of care. Still, there are many limitations associated with drug therapies, such as side effects from drugs, contraindications, and individual patient response variability. These limitations, in combination with a growing worldwide interest in natural and integrative health alternatives, have placed botanicals—bioactive compounds from plants–at center stage in terms of possible adjuncts in cardiovascular care. Many botanicals have multiple mechanisms of therapeutic action, including lipid-modulating properties, blood pressure-lowering effects, antioxidant potential, antiinflammatory effects, and endothelial protection. Just as the majority of cultures have traditional medicine systems that incorporate plant-based interventions for heart health, modern scientific research is only now guiding the mechanisms of action. By modulating cholesterol metabolism, increasing nitric oxide (NO) bioavailability, improving vascular tone, and reducing oxidative stress, botanicals may offer multilayered strategies to lower cardiovascular risk. Alternative methods to incorporate botanicals alongside evidence-based preventive and therapeutic options, as well as lifestyle replacements and pharmacological sources, may improve treatment effectiveness, adherence, and healthcare burden (Rastogi et al., 2016; Biesinger et al., 2016; Lipert et al., 2022).

8.2 Overview of botanicals in cardiovascular health

8.2.1 Definition and classification of botanicals

For medicinal, preventive, or health-promoting uses, botanicals are preparations made from various plant parts, such as leaves, roots, seeds, bark, flowers, and fruits. In terms of cardiovascular health, they are useful in treating modifiable risk factors, such as dyslipidemia, hypertension, oxidative stress, and inflammation. Botanicals may be classified as crude dietary plant materials (e.g., whole dried herbs, teas, powders), concentrated dietary plant preparations (e.g., extracts, tinctures, essential oils, phytopharmaceuticals), and purified bioactive materials resulting from the isolation of an active ingredient from a dietary plant. Botanicals can generally be grouped based on which part of the plant is used: leaf-based (green tea, hawthorn), root-based (ginseng), seed-based (flaxseed, fenugreek), bark-based (cinnamon), and flower-based (hibiscus), or by preparation type, including crude material, concentrated extract, and purified bioactive material. In pharmacological classification, botanicals are grouped by major function (e.g., hypocholesterolemic, antihypertensive, vasodilatory, antioxidant). They may also be classified into whole plants and standardized extracts, where whole plant formulations contain all the phytochemicals present in the particular plant, and standardized extracts contain consistent levels of constituents determined to be bioactive, as shown in Fig. 8.1 The various therapeutic potentials of botanicals are reflected in variations in their composition, production, and function, which also contribute to their growing clinical significance as supplemental treatments for the prevention and management of CVD (Durazzo et al., 2018; Nicoletti, 2012).

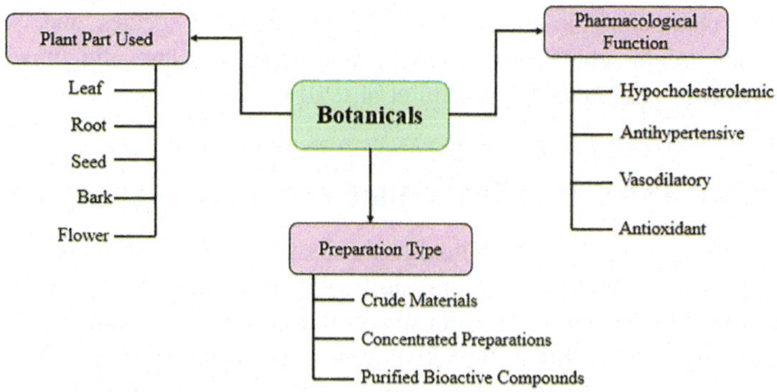

FIGURE 8.1

Classification of botanicals.

8.2.2 Bioactive phytochemicals relevant to cardiovascular disease

Botanicals confer benefits to cardiovascular physiologic function through a variety of bioactive phytochemicals, each with unique molecular targets that modulate metabolic pathways related to lipid metabolism and vascular tone, as well as inflammatory pathways. Polyphenols, flavonoids, phenolic acids, stilbenes, and lignans are the phytochemicals that have attracted the most interest for their heart-protective properties. Green tea, chocolate, berries, and citrus fruits contain flavonoids that increase endothelial nitric oxide synthase (eNOS) activity, promote vasodilation, reduce oxidative stress, and improve lipid markers by preventing the oxidation of low-density lipoprotein cholesterol (LDL-C). Whole grains and coffee contain phenolic acids, including caffeic and chlorogenic acids, which have beneficial antiinflammatory and antioxidant properties that enhance cardiovascular health. The stilbenes, and particularly resveratrol from grapes and berries, activate sirtuin pathways, improve mitochondrial function, and suppress vascular inflammation. The lignans from the seeds (flax and sesame) can have estrogenic and antioxidant functional activity, supporting endogenous regulation of lipids and vascular integrity. Another important group is plant-based sterols and stanols, which are structurally similar to cholesterol and competitively block its intestinal absorption, leading to a reduction in serum LDL-C without impacting HDL-C levels. Saponins, found in legumes, fenugreek, and ginseng, bind to bile acids, facilitating increased, concurrent bile acid excretion and hepatic catabolism of cholesterol. Organosulfur compounds derived from garlic, onions, and leeks can inhibit cholesterol synthesis by modulating HMG-CoA reductase, reduce platelet aggregation, and improve vascular function. Beta-carotene from carrots and lycopene from tomatoes are carotenoids providing potent antioxidant defense against lipid peroxidation and vascular injury. Omega-3 fatty acids found in flaxseed, chia seeds, and algae can also provide potent antiinflammatory effects, enhance lipid metabolism, and improve arterial compliance. Collectively, these phytochemicals offer a broad spectrum of pharmacologic properties found in botanical species and, when used synergistically, inhibit cholesterol levels, lower blood pressure, and prevent atherosclerosis (Pagliaro et al., 2015; Vasanthi et al., 2012).

8.2.3 Mechanisms of action in lipid and blood pressure regulation

Botanicals affect cardiovascular health through multifactorial mechanisms that target lipid metabolism, vascular physiology, and hemodynamic regulation. By competing with dietary cholesterol for absorption into mixed micelles, plant sterols and stanols, as well as other bioactive phytochemicals, help prevent intestinal cholesterol absorption in lipid management. Other plant bioactives enhance bile acid excretion, assisting the liver in converting cholesterol to bile salts, which likely decreases serum LDL-C. By lowering endogenous cholesterol production and modifying the enzymatic activity of lipid-regulating enzymes such as acyl-CoA: cholesterol acyltransferase and HMG-CoA reductase, plant polyphenols and organosulfur compounds reduce in vivo oxidative modification of LDL, a prerequisite for atherogenicity. In addition, antioxidant-enriched bioactive flavonoids and carotenoids reduce lipid oxidation, which decreases HDL-C functionality. Botanicals exert their effects on blood pressure through multiple molecular pathways. Polyphenols positively influence endothelial NO production by stimulating eNOS, producing a vasodilating effect, and reducing vascular resistance. Some peptides derived from plant proteins inhibit angiotensin-converting enzyme (ACE), thereby altering angiotensin II levels and lowering arteriolar systemic vascular tone. Botanicals rich in potassium enhance the sodium–potassium balance in the body that contributes to blood pressure regulation. Magnesium-rich plants help relax vascular smooth muscle, possess antiinflammatory phytochemicals, and facilitate vascular antistiffness. Omega-3 fatty acids also improve arterial compliance and attenuate sympathetic overactivity, as reflected in heart rate and blood pressure. The combined effect of these actions leads to a healthy lipid profile, improved vascular function, and immediate and long-term protection against atherosclerosis, hypertension, and cardiovascular comorbidities, as shown in Fig. 8.2 (Al Disi et al., 2016; Chrysant and Chrysant, 2017; Wong et al., 2014).

8.3 Hypocholesterolemic botanicals

Botanicals with lipid-lowering properties have been identified as potential complementary therapies for dyslipidemia, as mentioned in Table 8.1. These botanicals may act by inhibiting cholesterol synthesis, facilitating bile acid synthesis/excretion, modulating lipid absorption, and providing antioxidant protection against oxidative modification of LDL. The botanicals with the most established and most studied clinically relevant hypocholesterolemic properties include garlic (*Allium sativum*), fenugreek (*Trigonella foenum-graecum*), the

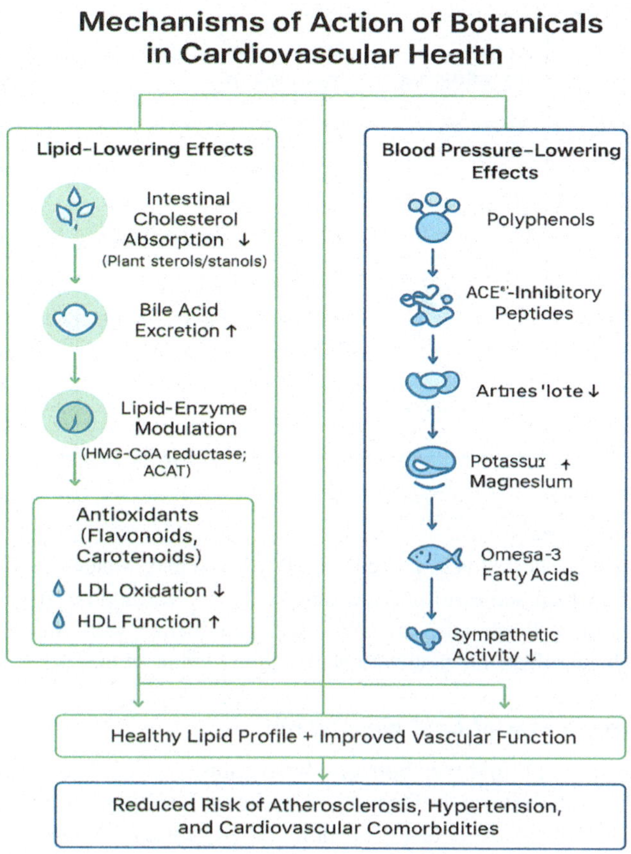

FIGURE 8.2

Mechanism of action of botanicals in cardiovascular health.

psyllium plant (*Plantago ovata*), and green tea (*Camellia sinensis*). According to studies and recommendations, regularly consuming these botanicals in the diet may help lower triglycerides, LDL-C, and total cholesterol while raising HDL-C and lowering the risk of cardiovascular events.

8.3.1 Garlic (*Allium sativum*)

8.3.1.1 *Active constituents*

Garlic contains a unique mixture of organosulfur compounds, phenolics, and bioactive metabolites that confer lipid-lowering properties. The main sulfur-containing constituents of garlic are allicin, ajoene, S-allylcysteine (SAC), diallyl disulfide, and diallyltrisulfide, produced during the enzymatic conversion of alliin (by alliinase) during

Table 8.1 Hypocholesterolemic botanicals and evidence.

Botanical	Active constituents	Mechanism of lipid modulation	Clinical evidence
Garlic (*Allium sativum*)	Allicin, S-allylcysteine	Inhibits HMG-CoA reductase, enhances bile acid excretion	↓ LDL and total cholesterol in RCTs
Green tea (*Camellia sinensis*)	Catechins (EGCG)	Inhibits cholesterol absorption, increases LDL clearance	Modest ↓ in LDL, improved lipid profile
Psyllium (*Plantago ovata*)	Soluble fiber	Inhibits the absorption of cholesterol and binds bile acids.	FDA-approved for cholesterol reduction
Fenugreek (*Trigonella foenum-graecum*)	Saponins, galactomannan	Reduces intestinal cholesterol absorption	Significant ↓ in LDL and triglycerides

crushing. Aged garlic extracts are composed of a stable, water-soluble compound, namely, SAC, which serves to provide antioxidant and cardioprotective action. Other compounds, such as flavonoids, saponins, and trace minerals, work together to produce garlic's beneficial effects on cardiovascular health (Martins et al., 2016).

8.3.1.2 Mechanism of lipid modulation

Garlic affects several lipid metabolism-related pathways. The rate-limiting enzyme in endogenous cholesterol production, 3-hydroxy-3-methylglutaryl coenzyme A reductase (HMG-CoA reductase), is inhibited by allicin, thereby reducing total cholesterol and LDL-C. Organosulfur compounds promote the excretion of bile acids, which in turn promote the utilization of hepatic cholesterol. Garlic also has antioxidant activity that inhibits the oxidative modification of LDL particles, thereby reducing their atherogenic potential. The modulation of fibrinolytic activity and platelet aggregation by ajoene and similar compounds indirectly benefits vascular health and reduces the likelihood of atherosclerosis (Ha et al., 2015).

8.3.1.3 Clinical evidence

The vast majority of studies endorse the hypocholesterolemic effects of garlic, and multiple metaanalyses have shown benefit. Still, findings remain highly variable and depend on factors such as preparation type, dosage, and duration of use. A large number of placebo-controlled clinical trials conducted on garlic powder supplements (600–900 mg/day) have shown reductions in total cholesterol of approximately 10%–15% over 8–12 weeks when administration was standardized. It has been demonstrated that aged garlic extract enhances the cholesterol-lowering benefits of garlic powder and lowers triglyceride and LDL-C levels while slightly raising HDL-C. Variability in study findings is generally attributed to differences in allicin bioavailability and patient compliance. Within the scope of well-controlled evidence, the totality of the data supports garlic as a useful adjunct treatment for managing hypercholesterolemia, especially when part of a broader lifestyle and dietary modification (Chan et al., 2020).

8.3.2 Green tea (*Camellia sinensis*)

8.3.2.1 *Active ingredients*

Green tea, made from nonfermented leaves of *C. sinensis*, is a complex mixture of bioactive constituents that contribute to its cardioprotective properties. Among these, polyphenolic catechins are the most prominent, with epigallocatechin gallate (EGCG) being the most abundant and biologically active, along with epicatechin, epigallocatechin, and epicatechin gallate. Other active green tea constituents include caffeine, theanine, flavonols, flavoring, and trace minerals, which combine to enhance therapeutic activity. These catechins are strong antioxidants that neutralize free radicals, lower oxidative stress, and preserve vascular health, all of which are essential defenses against the development of atherosclerosis (Basnet et al., 2015).

8.3.2.2 *Mechanism of lipid modulation*

A variety of actions working together mediates green tea's effects on cholesterol reduction. Catechins inhibit intestinal cholesterol absorption by disrupting its micellar solubilization, thereby reducing dietary cholesterol uptake. Furthermore, EGCG has been shown to inhibit hepatic cholesterol production by downregulating HMG-CoA reductase, the rate-limiting enzyme in cholesterol biosynthesis. Green tea polyphenols also upregulate LDL receptor expression in hepatocytes, leading to more efficient clearance of circulating LDL cholesterol. Catechins also exhibit potent antioxidant activity, inhibiting the oxidative modification of LDL particles and thereby decreasing their atherogenic potential. Green tea also improves endothelial function by altering NO bioavailability, thereby increasing vasodilation, adding another layer of cardiovascular protection (Macêdo et al., 2022; Huang et al., 2018).

8.3.2.3 *Clinical evidence*

Green tea has been repeatedly linked to slightly significant lipid-lowering effects in clinical studies. Randomized controlled clinical trials have assessed standardized green tea extracts, such as 400–800 mg daily, which have demonstrated small improvements in HDL-C and decreases in total cholesterol and LDL-C. Habitual consumption of green tea beverages is associated with atherogenic lipid profiles in healthy people and those with hyperlipidemia, and metaanalyses of these trials suggest reductions in LDL cholesterol of approximately 7–10 mg/dL, again confirming its clinical importance as an adjunctive therapy. In addition, epidemiological studies that have assessed populations known to consume green tea, especially in East Asia, frequently report a decreased incidence of coronary artery disease and stroke. Variance in trial outcomes is mostly explained by catechins' bioavailability, mode of preparation, and noncompliance with supplementation. Overall, these findings support the position that green tea is a safe, natural, and effective treatment that could be implemented as an adjunct to beneficial lifestyle changes and/or adjunctive pharmacological therapy to manage hypercholesterolemia while reducing cardiovascular risk (Brimson et al., 2023; Zamani et al., 2023).

8.3.3 Psyllium (*Plantago ovata*)

8.3.3.1 *Active constituents*

Psyllium, which comes from the husks of *P. ovata* seeds, is a natural source of soluble dietary fiber and plays important roles in lipid regulation and cardiovascular health. The primary bioactive constituent of psyllium is a complex mixture of arabinoxylans and hemicelluloses, which are capable of forming a viscous gel through hydration once ingested. The gel-forming properties of psyllium have direct implications for its physiological effects (e.g., delayed gastric emptying, reduced cholesterol absorption, and improved glycemic control) and for its overall health benefits. In addition to soluble fiber, psyllium also contains very small amounts of trace phytochemicals, including flavonoids and sterols (phytosterols), that contribute to the cardioprotective characteristics of psyllium (Madgulkar et al., 2015).

8.3.3.2 *Mechanism of lipid modulation*

The hypocholesterolemic effect of psyllium is driven primarily by its fiber-induced effects. When ingested, psyllium forms a thick, viscous gel that binds bile acids in the intestine and increases their excretion in the feces. Bile acids are synthesized in the liver from cholesterol that is found in the circulation, and depletion of bile acids causes the liver to use more circulating cholesterol than to create bile acids, which causes a reduction in total cholesterol and LDL cholesterol. Psyllium also decreases intestinal absorption of glucose and lipids, an indirect effect on metabolic control. Furthermore, the beneficial effects of psyllium on satiety and postprandial lipid absorption may provide additional cardiovascular protection. Psyllium does not directly inhibit cholesterol biosynthesis, as some other botanicals do; rather, it alters cholesterol balance by increasing elimination and improving lipid metabolism (Przybyszewska et al., 2024).

8.3.3.3 *Clinical evidence*

Psyllium's ability to decrease cholesterol is supported by strong clinical evidence. Supplementing with 7–10 g of soluble psyllium fiber daily, taken consistently, has been shown in numerous randomized controlled trials and metaanalyses to lower total and LDL cholesterol by 5%–10%. These effects are greatest in patients with baseline hypercholesterolemia and when psyllium is taken with a low-fat or cholesterol-lowering diet. Multiple long-term studies indicate that psyllium is safe and well tolerated, with common side effects primarily mild GI symptoms, such as bloating or flatulence. Most notably, psyllium has been categorized by regulatory bodies like the U.S. Food and Drug Administration (FDA) as an additional treatment for lowering cholesterol levels and cardiovascular risk when used with a low-fat or cholesterol-reduction diet.

Collectively, evidence indicates that psyllium is an effective, safe, and available botanical for dyslipidemia management and promoting cardiovascular health (Jovanovski et al., 2018; Chen et al., 2022).

8.3.4 Fenugreek (*Trigonella foenum-graecum*)

8.3.4.1 Active constituents

T. foenum-graecum seeds are the source of fenugreek, which has long been used as a food and a medicinal. Traditionally, fenugreek has been used for metabolic and cardiovascular health. The seeds of fenugreek are comprised of numerous bioactive constituents, including steroidal saponins (notably diosgenin), galactomannan fiber, flavonoids, alkaloids, such as trigonelline, and a suite of other polyphenolic compounds found in the seeds and sprouted seeds. Fenugreek's soluble fibers form a viscous gel when hydrated. This is important because both glucose and lipid metabolism are strongly influenced by viscous gel fiber. Diosgenin and other saponins are well recognized for affecting cholesterol homeostasis; however, fenugreek fiber and polyphenols also exert hypocholesterolemic and antioxidant effects (Arya et al., 2022).

8.3.4.2 Mechanism of lipid modulation

The cholesterol-lowering effect of fenugreek comes from the combined effect of the saponins, fibers, and polyphenols in fenugreek. By delaying stomach emptying, reducing intestinal absorption of dietary cholesterol, and encouraging bile acid excretion in the feces, the soluble fiber fraction facilitates improved hepatic conversion of cholesterol into bile acids. It lowers LDL cholesterol levels in the blood. Saponins (such as diosgenin) can directly interfere with cholesterol absorption in the intestine by forming insoluble complexes with bile salts, thereby reducing cholesterol absorption. Fenugreek also possesses antioxidant properties, which protect LDL particles from oxidative modification and, therefore decrease their atherogenicity. Moreover, fenugreek's impact on glucose metabolism and insulin sensitivity indirectly supports lipid regulation, making it a valuable botanical support for patients with dyslipidemia and metabolic syndrome (Syed et al., 2020; Heshmat-Ghahdarijani et al., 2020).

8.3.4.3 Clinical evidence

Clinical studies have shown significant changes in lipid profile following fenugreek supplementation, particularly in hyperlipidemic subjects. Several randomized controlled trials using doses between 10 and 25 g of fenugreek (*T. foenum-graecum*) seed powder per day reported reductions of total cholesterol and reduced LDL cholesterol of 10%–15% and moderate increases in HDL-C. The remaining studies evaluated standardized fenugreek extracts that reported reduced serum triglycerides and improved lipid ratios. Long-term use of fenugreek in humans is safe, with only mild gastrointestinal side effects reported. Additionally, clinical studies suggest that fenugreek may be particularly beneficial for those with Type II diabetes, as it can improve lipid metabolism and glycemic control. All things considered, the data point to fenugreek as a safe, readily available herbal supplement that can be used in conjunction with dietary and lifestyle modifications to help control hypercholesterolemia and reduce the risk of CVD (Yousefi et al., 2017).

8.4 Antihypertensive botanicals

Several botanicals have demonstrated blood pressure-lowering activity through multiple mechanisms of action, including vasodilation, diuresis, inhibition of the renin-angiotensin system, and improved endothelial function, as summarized in Table 8.2. For instance, clinical studies have shown that hibiscus (*Hibiscus sabdariffa*) is an ACE inhibitor and diuretic, lowering both systolic and diastolic blood pressure. Hawthorn (*Crataegus monogyna*) has been shown to improve coronary blood flow and lower vascular resistance. Flaxseed (*Linum usitatissimum*) increases endothelial protection due to its omega-3 fatty acids and lignans. Beetroot

Table 8.2 Antihypertensive botanicals and evidence.

Botanical	Active constituents	Mechanism of BP reduction	Clinical findings
Hibiscus (*Hibiscus sabdariffa*)	Anthocyanins, hibiscus acid	ACE inhibition, vasodilation, and diuretic effect	↓ SBP/DBP by 7–10 mmHg in RCTs
Hawthorn (*Crataegus monogyna*)	Flavonoids, oligomeric procyanidins	Improves coronary blood flow, vasodilatory	Improved BP and cardiac function in mild HTN
Flaxseed (*Linum usitatissimum*)	α-linolenic acid, lignans	Improves endothelial function, reduces vascular stiffness	↓ SBP by ~10 mmHg in hypertensives
Beetroot (*Beta vulgaris*)	Dietary nitrates	Increases NO bioavailability, vasodilation	↓ SBP by 4–8 mmHg, improved vascular health

(*Beta vulgaris*) is high in dietary nitrates, which impact NO bioavailability and promote vasodilation. These botanicals are used as natural and complementary interventions for the management of hypertension; however, further research is needed regarding long-term safety and standardization of dosing (Baharvand-Ahmadi et al., 2016).

8.4.1 Hibiscus (*Hibiscus sabdariffa*)

8.4.1.1 Active constituents

Hibiscus, a tea made from the calyces of *H. sabdariffa*, has been consumed for centuries as an herbal tea and has enjoyed a long history of traditional use for cardiovascular health. Hibiscus has beneficial effects attributed to its chemical ingredients: anthocyanins (delphinidin and cyanidin derivatives), flavonoids, organic acids such as hibiscus acid, minor amounts of vitamin C, and polysaccharides. Hibiscus's phytochemicals provide potent antioxidant, antiinflammatory, and vasoprotective properties, making it a useful botanical for regulating blood pressure (Jabeur et al., 2017).

8.4.1.2 Mechanism of blood pressure modulation

Hibiscus lowers blood pressure through several complementary mechanisms. As natural ACE inhibitors, anthocyanins and hibiscus acid reduce the production of angiotensin II, which in turn causes vasodilation. There are increases in endothelial NO bioavailability, which facilitate vascular relaxation (constriction and arterial compliance). Moreover, potassium and the flavonoids have a mild diuretic effect, lowering blood volume and systemic vascular resistance. Hibiscus antioxidant activity protects the vascular endothelium from oxidative damage, which is often a concern in hypertension and atherosclerosis (Liu et al., 2020).

8.4.1.3 Clinical evidence

Several randomized clinical trials have supported the antihypertensive effects of hibiscus tea and its standardized extracts. In persons with prehypertension and moderate hypertension, daily consumption of hibiscus tea (2–3 cups =1–2 L) has been demonstrated to reduce systolic blood pressure by 7–10 mmHg and diastolic blood pressure by 3–5 mmHg. The hibiscus tea or hibiscus extract doses employed in the clinical research used to make this determination were equivalent to two to three cups of tea daily. Clinical studies comparing hibiscus to low-dose antihypertensive medications showed similar effects on lowering baseline blood pressure. In most cases, effects were strongest with hibiscus, especially when taken consistently for 6–12 weeks. Evidence from metaanalysis on the antihypertensive effect of hibiscus supports the potential to be a safe and effective adjunct in hypertension management. Mild and transient gastrointestinal upset would be the most common side effect, but most would find hibiscus well-tolerated and a simple addition to their dietary interventions (Al-anbaki et al., 2021; Obu, 2025).

8.4.2 Hawthorn (*Crataegus monogyna*)

8.4.2.1 *Active ingredients*

Traditional European medicine frequently uses hawthorn, which is derived from the berries, leaves, and flowers of *C. monogyna* and closely related species, to treat cardiovascular conditions. The bioactive components of hawthorn are primarily flavonoids like hyperoside, vitexin, and rutin, as well as oligomeric procyanidins. The flavonoids have cardiovascular health benefits, including vasodilation, antioxidant activity, and cardiotonic activity. Thus hawthorn provides an acceptable herbal medicine for managing hypertension and heart function (Nabavi et al., 2015).

8.4.2.2 *Mechanism of action for blood pressure modification*

Hawthorn operates primarily by enhancing vascular function. It has been demonstrated that hawthorn's flavonoids enhance endothelial cells' NO release, resulting in vasodilation and decreased peripheral resistance. Procyanidins and other polyphenols found in hawthorn have been shown to improve coronary blood flow, reduce arterial stiffness, and support myocardial contractility, thereby improving overall cardiac work. Hawthorn has also been shown to have some ACE-inhibition and calcium channel-blocking activity, which will also help lower systemic blood pressure. Most impressively, hawthorn is a strong antioxidant, which protects vascular endothelial cells and helps slow the progression of atherosclerosis (Cloud et al., 2020).

8.4.2.3 *Clinical evidence*

Clinical research has shown that supplementing with hawthorn extract lowers blood pressure somewhat but significantly, particularly in individuals with moderate hypertension and heart failure. Hawthorn extracts under control, up to 500–1300 mg daily, have demonstrated reductions in systolic and diastolic pressures, and improvements in exercise tolerance and cardiac output in heart failure patients. Variability exists between trials, partly due to differences in extract standardization and between trial populations. Overall, the evidence indicates that hawthorn is well tolerated and can be safely and effectively used to treat hypertension (Szikora et al., 2025).

8.4.3 Flaxseed (*Linum usitatissimum*)

8.4.3.1 *Active constituents*

Alpha-linolenic acid (ALA), lignans including secoisolariciresinol diglucoside (SDG), and soluble fiber are abundant in flaxseed, a seed derived from *L.*

usitatissimum. Its active constituents potentially have antioxidant, antiinflammatory, and vasoprotective properties. The lignans function as phytoestrogens and have additional benefits, including preventing CVD; ALA is a long-chain omega-3 fatty acid whose cardiovascular and vascular health benefits and capacity to modulate lipids and blood pressure are well established (Yang et al., 2021).

8.4.3.2 Mechanism of blood pressure regulation

The antihypertensive effect of flaxseed results from multiple actions on the vascular environment and lipids. ALA improves endothelial function by increasing NO availability and reducing arterial stiffening, facilitating vasodilation. Lignans exhibit antioxidant effects, reducing vascular inflammation and increasing arterial compliance. The soluble fiber fraction improves metabolic control and, indirectly, contributes to vascular health by lowering cholesterol levels. All these actions ultimately lower systemic vascular resistance and affect blood pressure levels (Khalesi et al., 2015; Ursoniu et al., 2016).

8.4.3.3 Clinical evidence

Numerous clinical studies have shown that supplementing with flaxseed significantly lowers blood pressure. In a seminal randomized controlled trial, for example, individuals with hypertension who consumed 30 g of ground flaxseed daily for 6 months saw decreases in both systolic (10 mmHg) and diastolic (7 mmHg) blood pressure. Consistent drops in both systolic and diastolic blood pressure are also supported by metaanalyses of flaxseed therapies, with longer-term consumption showing the strongest effects. Although consuming flaxseed is safe and well-tolerated, excessive intake may cause mild gastrointestinal distress (Ansari et al., 2019).

8.4.4 Beetroot (*Beta vulgaris*)

8.4.4.1 Active ingredients

Beetroot, a potent food product derived from the root of *B. vulgaris*, is well-known for its extremely high dietary nitrate levels, which serve as precursors to NO. Based on their high levels of nitrates, betalains, polyphenols, vitamins, and minerals (potassium and magnesium), they all help to provide cardioprotective effects. The nitrates, in particular, make beetroot a special functional food for vascular health (Thiruvengadam et al., 2024).

8.4.4.2 Mechanism of blood pressure regulation

The antihypertensive effect of beetroot is primarily mediated through the nitrate–nitrite–nitric oxide pathway. In the oral cavity, dietary nitrates are converted by bacteria into nitrites, which are subsequently reduced to nitric oxide in the circulation. Nitric oxide promotes vasodilation, improves endothelial function, and enhances arterial compliance. Beetroot has also been shown to reduce arterial stiffness, reduce systemic vascular resistance, improve endothelial function and vessel structure—potassium and magnesium further control vascular tone and water balance. Betalains and polyphenols exhibit antioxidant effects that safeguard vascular endothelial cells from free radical damage (Bahadoran et al., 2017).

8.4.4.3 Clinical evidence

Robust clinical evidence exists that supports beetroot as a reliable antihypertensive botanical. In a study of beetroot juice consumption (250–500 mL/day), systolic blood pressure decreased by 5–10 mmHg, with effects generally observed within hours, whereas longer-term interventions demonstrated persistence. Randomized controlled trials in hypertensive subjects have all demonstrated significant reductions in both systolic and diastolic blood pressure, indicating that beetroot is a viable nonpharmacological option. Beetroot's effects are stronger in people with uncontrolled hypertension. While most people can safely consume beetroot, excessive consumption may result in gastrointestinal distress or beeturia (reddish urine) (Grönroos et al., 2024).

8.5 Safety, dosage, and standardisation

The clinical application of botanicals for CVD requires careful consideration of safety, dosage, and standardization to ensure therapeutic efficacy while limiting risk. Generally, botanicals typically have a history of safety; their pharmacological activity is affected by dosage, preparation, and patient-specific factors, including comorbidities and medications.

8.5.1 Recommended dosage ranges

Appropriate botanical product formulations that deliver the best possible results in treating CVD are essential to achieving high effectiveness in managing this condition. Experimental and clinical trials have established the potency dosage ranges, as shown in Table 8.3. Unlike most conventional pharmaceutical drugs, botanical products may exhibit greater variability in the concentration of their active ingredients (bioactive constituents); therefore to achieve consistent therapeutic outcomes, dosage recommendations based on evidence-based medicine should be adhered to. Garlic (*A. sativum*) is one of the most widely researched botanicals and has been shown in studies to provide both lipid-lowering and antihypertensive benefits. The highest daily doses of garlic extracts used in clinical studies to obtain these benefits range from 600 to 1200 mg of standardized garlic extract (equivalent to approximately 2 to 5 g of whole, crushed garlic cloves). They are sufficient to yield the bioactive substances needed to effect measurable cardiovascular benefits. Psyllium (*P. ovata*), a soluble fiber obtained from flowers, is frequently

Table 8.3 Recommended dosage ranges of botanicals.

Botanical	Common dosage form	Suggested dose range	Duration of use
Garlic	Standardized extract	600–1200 mg/day	4–24 weeks
Green tea	Capsules/tea infusion	300–600 mg EGCG/day	12 weeks–6 months
Psyllium	Powder/capsules	5–10 g/day	Long-term safe
Fenugreek	Powdered seeds/ extract	5–25 g/day	4–12 weeks
Hibiscus	Tea/extract	2–3 cups/day	4–12 weeks
Beetroot	Juice/extract	250–500 mL/day	2–6 weeks

recommended to reduce the risk of CVD because it reduces cholesterol levels. Clinical research shows that taking 5–10 g of psyllium twice daily, while drinking enough water, can help lower total and LDL cholesterol substantially and is very well tolerated by the digestive system. Similarly, hibiscus (*H. sabdariffa*) has traditionally been used as a herbal tea, being touted for its ability to lower blood pressure when 2–3 g of dried calyces (the floral parts that form a cup-like structure) are consumed daily in the form of 2–3 cups of tea. This is consistent with both traditional uses and recent clinical studies showing reductions in systolic and diastolic blood pressure when consuming hibiscus. Beetroot (*B. vulgaris*) juice is typically prescribed to the general public, with regulations requiring daily consumption of 250–500 mL. This quantity of beetroot juice is believed to contain enough nitrates to enhance NO production, promote vasodilation, and reduce blood pressure. Various botanicals, such as fenugreek and flaxseed, have also shown therapeutic cardiovascular effects when consumed at adequate clinical doses in clinical studies. Many botanicals, such as fenugreek (*T. foenum-graecum*), have been supported by clinical trials; both fenugreek and flaxseed (*L. usitatissimum*) have been shown to influence cardiovascular health when provided at clinically effective dosages. In general, fenugreek seed powder is taken at 5–25 g daily and has been shown to improve lipid and glucose metabolism. Flaxseed is usually used as a nutritional supplement at doses of 30–50 g daily to reduce cholesterol and blood pressure.

When determining the appropriate use of prescription medicine, it is important to take into consideration the potential side effects associated with administering doses that are either below the clinically effective range or above the established therapeutic range; in general, an inadequate or inappropriate dose may potentially have an adverse, nonbeneficial effect on the heart and reduce the ability of the drug to be effective. Conversely, excessive use of prescription drugs may create an increased risk of developing adverse effects, such as gastrointestinal discomfort, nausea, or low blood pressure. Thus botanical selections for the treatment of CVD should follow clinically validated dosage guidelines that ensure optimal therapeutic effectiveness while minimizing the risk of unwanted side effects (Liperoti et al., 2017; Shaito et al., 2020).

8.5.2 Potential herb-drug interactions

Botanicals integrated with current CVD therapies have demonstrated potential efficacy as new adjuncts. However, there are important issues related to herb-drug interactions that need to be addressed, particularly for patients taking long-term cardiovascular medications (e.g., antilipemic agents, antihypertensive drugs, etc.). As described in Table 8.4, it provides a comprehensive list of herb-drug interactions to consider for each botanical product used in the treatment of CVD. One example of this interaction is the use of garlic (*A. sativum*). This natural antiplatelet agent can increase the effects of antiplatelet and anticoagulant agents such as aspirin, clopidogrel, and warfarin. The potential for increased bleeding complications exists when using combination agents for antithrombotic therapy and for patients undergoing surgical procedures. The other plant is hibiscus (*H. sabdariffa*), a mild diuretic with modest potential to block the ACE. The effect of hibiscus is to enhance the blood pressure-lowering effects of standard antihypertensive agents, potentially resulting in excessive hypotension, dizziness, and/or electrolyte imbalance if not well-monitored.

Due to the high content of dietary nitrates, beetroot (*B. vulgaris*) increases the availability of NO. Beetroot will interact with medicines containing nitrate, such as nitroglycerin, isosorbide dinitrate, and phosphodiesterase-5 inhibitors. This can lead to excessive vasodilation and a hazardous decline in blood pressure. Additionally, fiber-rich plants such as Psyllium (*P. ovata*) can block the intestines' ability to fully absorb some medications, such as statins, beta-blockers, and oral antidiabetic agents, when consumed together without adequate spacing between doses. As a result, this can lead to decreased drug bioavailability and reduced therapeutic effect.

Table 8.4 Potential herb–drug interactions.

Botanical	Interacting drug/ class	Mechanism of interaction	Clinical implication
Garlic	Warfarin, aspirin	Antiplatelet effect	↑ Bleeding risk
Green tea	β-blockers, warfarin	Alters drug metabolism, vitamin K interference	Reduced efficacy of anticoagulants
Psyllium	Oral hypoglycemics	Delays drug absorption	Reduced bioavailability
Fenugreek	Insulin, sulfonylureas	Hypoglycemic effect	↑ Risk of hypoglycemia
Hibiscus	Hydrochlorothiazide, ACE inhibitors	Additive BP lowering	Hypotension

Fenugreek (*T. foenum-graecum*), which is often utilized for its lipid-lowering and hypoglycemic effects, has been found to have additive effects when combined with insulin and oral hypoglycemic agents, increasing the risk of hypoglycemia. In the same way, Flaxeed (*L. usitatissimum*), known for its high lignan and omega-3 content, can modify hormonal metabolism, thereby enhancing or augmenting the anticoagulation/antiplatelet activity of other medications. As a result, this effect can make bleeding much more likely in certain patients than before. These interactions emphasize the importance of performing a full risk assessment on each patient before adding any herbal supplements to their current cardiology treatment regimen. Physicians should always supervise the use of botanicals when combined with standard cardiovascular drugs, provide clinical follow-up to ensure patient safety, and provide thorough instructions on their use. Healthcare providers must evaluate the need for dose adjustments or changes in product administration and be open to disclosing the use of any herbal product. These precautions will reduce the risk of adverse events while increasing the likelihood of achieving optimal benefit from integrated cardiovascular therapy (Shaikh et al., 2020; Jyothi et al., 2025).

8.5.3 Quality control and standardisation challenges

One of the main barriers to the adoption of botanicals for cardiovascular health is a lack of quality control and standardization. Whereas manufactured pharmaceuticals follow strict guidelines for consistency, herbal products can exhibit significant variability in their chemical composition (e.g., from the same plant species, grown under the same conditions, harvested at different times, prepared using different methods, etc.). Within herbal products, major variability exists, for example, when determining allicin content in garlic and anthocyanin content in hibiscus. It has been harder to determine consistent therapeutic benefit among unsolicited herbal botanical products. Products have also been compromised when convenience led to adulteration (e.g., with heavy metals, pesticides, or microbial toxins). Patients are susceptible to poor-quality products when produced without scrutiny at any level of governance. Standardization of botanicals is defined as the identification, quantification, and consistency of the bioactive phytochemicals involved in each preparation's function. In this case, botanicals are particularly challenging to standardize because many stimulate effects through multiple synergistic compounds rather than a single active constituent. For instance, the cardiovascular benefits attributed to hawthorn (*Crataegus* spp.) are thought to result from a combination of flavonoids and oligomeric proanthocyanidins. Therefore standardization of hawthorn requires it to include all desirable cofactors, not just to quantify them to define the level of standardization. Advanced analytical technologies, including HPLC, LC-MS, and DNA barcoding, are increasingly being applied in the industry to improve quality assurance protocols, but global harmonization of standards does not exist! Furthermore, to ensure patient safety and guarantee reliable therapeutic benefits, an interdisciplinary collaboration among researchers, industry, and regulatory agencies will be necessary to implement validated protocols for the production, labeling, and clinical testing of botanicals for cardiovascular care (Srivastava and Misra, 2018).

8.6 Integrating botanicals into cardiovascular care

Botanicals as adjunctive strategies for heart health have gained recognition as part of a multifaceted approach to managing cardiovascular risk factors, including dyslipidemia, hypertension, and oxidative stress. The incorporation of botanical products into usual care enables evidence-based application, synergistic use with other lifestyle modifications, and approaches to patient-centered care, thereby improving therapeutic recommendations, as shown in Fig. 8.3.

8.6.1 Evidence-based clinical applications

An increasing body of research supports the clinical use of botanical products for the treatment of CVD, including observational studies, metaanalyses, and randomized controlled trials. The findings from these types of studies collectively suggest that many botanical products offer possible ways to support the management of cardiovascular risk factors, including dyslipidaemia and hypertension. Garlic (*A. sativum*) is one of the botanical products that scientists have studied the most; in individuals with mild to moderate hypercholesterolemia, it has been demonstrated to produce a small but clinically significant reduction in both total cholesterol and LDL cholesterol. Investigators believe that garlic's cholesterol- and lipid-lowering effects are due to its sulfur-containing bioactive compounds, which inhibit cholesterol synthesis and promote lipid metabolism. Therefore in addition to

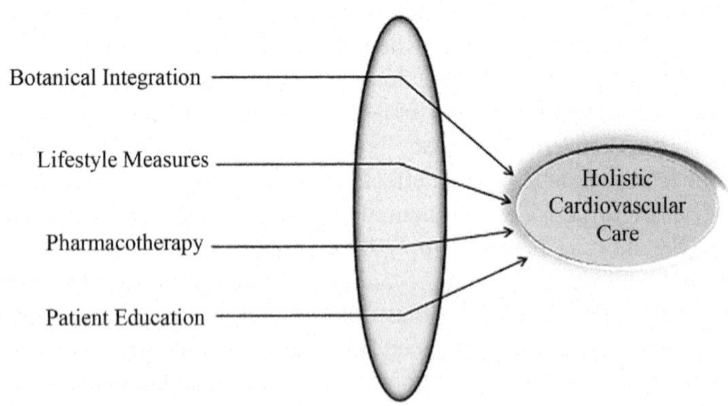

FIGURE 8.3

Pathways to heart health.

cholesterol-lowering medication, garlic is a viable adjunct therapy. In addition to garlic, *B. vulgaris* (beetroot), and *H. sabdariffa*, two other botanicals have been shown to lower blood pressure in both prehypertensive and hypertensive individuals significantly. By enhancing vascular endothelial function, inducing vasodilation, increasing NO bioavailability, and lowering oxidative stress, these botanicals primarily improve cardiovascular performance and reduce vascular resistance.

While many clinical studies have shown that psyllium husk is an effective soluble dietary fiber obtained from *P. ovata* and reduces LDL cholesterol and total cholesterol through mechanisms of reducing the absorption of cholesterol into the body from the intestines and enhancing the excretion of bile acids, it has become widely accepted in medicine since many regulatory agencies, including the FDA, have recognized its effectiveness as an evidence-based dietary intervention for reducing cardiovascular risks. However, there are still many challenges in incorporating botanical therapies into standard clinical practice due to significant differences in design, size, time frame, and formulation across randomized controlled trials (RCTs). In addition, different methods of plant extraction, the degree to which the active constituents were standardized, and the amount and frequency of each dosage have produced inconsistent results regarding the health benefits reported in the trials, with some trials reporting significant benefits. In contrast, others reported little or no effect. Nonetheless, botanicals based on solidclinical evidence and complying with rigorous quality control systems and standards can contribute to the safe use of botanicals as part of an integrative model of cardiovascular care. Botanicals may improve the efficacy of conventional pharmacotherapy, increase patient compliance with medication regimens, and decrease adverse drug reactions associated with pharmaceutical agents. The use of specific botanicals in the treatment of CVD is currently supported by sufficient clinical evidence, provided rigorous monitoring is carried out to ensure product safety, dosage, and quality (Dog, 2022).

8.6.2 Combining botanicals with lifestyle and pharmacotherapy

Cardiovascular care is being provided through a multifaceted and integrative approach to address the modifiable risk factors of CVD, including dyslipidemia, hypertension, obesity, tobacco use, and lack of physical activity. The use of botanicals may enhance the effects of pharmaceutical agents and lifestyle changes (diet and exercise) by synergizing with them. For example, Psyllium has been shown in clinical studies to lower cholesterol when taken in conjunction with a diet rich in whole grains, fruits, and vegetables. This is because it prevents cholesterol from being absorbed in the intestines and increases the body's capacity to metabolize fats through increased lipid metabolism, thereby supporting and enhancing the effects of dietary changes. Additionally, beetroot juice is a great source of nitrates, which can improve endothelial function and NO-mediated vasodilation, allowing for longer-lasting blood pressure control even during regular physical activity. These examples demonstrate how combinations of botanicals and established lifestyle habits can have a synergistic effect.

The addition of botanicals to existing treatment regimens can also be useful for certain groups of patients, such as those with statin intolerance, those who have developed some form of drug resistance, or those experiencing adverse drug effects associated with more traditional cardiovascular therapies. For these patients, incorporating botanicals into their care regimen may provide modest to clinically significant improvements in lipid levels and blood pressure, allowing a decrease in medication dose while maintaining similar clinical benefit and increasing adherence. Additionally, this method will help decrease the patient's pill load, which is a very important determinant for complying with long-term management of chronic CVD. However, caution must be exercised when introducing botanicals into the pharmacological regimen, as there is a potential for herb-drug interactions, especially among patients on anticoagulant, antihypertensive, and lipid-lowering therapy. Therefore it is critical to assess each patient's profile accurately, closely monitor patients throughout treatment, and supervise all patients receiving botanicals alongside other prescribed medications.

Medicinal herbs, including garlic (*A.sativum*) and Hawthorn (*Crataegus* spp.), should always be considered as adjunctive therapies to pharmacologic agents used to treat critical or serious CVDs rather than as replacements for necessary pharmacologic therapies for these serious conditions. Only a qualified health professional should prescribe these herbs to patients as part of a larger treatment plan that incorporates both pharmaceutical and lifestyle changes to assist the patient in managing their lipid levels, controlling their blood pressure, and maximizing their overall cardiovascular health. Together, the addition of these herbals to traditional lifestyle and pharmacologic therapies MAY help improve cardiovascular biomarkers and improve patient-reported quality of life, satisfaction with treatment, and other benefits. The use of herbals along with lifestyle changes and pharmacological therapy is consistent with the foundation of holistic medicine and patient-centric approaches. Current trends in CVD prevention promote the adoption of individualized, scientifically validated treatment plans (Sharifi-Rad et al., 2020).

8.6.3 Patient education and adherence

To improve the integration of botanicals into the treatment of CVD, individuals must be educated by their provider about botanicals and continue compliance with their prescribed treatments. Education on botanicals is necessary, as many patients who are interested in using botanical medicine also hold beliefs and/or misconceptions regarding the safety, efficacy, dosing, and place in therapy of botanical products. There is a difference between patients who are informed about the research supporting a botanical and those who have unrealistic expectations of what the botanical can do, believing that using it will allow them to stop taking the medications prescribed for their CVD. Therefore healthcare providers should provide evidence-based education on the role of botanical products, specifically that botanicals are beneficial adjunctive treatment options for cardiovascular conditions. Healthcare professionals play an important role in helping their patients determine which botanicals to use, how to dose them, and for how long, and have access to clinical studies to support their recommendations. Providing patients with information on several commonly studied botanicals (garlic, hibiscus, and flaxseed), including their mechanisms of action, potential benefits, and limitations, will help build confidence among patients and their families. Additionally, providing patients with the information they need about certain botanicals equips them to make informed decisions about their treatment and to reduce any skepticism or inflated expectations about the results of their use.

Addressing practical and psychological barriers to consistent use of botanicals will enhance patient compliance. Examples of these barriers include inconsistent quality among herbal products, concerns about possible negative interactions between prescribed pharmaceuticals and herbal products, and conflicting information from different sources about herbal products or supplements. Patients should be prompted to discuss their use of any herbal or dietary supplements with their healthcare provider, and healthcare providers must also actively encourage patients to raise the topic during appointments. Patients will benefit from this conversation because it enables providers to monitor patient progress, identify potential interactions sooner rather than later, and adjust therapies as needed to reduce risk. Providers may also utilize patient engagement techniques, including incorporating botanicals into patients' usual eating habits, creating simple dosing schedules, and using standardized, evidence-based formulations.

All things considered, patient and family education will provide people with the confidence they need to actively manage their healthcare. When patients have the knowledge to understand and see the evidence for using herbal remedies appropriately, along with confidence in using them safely and under a doctor's supervision, their adherence will increase. It will also affect their overall treatment success rate. Informed decision-making, Shared Responsibility, and personalized care are what drive patient-centered care and holistic medicine. From this, we can see that botanicals can safely and effectively be incorporated into a comprehensive prevention and management plan for CVD.

8.7 Policy, research, and future perspectives

8.7.1 Regulatory considerations for botanicals

Regulating botanicals for cardiovascular health is often difficult because they fall somewhere between traditional medicines and therapeutics, and the regulatory frameworks are not quite the same. Pharmaceutical agents undergo extensive preclinical and clinical evaluation before marketing authorization; however, much like pharmaceuticals, botanicals are often marketed as dietary supplements or herbal products that do not undergo extensive regulatory oversight. Due to a lack of study data or regulatory requirements, botanicals lack consistent testing during production, resulting in serious challenges such as variability in potency, contamination with heavy metals or pesticides, adulteration with synthetic drugs or botanical components, and inconsistent labeling of active ingredients. These challenges are concerning, patient-safety-related, and are barriers to the acceptance of botanicals and their bioactive constituents in cardiovascular care. In general, for botanicals to be reliable, validated analytical methods, standardized methods of writing source material and extraction procedure, and labeled contents or active constituents will need to be established. There is a wide range of regulatory approaches to botanicals worldwide. Botanicals are regulated by the FDA in the United States, which views them as "dietary supplements" under the Dietary Supplement Health and Education Act. That does not require the same level of evidence as drugs do. For example, in Europe, the European Medicines Agency has stricter frameworks through the Committee on Herbal Medicinal Products. In India, the Ministry of AYUSH plays an important role in approving and promoting traditional herbal medicines. However, differences between countries create challenges for international trade and collaborative research aimed at understanding the therapeutic benefits of botanical products. A harmonized regulatory system, supported by pharmacovigilance databases and postmarket surveillance, would improve the quality control of botanicals, enhance acceptance worldwide, and encourage healthcare providers to have greater confidence in using botanicals for CVD prevention, management, and treatment (Parihar et al., 2023).

8.7.2 Emerging research trends

Botanicals have attracted growing interest from researchers and healthcare professionals for their potential to promote cardiovascular health, leading to more studies and investigations on the topic. This renewed interest has also allowed researchers to combine historical/traditional uses of botanicals with contemporary methods to understand how they work, where they act, and so on. Past research focused solely on ethnobotanical studies, observational studies, and anecdotal evidence. The mechanistic implications of botanical products weren't thoroughly researched until modern times. The use of new techniques, such as genomic and metabolomic testing, has generated significant insights into how botanicals affect the body. By better understanding the biology of botanical actions, we can now more accurately define their mechanisms of action using modern scientific instrumentation. For example, polyphenols, flavonoids, and alkaloids are being studied for their ability to modulate the effects of oxidative stress, endothelial function, inflammation, and lipid metabolism through genomic, epigenetic, and metabolic regulation of the cardiovascular system, areas not included in earlier research. The latest trend is using nanotechnology to create botanical products with better bioavailability, stability, and targeted delivery of their active ingredients. By improving the solubility of these substances and protecting them from breakdown due to their delicate nature, these nanoformulations will enhance the effectiveness of botanical medicine, allow lower dosages, and result in fewer side effects; thus they may be used in conjunction with conventional treatments for heart disease. With this increased methodological rigor, botanicals can have greater scientific credibility as interventions and, therefore may be incorporated into evidence-based practices for heart health. Utilizing artificial intelligence, machine learning, and big data analytics as tools in botanical research is a new development for researchers in this field.

The use of these technologies enables researchers to analyze large datasets derived from controlled clinical trials, observational clinical data, and various pharmacovigilance sources. This type of analysis will help researchers better predict potential interactions between herbs and medications, determine how to tailor herbal remedies (and to whom), and, based on an individual's risk profile, which therapies should be used. In addition, the growing collaboration between practitioners of traditional medicine and biomedical researchers is not only validating the historical uses of herbs but also creating an environment where the knowledge accumulated over the centuries about botanicals is systematically reviewed, studied, and applied to real-world therapeutic approaches. Taken together, the many developments in current botanical research indicate a clear move away from anecdotal and observational evidence toward more scientific, mechanistically based, and clinically relevant evaluations of botanicals. As researchers combine emerging advanced omics technologies with nanotechnology, AI-fueled analytics, and an increased emphasis on collaborative translational research, more high-quality evidence will be available to demonstrate the efficacy and safety of botanicals as an adjunct or alternative to cardiovascular drugs and to assist in meeting the individualized treatment needs of patients around the world.

8.7.3 Path forward for clinical adoption

To successfully integrate botanical products into cardiovascular healthcare, we must find common ground between traditional knowledge and evidence-based clinical practice. While there has been some controlled clinical trial evidence that supports certain botanicals like garlic (*A. sativum*), hibiscus (*H. sabdariffa*), flaxseed (*L. usitatissimum*), and psyllium (*P. ovata*) as having potential lipid-lowering or antihypertensive or other cardiovascular benefits, more extensive, well-designed clinical research on a larger scale is needed to establish this credibility before they will be widely accepted within traditional Cardiology. Establishing botanical product standardization is one of the first steps towards providing botanical products with clinical credibility. Evidence-based clinical practice requires establishing the active ingredients and their dosages, delineating the standardization of each batch of product, and validating extraction protocols. Current regulatory practices must also evolve as new scientific evidence emerges to create a regulatory framework that clearly defines dosages, labeling, safety monitoring, and postmarket surveillance. Once there is sufficient regulatory evidence, botanical products will be viewed like prescription drugs and as prescription-drug-compatible adjunct therapies within the field of Cardiovascular Care. Additionally, establishing standards for quality, safety, and reproducibility will enable both clinicians and patients to feel confident in using botanical products.

In conjunction with developing standards and regulations, the implementation of botanical products in the treatment of CVD is important to ensure that these botanicals can be used alongside the lifestyle changes listed above or added to lifestyle interventions. Botanical products are a means of providing individualized therapy for the patient in addition to the prescribed medical therapy. This, in turn, can enhance patient satisfaction. If we can develop professionals, such as doctors, nurses, mid-level providers, and pharmacists, who have a good understanding of the therapeutic effects of botanical products and the mechanisms by which they operate, we will promote the use of botanical products by more patients. Furthermore, by including botanical medicine in medical education programs and creating opportunities for interprofessional collaboration between those who specialize in herbal medicine and those who specialize in biomedicine, we will continue to develop professionals who are knowledgeable about the therapeutic value of botanicals.

The recent developments in pharmacogenomics (personalized medicine) will provide opportunities to optimize the use of botanical therapies by taking into account each patient's individualized characteristics, such as metabolism, drug response, and genetic predispositions, and create an individualized botanical program that can maximize the potential benefits of botanical therapy while minimizing any potential adverse effects or interactions with other medications used. The proposed methods, by taking into account individualized characteristics, may shift away from using botanicals solely as supportive therapies and position them as an integral part of the personalized cardiovascular treatment approach. Overall, the path forward to the clinical use and acceptance of botanicals will require a collaboration of many components, including robust scientific evidence regarding safety and efficacy, standardization and regulation of the formulations of the botanical products, integration into comprehensive care models, and professional education for healthcare professionals, and an integration of the use of botanicals as part of the overall personalized medicine approach to cardiovascular health. By combining these elements, botanicals could move from complementary therapies to more widely recognized, evidence-based therapeutic agents that can achieve safe and effective cardiovascular health outcomes at the individual, national, and international levels. They would also play a significant role in the management and prevention of CVD.

8.8 Conclusion

The leading cause of morbidity and mortality around the world is CVD, and the major contributors to CVD are modifiable risk factors such as hypertension, hypercholesterolemia, and physical inactivity. Despite the large number of cardiovascular events and improved survival rates associated with conventional evidence-based medication therapy (including antihypertensives, statins, and antiplatelet medications), cardiovascular events are still occurring due to the limitations of the medications. Adverse effects, challenges with long-term adherence, variability in individuals' responses to medications, and the presence of residual cardiovascular risk have led to a growing need to investigate adjunctive/integrative techniques that both enhance cardiovascular risk reduction and support patients' adherence to treatment. Using evidence-based botanicals in conjunction with standard pharmacologic therapy can result in clinically significant improvements in health status. Integrative Health Approaches improve cardiovascular risk factors, reduce the need for medications or lower pill burden for patients, prevent adverse reactions related to medications for some patients, and help you achieve long-term treatment compliance. Importantly, although botanicals may provide some supportive benefits for certain patients, they should not be considered a substitute for reasonable medication use. Rather, they should be considered adjunct treatment options in the context of providing more holistic, patient-focused care.

Many issues must be addressed before botanicals can be incorporated into mainstream cardiovascular care, even though a large body of evidence already exists and continues to grow. Now, a few key concerns include: no standardized botanical preparation exists; the concentration of active constituents is not consistent; large-scale, carefully planned clinical trials are required to ensure the long-term safety and effectiveness of botanical use are known, as the dosage schedules for each botanical are yet unknown. Also, the risk of herb-drug interactions must be carefully monitored, as it may be particularly significant in patients who take multiple cardiovascular medications. To enable safe and effective use, the establishment of a sound regulatory harmonization system, a rigorous quality control process, and a framework of clinical practice guidelines will be critical in facilitating the wider acceptance of botanicals in clinical practice. In the future, the combination of botanicals, lifestyle modifications, and drug therapy will be defined through a holistic, evidence-based approach. It will benefit from advancing interdisciplinary research, supportive regulatory policies, and further improvements in clinician and patient education regarding the use of botanicals for the prevention and management of CVD. Through ongoing scientific validation and clinical oversight, it is anticipated that botanicals will transition from the category of complementary therapy to that of fully accepted, evidence-based components of a comprehensive, integrative cardiovascular clinical practice model.

References

Al Disi SS, Anwar MA, Eid AH. Anti-hypertensive herbs and their mechanisms of action: part I. Frontiers in Pharmacology 2016;6 Available from https://doi.org/10.3389/fphar.2015.00323, http://journal.frontiersin.org/article/10.3389/fphar.2015.00323/full.

Al-anbaki M, Cavin AL, Nogueira RC, Taslimi J, Ali H, Najem M, et al. Hibiscus sabdariffa, a treatment for uncontrolled hypertension. Pilot comparative intervention. Plants 2021;10(5) Available from https://doi.org/10.3390/plants10051018, https://www.mdpi.com/2223-7747/10/5/1018/pdf.

Ansari R, Zarshenas MM, Dadbakhsh AH. A review on pharmacological and clinical aspects of *Linum usitatissimum* L. Current Drug Discovery Technologies 2019;16(2):148–58. Available from https://doi.org/10.2174/1570163815666180521101136, http://www.eurekaselect.com/162303/article.

Arya P, Kumari N, Ahmad Wani S, Kumar P. Trigonella foenum graecum. Herbs, Spices and Their Roles in Nutraceuticals and Functional Foods. India: Elsevier; 2022. p. 133–48. Available from https://www.sciencedirect.com/book/9780323907941, https://doi.org/10.1016/B978-0-323-90794-1.00001-6.

Bahadoran Z, Mirmiran P, Kabir A, Azizi F, Ghasemi A. The nitrate-independent blood pressure-lowering effect of beetroot juice: a systematic review and meta-analysis. Advances in Nutrition 2017;8(6):830–8. Available from https://doi.org/10.3945/an.117.016717, http://advances.nutrition.org/content/8/6/830.full.pdf#page=1&view=FitH.

Baharvand-Ahmadi B, Bahmani M, Tajeddini P, Rafieian-Kopaei M, Naghdi N. An ethnobotanical study of medicinal plants administered for the treatment of hypertension. Journal of Renal Injury Prevention 2016;5(3):123–8. Available from https://doi.org/10.15171/jrip.2016.26.

Basnet S, Otsuka M, Sasaki C, Asada C, Nakamura Y. Functionalization of the active ingredients of Japanese green tea (*Camellia sinensis*) for the synthesis of bio-based epoxy resin. Industrial Crops and Products 2015;73:63–72. Available from https://doi.org/10.1016/j.indcrop.2015.03.091.

Biesinger S, Michaels HA, Quadros AS, Qian Y, Rabovsky AB, Badger RS, et al. A combination of isolated phytochemicals and botanical extracts lowers diastolic blood pressure in a randomized controlled trial of hypertensive subjects. European Journal of Clinical Nutrition 2016;70(1):10–6. Available from https://doi.org/10.1038/ejcn.2015.88.

Brimson JM, Prasanth MI, Kumaree KK, Thitilertdecha P, Malar DS, Tencomnao T, et al. Tea plant (*Camellia sinensis*): a current update on use in diabetes, obesity, and cardiovascular disease. Nutrients 2023;15(1) Available from https://doi.org/10.3390/nu15010037, http://www.mdpi.com/journal/nutrients/.

Chan WJJ, McLachlan AJ, Luca EJ, Harnett JE. Garlic (*Allium sativum* L.) in the management of hypertension and dyslipidemia – a systematic review. Journal of Herbal Medicine 2020;19 Available from https://doi.org/10.1016/j.hermed.2019.100292, http://www.elsevier.com/wps/find/journaldescription.cws_home/725778/description#description.

Chen C, Shang C, Xin L, Xiang M, Wang Y, Shen Z, et al. Beneficial effects of psyllium on the prevention and treatment of cardiometabolic diseases. Food and Function 2022;13(14):7473–86. Available from https://doi.org/10.1039/d2fo00560c, http://pubs.rsc.org/en/journals/journal/fo.

Chong B, Jayabaskaran J, Jauhari SM, Chan SP, Goh R, Kueh MTW, et al. Global burden of cardiovascular diseases: projections from 2025 to 2050. Journal of Preventive Cardiology 2025;32(11):1001–15. Available from https://doi.org/10.1093/eurjpc/zwae281, https://academic.oup.com/eurjpc.

Chrysant SG, Chrysant GS. Herbs used for the treatment of hypertension and their mechanism of action. Current Hypertension Reports 2017;19(9) Available from https://doi.org/10.1007/s11906-017-0775-5, www.springer.com.

Cloud A, Vilcins D, McEwen B. The effect of hawthorn (*Crataegus* spp.) on blood pressure: a systematic review. Advances in Integrative Medicine 2020;7(3):167–75. Available from https://doi.org/10.1016/j.aimed.2019.09.002.

Dog TL. The Role of Botanicals in Cardiovascular Health. Informa UK Limited; 2022. p. 369–79. Available from https://doi.org/10.1201/9781003137849-18.

Durazzo A, D'Addezio L, Camilli E, Piccinelli R, Turrini A, Marletta L, et al. From plant compounds to botanicals and back: a current snapshot. Molecules 2018;23(8):1844. Available from https://doi.org/10.3390/molecules23081844.

Grönroos R, Eggertsen R, Bernhardsson S, Praetorius Björk M. Effects of beetroot juice on blood pressure in hypertension according to European society of hypertension guidelines: a systematic review and meta-analysis. Nutrition, Metabolism and Cardiovascular Diseases 2024;34(10):2240–56. Available from https://doi.org/10.1016/j.numecd.2024.06.009.

Ha AW, Ying T, Kim WK. The effects of black garlic (*Allium satvium*) extracts on lipid metabolism in rats fed a high fat diet. Nutrition Research and Practice 2015;9(1):30–6. Available from https://doi.org/10.4162/nrp.2015.9.1.30, http://www.ncbi.nlm.nih.gov/pmc/articles/PMC4317477/pdf/nrp-9-30.pdf.

Heshmat-Ghahdarijani K, Mashayekhiasl N, Amerizadeh A, Teimouri Jervekani Z, Sadeghi M. Effect of fenugreek consumption on serum lipid profile: a systematic review and meta-analysis. Phytotherapy Research 2020;34(9):2230–45. Available from https://doi.org/10.1002/ptr.6690.

Huang J, Feng S, Liu A, Dai Z, Wang H, Reuhl K, et al. Green tea polyphenol EGCG alleviates metabolic abnormality and fatty liver by decreasing bile acid and lipid absorption in mice. Molecular Nutrition & Food Research 2018;62(4). Available from https://doi.org/10.1002/mnfr.201700696.

Jabeur I, Pereira E, Barros L, Calhelha RC, Soković M, Oliveira MBPP, et al. *Hibiscus sabdariffa* L. as a source of nutrients, bioactive compounds and colouring agents. Food Research International 2017;100:717–23. Available from https://doi.org/10.1016/j.foodres.2017.07.073, www.elsevier.com/inca/publications/store/4/2/2/9/7/0.

Jovanovski E, Yashpal S, Komishon A, Zurbau A, Mejia SB, Ho HVT, et al. Effect of psyllium (*Plantago ovata*) fiber on LDL cholesterol and alternative lipid targets, non-HDL cholesterol and apolipoprotein B: A systematic review and meta-analysis of randomized controlled trials. American Journal of Clinical Nutrition 2018;108(5):922–32. Available from https://doi.org/10.1093/ajcn/nqy115, http://www.ajcn.org/contents-by-date.2005.shtml.

Jyothi MV, VL AB, Wagh VD, Rasheed A, Dayaramani R, Panigrahy UP, et al. Herbal interactions with cardiac medications: a comprehensive review of potential interactions between herbal drugs and commonly prescribed cardiac medications. Current Drug Safety 2025;20(2):94–119. Available from https://doi.org/10.2174/0115748863289321240424063819, www.eurekaselect.com/588/journal/current-drug-safety.

Khalesi S, Irwin C, Schubert M. Flaxseed consumption may reduce blood pressure: a systematic review and meta-analysis of controlled trials. Journal of Nutrition 2015;145(4):758–65. Available from https://doi.org/10.3945/jn.114.205302, http://jn.nutrition.org/content/145/4/758.full.pdf.

Liperoti R, Vetrano DL, Bernabei R, Onder G. Herbal medications in cardiovascular medicine. Journal of the American College of Cardiology 2017;69(9):1188–99. Available from https://doi.org/10.1016/j.jacc.2016.11.078.

Lipert A, Szadkowska I, Matusiak-Wieczorek E, Kochan E. The effect of herbal supplements on blood pressure: systematic review and meta-analysis. Antioxidants 2022;11(8):1419. Available from https://doi.org/10.3390/antiox11081419.

Liu, H., Liang, Z.M., Li, R.T., Yu, Y.G., 2020. 2 6 Advances in the mechanisms of Hibiscus sabdariffa L. on hypertension. E3S Web of Conferences EDP Sciences China www.e3s-conferences.org/ 145 10.1051/e3sconf/202014501039 .

Macêdo APA, Gonçalves MdS, Barreto Medeiros JM, David JM, Villarreal CF, Macambira SG, et al. Potential therapeutic effects of green tea on obese lipid profile – a systematic review. Nutrition and Health 2022;28(3):401–15. Available from https://doi.org/10.1177/02601060211073236, https://journals.sagepub.com/home/NAH.

Madgulkar AR, Rao MRP, Warrier D. Characterization of psyllium (*Plantago ovata*) polysaccharide and its uses. Polysaccharides: Bioactivity and Biotechnology. India: Springer International Publishing; 2015. p. 871–90. Available from http://doi.org/10.1007/978-3-319-16298-0, https://doi.org/10.1007/978-3-319-16298-0_49.

Martins N, Petropoulos S, Ferreira ICFR. Chemical composition and bioactive compounds of garlic (*Allium sativum* L.) as affected by pre- and post-harvest conditions: a review. Food Chemistry 2016;211:41–50. Available from https://doi.org/10.1016/j.foodchem.2016.05.029, www.elsevier.com/locate/foodchem.

Nabavi SF, Habtemariam S, Ahmed T, Sureda A, Daglia M, Sobarzo-Sánchez E, et al. Polyphenolic composition of *Crataegus monogyna* jacq.: from chemistry to medical applications. Nutrients 2015;7(9):7708–28. Available from https://doi.org/10.3390/nu7095361, http://www.mdpi.com/2072-6643/7/9/5361/pdf.

Nicoletti M. Nutraceuticals and botanicals: overview and perspectives. International Journal of Food Sciences and Nutrition 2012;63(sup1):2–6. Available from https://doi.org/10.3109/09637486.2011.628012.

Obu RN. Systematic review and meta-analysis of hibiscus sabdariffa in hypertension management. Journal of Complementary and Alternative Medical Research 2025;26(1):37–58. Available from https://doi.org/10.9734/jocamr/2025/v26i1613.

Pagliaro B, Santolamazza C, Simonelli F, Rubattu S. Phytochemical compounds and protection from cardiovascular diseases: a state of the art. BioMed Research International 2015;2015 Available from https://doi.org/10.1155/2015/918069, http://www.hindawi.com/journals/biomed/.

Parihar N, Barve K, Bhatt LK. Clinical, ethical, and regulatory issues of botanicals. Phytopharmaceuticals and Herbal Drugs: Prospects and Safety Issues in the Delivery of Natural Products. India: Elsevier; 2023. p. 265–79. Available from https://www.sciencedirect.com/book/9780323991254, https://doi.org/10.1016/B978-0-323-99125-4.00018-4.

Przybyszewska J, Kuźmiński A, Przybyszewski M. Importance of plantago ovata seed husk (psyllium husk) in the prevention and non-pharmacological treatment of hypercholesterolemia, postprandial hyperglycemia and hypertension. Health Problems of Civilization 2024. Available from https://doi.org/10.5114/hpc.2024.142911.

Rastogi S, Pandey MM, Rawat AKS. Traditional herbs: a remedy for cardiovascular disorders. Phytomedicine 2016;23(11):1082–9. Available from https://doi.org/10.1016/j.phymed.2015.10.012, www.urbanfischer.de/journals/phytomed.

Shaikh AS, Thomas AB, Chitlange SS. Herb–drug interaction studies of herbs used in treatment of cardiovascular disorders—a narrative review of preclinical and clinical studies. Phytotherapy Research 2020;34(5):1008–26. Available from https://doi.org/10.1002/ptr.6585, http://onlinelibrary.wiley.com/journal/10.1002/(ISSN)1099-1573.

Shaito A, Thuan DTB, Phu HT, Nguyen THD, Hasan H, Halabi S, et al. Herbal medicine for cardiovascular diseases: efficacy, mechanisms, and safety. Frontiers in Pharmacology 2020;11 Available from https://doi.org/10.3389/fphar.2020.00422, http://www.frontiersin.org/Pharmacology.

Sharifi-Rad J, Rodrigues CF, Sharopov F, Docea AO, Karaca AC, Sharifi-Rad M, et al. Diet, lifestyle and cardiovascular diseases: linking pathophysiology to cardioprotective effects of natural bioactive compounds. International Journal of Environmental Research and Public Health 2020;17(7) Available from https://doi.org/10.3390/ijerph17072326, https://www.mdpi.com/1660-4601/17/7/2326/pdf.

Srivastava S, Misra A. Quality Control of Herbal Drugs: Advancements and Challenges. Springer Science and Business Media LLC; 2018. p. 189–209. Available from https://doi.org/10.1007/978-981-10-8291-7_10.

Syed QA, Rashid Z, Ahmad MH, Shukat R, Ishaq A, Muhammad N, et al. Nutritional and therapeutic properties of fenugreek (*Trigonella foenum*-graecum): a review. International Journal of Food Properties 2020;23(1):1777–91. Available from https://doi.org/10.1080/10942912.2020.1825482, https://www.tandfonline.com/loi/ljfp20.

Szikora Z, Mátyus RO, Szabó BV, Csupor D, Tóth B. Hawthorn (*Crataegus* spp.) clinically significantly reduces blood pressure in hypertension: a meta-analysis of randomized placebo-controlled clinical trials. Pharmaceuticals 2025;18(7):1027. Available from https://doi.org/10.3390/ph18071027.

Thiruvengadam M, Chung IM, Samynathan R, Chandar SRH, Venkidasamy B, Sarkar T, et al. A comprehensive review of beetroot (*Beta vulgaris* L.) bioactive components in the food and pharmaceutical industries. Critical Reviews in Food Science and Nutrition 2024;64(3):708–39. Available from https://doi.org/10.1080/10408398.2022.2108367, https://www.tandfonline.com/loi/bfsn20.

Ursoniu S, Sahebkar A, Andrica F, Serban C, Banach M. Effects of flaxseed supplements on blood pressure: a systematic review and meta-analysis of controlled clinical trial. Clinical Nutrition 2016;35(3):615–25. Available from https://doi.org/10.1016/j.clnu.2015.05.012.

Vasanthi HR, ShriShriMal N, Das DK. Retraction notice: phytochemicals from plants to combat cardiovascular disease. Current Medicinal Chemistry 2012;19(14):2242–51. Available from https://doi.org/10.2174/092986712800229078.

Wong CP, Kaneda T, Morita H. Plant natural products as an anti-lipid droplets accumulation agent. Journal of Natural Medicines 2014;68(2):253–66. Available from https://doi.org/10.1007/s11418-014-0822-3.

Yang J, Wen C, Duan Y, Deng Q, Peng D, Zhang H, et al. The composition, extraction, analysis, bioactivities, bioavailability and applications in food system of flaxseed (*Linum usitatissimum* L.) oil: a review. Trends in Food Science & Technology 2021;118:252–60. Available from https://doi.org/10.1016/j.tifs.2021.09.025.

Yousefi E, Zareiy S, Zavoshy R, Noroozi M, Jahanihashemi H, Ardalani H. Fenugreek: a therapeutic complement for patients with borderline hyperlipidemia: a randomised, double-blind, placebo-controlled, clinical trial. Advances in Integrative Medicine 2017;4(1):31–5. Available from https://doi.org/10.1016/j.aimed.2016.12.002.

Zamani M, Kelishadi MR, Ashtary-Larky D, Amirani N, Goudarzi K, Torki IA, et al. The effects of green tea supplementation on cardiovascular risk factors: a systematic review and meta-analysis. Frontiers in Nutrition 2023;9 Available from https://doi.org/10.3389/fnut.2022.1084455, journal.frontiersin.org/journal/nutrition.

PART

4

Safety, efficacy, and regulatory considerations

CHAPTER

9 Ensuring safety and efficacy: clinical trials and quality control

Ashish Kumar Sharma[1], Shivam Singh[1], Arindam Kolay[1], G.L. Viswanatha[2], Minakshi Kommoju[3] and Anasuya Tanneeru[4]

[1]NIMS Institute of Pharmacy, NIMS University, Jaipur, Rajasthan, India,
[2]Clinchoice Pvt Ltd & Department of Pharmacology, Manipal College of Pharmaceutical Sciences, Manipal, Karnataka, India,
[3]Department of Pharmaceutics and Biotechnology, KVSR Siddhartha College of Pharmaceutical Sciences, Vijayawada, Andhra Pradesh, India,
[4]University of Glasgow, Glasgow, G12 8QQ, United Kingdom

9.1 Introduction

Good clinical practice (GCP) training is a fundamental requirement for all individuals involved in the conduct and oversight of clinical research. GCP represents a universally accepted standard that governs the design, conduct, monitoring, recording, and reporting of clinical studies involving human participants. It encompasses a comprehensive set of internationally recognized ethical principles and scientific quality requirements that must be followed at every stage of a clinical trial.

Following extensive laboratory investigations and preclinical studies in animal models, only the most promising therapeutic interventions progress to clinical trials involving human subjects. A clinical trial, also referred to as a clinical study, is a structured research activity designed to evaluate the safety, efficacy, and effectiveness of an intervention in a defined group of participants. Clinical trials are conducted to assess new or improved methods for disease screening, prevention, diagnosis, or treatment, and are typically carried out in a series of well-defined phases (Dorsey et al., 2015).

Throughout a clinical trial, critical information is generated about the investigational intervention, including its potential risks, benefits, and therapeutic value. Importantly, clinical trials are ethically justifiable only when genuine scientific uncertainty (equipoise) exists regarding the outcome; studies cannot be conducted when the outcome is already conclusively known from prior research. Milestones were given in Table 9.1.

Healing Botanicals. DOI: https://doi.org/10.1016/B978-0-443-44295-7.00008-2

Table 9.1 Historical milestones in research ethics and clinical trials.

S. No.	Year	Document/milestone	Significance and key contributions
1.	1947	Nuremberg code	Established foundational ethical principles for conducting research involving human participants, emphasizing voluntary consent.
2.	1948	Universal declaration of human rights	Defined basic human rights and freedoms, forming the ethical backbone for protections in biomedical research.
3.	1960	Emergence of pharmacovigilance	Introduced systematic monitoring of medicine safety, focusing on adverse drug reaction reporting.
4.	1964	Declaration of Helsinki	Provided ethical principles for medical research involving human subjects; periodically revised to reflect evolving standards (latest major revision 2008).
5.	1965	Institutional Review Boards	Formalized independent ethical review of research studies in the USA under 45 CFR Part 46.
6.	1968	WHO Scientific Group recommendations	Issued early guidance on the scientific and ethical evaluation of pharmaceutical products.
7.	1974	Belmont report (USA)	Articulated three core ethical principles: respect for persons, beneficence, and justice.
8.	1975	WHO drug development guidelines	Provided structured guidance on ethical and scientific aspects of drug research and development.
9.	1979	FDA regulations (21 CFR Part 50)	Established legally binding requirements for the protection of human research participants in the USA.
10.	1980	International ethical guidelines for biomedical research	Promoted globally accepted ethical standards for biomedical research involving human subjects.
11.	1995	WHO good clinical practice guidelines	Introduced good clinical practice (GCP) principles specifically for pharmaceutical clinical trials worldwide.
12.	1996	ICH E6 GCP	Harmonized GCP standards across the US, Europe, and Japan, becoming the global benchmark for clinical trials.
13.	1997	UNESCO declaration on the human genome	Addressed ethical issues related to genetics, genomics, and human rights.
14.	1998	Data Protection Act	Strengthened legal safeguards for personal and health-related data used in research.
15.	2000	WHO ethics committee standards	Defined operational and ethical standards for independent ethics committees.
16.	2001	ISO standards for clinical investigations	Established quality and safety standards for clinical investigations of medical devices.

(*Continued*)

Table 9.1 Historical milestones in research ethics and clinical trials. (*Continued*)

S. No.	Year	Document/milestone	Significance and key contributions
17.	2001	European Union (EU) clinical trials directive (2001/20/EC)	Harmonized clinical trial regulations across EU member states.
18.	2002	CIOMS ethical guidelines	Provided internationally applicable ethical guidance for biomedical research, especially in low- and middle-income countries.
19.	2004	Medicines for Human Use Regulations (UK)	Implemented EU clinical trial directives into UK law.
20.	2005	PAHO–PANDRH GCP guidelines	Promoted GCP standards across the Americas.
21.	2005	EU GCP Directive (2005/28/EC)	Detailed practical requirements for implementing GCP within the EU.
22.	2006	Human Tissue (Scotland) Act	Regulated ethical collection, storage, and use of human tissue for research purposes.
23.	2006	Medicines for Human Use (Amendment) Regulations	Updated and strengthened clinical trial regulatory requirements in the UK.
24.	2008	CIOMS epidemiological ethics guidelines	Established ethical standards specific to epidemiological and population-based research.

9.2 International Council for Harmonization, International Council for Harmonization-good clinical practice

GCP serves as a foundational framework for hospitals, academic researchers, and pharmaceutical and biotechnology companies engaged in clinical research. Established by the International Council for Harmonization (ICH), GCP provides an international quality standard for clinical trials involving human participants. Role of ICH is best described by Fig. 9.1. These guidelines ensure that investigational products are evaluated in a scientifically robust manner, while simultaneously safeguarding the rights, safety, and well-being of trial participants. In addition, GCP clearly defines the responsibilities and interactions of key stakeholders, including sponsors, investigators, clinical research associates, and ethics committees (Dixon, 1999). Details were described in Table 9.2.

One of the main ethical principles in GCP is that the interests of science and society should not override the rights, safety, and well-being of research participants. GCP strives to ensure that clinical research is scientifically sound and ethically correct, and that comprehensive reports on the clinical attributes of investigational products are well documented. These guidelines are founded on two fundamental principles: the safety of human subjects and the integrity and dependability of

FIGURE 9.1

Role of International Council for Harmonization (*ICH*) in ensuring ethical and quality clinical trials. ICH quality control measures.

biomedical data. GCP guarantees the public that the data from clinical research are accurate, credible, and verifiable, and that the rights, dignity, and confidentiality of participants are strictly upheld by imposing standard procedures for conducting and reporting the study. Therefore GCP is a vital tenet of contemporary clinical research, which fosters responsible conduct of ethical research, scientific innovation, and builds public confidence in clinical studies (Ojha and Bhargava, 2022).

The main goal of the ICH GCP guideline is to establish a common international standard for the design, conduct, performance, monitoring, auditing, recording, analysis, and reporting of clinical trials involving human subjects. The guideline was created through a joint effort by regulatory bodies and industry players in the European Union (EU), Japan, and the United States to ensure that various regulatory demands are combined into a single, universally acceptable standard. ICH GCP also enhances interregional acceptance of clinical trial data, thereby minimizing the likelihood of duplication in a study and shortening the time required for regulatory review. Such harmonization not only increases efficiency in drug development but also ensures that trials are conducted in a manner that is sensitive to participants' rights, safety, and well-being. Moreover, the guideline enhances data credibility and integrity by assigning responsibility to investigators, sponsors, and ethics committees. Ultimately, ICH GCP supports the ethical globalization of pharmaceutical research, facilitating timely access to safe, effective, and high-quality medicinal products intended for human use worldwide (Ravinetto, 2017).

Table 9.2 Brief introduction of International Council for Harmonization (*ICH*)-good clinical practice (*GCP*).

S. No.	Aspect	Description
1.	Purpose	To achieve global harmonization of regulatory requirements and guidelines governing the development of drugs, biologics, and medical devices.
2.	Stakeholders involved	Regulatory authorities and pharmaceutical industry representatives from Europe, Japan, and the United States, with contributions from the World Health Organization, Canada, Nordic countries, and Australia.
3.	Core objectives	To safeguard the rights, safety, and well-being of human participants involved in clinical research. To ensure the credibility, accuracy, and integrity of clinical trial data. To establish uniform standards for the ethical and scientific conduct of clinical studies.
4.	Concept of GCP	GCP represents the integration of ethical principles with robust quality systems to generate reliable clinical evidence.
5.	Development and implementation process	GCP guidelines are formulated for pharmaceuticals, biologics, and medical devices, endorsed by ICH member organizations, and subsequently adopted or incorporated into national regulatory frameworks (Rana et al., 2024).
6.	Parties responsible for GCP compliance	Sponsors, clinical investigators, independent ethics committees, Institutional Review Boards, contract research organizations, clinical research coordinators, clinical research associates, research nurses, medical monitors, data management and entry personnel, and other trial-related professionals (Bunn, 2024).

The ICH of Technical Requirements for Pharmaceuticals for Human Use (ICH) plays a unique and pivotal role in uniting regulatory authorities and representatives from the pharmaceutical industry across Europe, Japan, and the United States. Through collaborative dialog, ICH addresses scientific and technical challenges related to drug development and registration. In developing the GCP guideline, careful consideration was given to existing GCP standards in the EU, Japan, and the United States, as well as those adopted in Australia, Canada, the Nordic countries, and by the World Health Organization (WHO). As a result, the guideline reflects broad international consensus on the ethical and scientific requirements for clinical research (Allin et al., 2004).

The ICH GCP guideline is intended to be followed when generating clinical trial data for submission to regulatory authorities. Its principles are also applicable to other types of clinical investigations that may influence the safety, rights, and well-being of human participants. One of the goals of this harmonization process is to facilitate the effective utilization of human, animal, and material resources and to reduce the number of redundant clinical studies. The guideline can help enhance timely access to new and effective medicines worldwide, without negatively affecting quality, safety, or effectiveness, and without undermining regulatory roles in ensuring population safety (Saltman and Figueras, 1998).

Clinical trials conducted in accordance with ICH and WHO GCP standards follow a common framework accepted across several regions, including the EU, Japan, the United States, Australia, Canada, and the Nordic countries. The success of countries that have implemented these guidelines is that they are all operating within the same internationally accepted standard. The reason this harmonized approach is desirable is to facilitate the interpretation and utilization of the technical and regulatory requirements for registering a product, thereby minimizing unnecessary testing processes in the research and development process and encouraging innovation in the development of new therapeutic agents.

9.3 Principles of good clinical practice

GCP is anchored in an elaborate set of ethical and scientific principles that aim to ensure secure, open, and accountable clinical research involving human subjects. These principles provide a clear framework for designing, conducting, monitoring, recording, and reporting clinical trials which was represented by Fig. 9.2. GCP assures the public and the scientific credibility of research findings by protecting the rights, dignity, and health of participants and ensuring data accuracy and reliability (Organization WH, 2005; Foroughi-Heravani et al., 2020).

9.3.1 Primacy of participant welfare

The rights, safety, and well-being of the trial participants should always come first ahead of the interests of science and society. Participant welfare is the highest priority in GCP and the ethical basis of any clinical research involving human subjects. This principle holds that the rights, safety, dignity, and well-being of research subjects should not be subordinated to the interests of science, society, or commercial interests. Clinical trials are meant to enhance medical knowledge, and this can never in any way justify exposing people to harm and exploitation. In the past, this principle was introduced due to unethical research practices that failed to respect human dignity. Influential models such as the Belmont Report and the Declaration of Helsinki codified the need to give precedence to participants over the aims of the research. In current GCP guidelines, especially those developed by the International Council for Harmonization (ICH E6), this requirement is implemented through stringent ethical protections (Idänpään-Heikkilä, 1994).

FIGURE 9.2

Principles of good clinical practice.

The well-being of the participants is safeguarded by a scientifically valid study design. Trials should be informed by sufficient preclinical and clinical evidence that the expected benefits outweigh the risks. Risk-benefit assessment is not an isolated process but one that continues throughout the study. Investigators need to follow up on safety information, report side effects promptly, and alter or discontinue the trials if the risks outweigh the benefits. Another main mechanism that protects the welfare is informed consent. Participants need to be given clear and detailed information about the purpose of the study, the research process, the risks, benefits, and alternatives. The consent should be voluntary, without any coercion, and recorded properly. Children, economically disadvantaged individuals, and those with impaired decision-making capacity are vulnerable people who need special protection. Protection of participants is further enhanced by independent ethical review by Institutional Review Boards (IRBs) or ethics committees. Such entities screen study protocols to ensure they are ethically acceptable before recruitment. Also, the confidentiality of personal data and the respect for privacy are key elements of welfare protection (Vijayananthan and Nawawi, 2008).

Ultimately, the primacy of participant welfare transforms clinical research from a purely scientific endeavor into a morally accountable practice. By placing human dignity at the center of research governance, GCP ensures that medical progress proceeds responsibly and ethically, with public trust.

9.3.2 Qualified research personnel

Everyone participating in the performance of a clinical trial should have the right education, training, and experience to discharge the expressed duties. Any ethical and scientifically sound clinical trial is conducted by qualified research personnel. As stipulated by the GCP guidelines, all people participating in a clinical study should have the necessary education and professional training, licensure where required, and experience to perform the duties allocated to them effectively. This study is necessary to guarantee the safety of the participants and the integrity of the research data (Shanley et al., 2017).

The principal investigator (PI) is in charge of the overall running of the trial in the study site. According to the guidelines provided by the International Council for Harmonization (ICH E6 R2/R3), investigators must be qualified based on education, training, and experience. They must also demonstrate the capacity to conduct the study properly. Qualification information should include a curriculum vitae, medical licenses, and GCP training certificates, which should be recorded and kept as important study papers (Vijayananthan and Nawawi, 2008).

The PI is not the only qualified person. The subinvestigators, study coordinators, research nurses, pharmacists, laboratory staff, and data managers should also be proficient in their areas. In the event of delegation of duties, the investigator should ensure that personnel are adequately trained and controlled. Delegation logs are expected to be written with responsibility, accountability, and transparency in mind. Another important element is continuous education. GCP training is not a one-time need; staff need to keep up with regulatory changes, protocol updates, and new best practices. Sponsors also share this responsibility by providing protocol-specific training and monitoring performance at sites to ensure compliance. Qualified personnel also have a direct influence on the quality of trials. Poorly trained personnel could undermine informed consent, mismanage investigational products, fail to record information correctly, or fail to report adverse events appropriately. Such gaps may jeopardize participants and nullify the study's results (Idänpään-Heikkilä, 1994).

9.3.3 Scientific and ethical integrity

Clinical trials should be scientifically rational, methodologically sound, and conducted in accordance with universally accepted ethical principles. Any responsible clinical trial is all about scientific and ethical integrity. This principle fundamentally acknowledges a straightforward yet powerful fact: individuals do not volunteer for research to produce statistics, but to contribute to knowledge that can help improve lives. No such trust should be lost. A clinical trial should be scientifically justified in the first place. It implies that the research question must be grounded in a real medical need, be supported by prior evidence, and have a methodologically sound research design. When a study is done improperly without proper controls, endpoints, or proper statistical planning, it stands the risk of giving unreliable results. Inflicting inconvenience or even harm on participants without the opportunity to generate valid knowledge breaches both the scientific responsibility and ethical obligation (Berghammer, 2014).

Scientific rigor goes hand in hand with ethical integrity. The Declaration of Helsinki and the ICH GCP are international standards that require minimizing risks to a level that is reasonable in relation to the expected benefits. An attentive risk-benefit evaluation should be done before the trial and kept under review as more information is obtained. Independent ethics committees (IECs) review study protocols to ensure that the rights, safety, and well-being of participants are protected. It is also critical to be transparent and honest. Reporting adverse events properly, correctly recording data, and publishing true results, whether positive or negative, reinforces public trust. Distorting statistics or distorting information by reporting what one wants to portray is detrimental to science and society (Towolawi, 2026).

There is also scientific and ethical integrity, with respect for participants as individuals. The principle of human dignity overriding the goals of research is strengthened by clear communication, free and informed consent, and the right to withdraw at any time.

9.3.4 Quality assurance and control

Effective systems and protocols should be established and adhered to to maintain the quality, reliability, and consistency of all events in the clinical trial. The silent protectors of all clinical trials are quality assurance (QA) and quality control (QC). Although the participants can only observe the study visits and procedures, off the record, well-designed systems ensure accuracy, consistency, and reliability of each step. Such systems safeguard both scientific outcomes and the individuals who put their confidence in research (Funning et al., 2009).

QA is the overarching framework of policies and procedures that ensures a clinical trial is conducted in accordance with GCP and regulatory requirements. It is initiated way before the enrollment of the first participant. The protocols are studied, risks are determined, duties are specified, and training is done. The solution is to design quality into the study from the very beginning, a concept sometimes called quality by design. Research teams avoid making mistakes rather than merely fixing them later (Brown, 2001).

QC, on the other hand, concentrates on operational methods applied to the trial. This incorporates source data verification, monitoring visits, data validation checks, lab calibration, accountability for investigational products, and the reporting of adverse effects promptly. The current GCP (ICH E6 R2/R3) encourages risk-based monitoring, which means focusing oversight efforts on the most significant aspects of the study participants' impact on their safety and data integrity. Effective QA and QC systems promote uniformity across study sites, proper documentation, and transparency. Electronic data capture systems, audit trail and secure data management processes also enhance reliability. In the event of any discrepancies, corrective and preventive actions (CAPAs) are implemented to prevent recurrence.

Humanizing these systems means acknowledging their greater object: each piece of data is a person's experience. An omitted record, an imprecisely set instrument, or an unexplained negative occurrence may alter results and jeopardize safety. Having solid quality systems will also ensure that every participant's contribution is not disrespected and is well represented (Rana BK, Yadav D, Yadav S. 2024).

9.3.5 Adequate supporting data

It should have adequate nonclinical and clinical information on the investigational medicinal product to support the initiation and design of the proposed clinical trial. Before a new medicine is tested on human beings, there must be strong scientific evidence that it is reasonably safe to do so. Adequate supporting data principle: no clinical trial should start without adequate nonclinical and, where available, prior clinical data to support the study design and its level of risk. This evidence is based on nonclinical studies. Lab tests and animal studies help the researcher understand how the investigational medicine product behaves in the body when it is taken, distributed, metabolized, and removed. Toxicology research determines potential toxic effects and identifies safe initial doses for first-in-human trials. Under ICH M3 (R2), the proposed studies should be based on the proposed clinical exposure, including the amount and timeframe of exposure. This cautious planning minimizes uncertainty and reduces unnecessary damage to participants (Grimes et al., 2005).

Early-stage clinical data are used to make decisions as the development continues. Findings from Phase I trials, such as safety profiles and pharmacokinetic results, are used to guide the design of subsequent studies. Each step is based on previous evidence, forming a structured, scientifically justified system. This principle entails acknowledging that each participant of a clinical trial is not merely a subject; he is a person who trusts science. When there is adequate supporting data, the researchers can easily justify potential risks and benefits during informed consent. In the absence of that evidence, it would be unwarranted participation (Noorzurani et al., 2009).

Ethical principles such as the Declaration of Helsinki and ICH GCP (E6) make it clear that any research involving human subjects should be grounded in a comprehensive understanding of prior scientific knowledge. Ethics committees are very keen to examine evidence before approving it, and the expected benefits should be reasonable compared to the expected risks.

9.3.6 Compliance with ethical declarations

The clinical trials should be conducted in accordance with the ethical principles of the Declaration of Helsinki and other relevant international ethical principles. Adherence to ethical statements implies that clinical trials are not conducted solely for scientific ambition, but that respect for human dignity drives them. The first version of the Declaration of Helsinki, adopted by the World Medical Association in 1964 and subsequently updated several times, serves as a moral compass for research involving human participants. It confirms that the interests of science and society should be put in the background in favor of the well-being of the individual (DeVito et al., 2020).

In reality, this principle takes more than a token acceptance. Clinical trials should adhere to the main commitments of Helsinki: voluntary informed consent, a positive risk-benefit analysis, protection of vulnerable populations, independent ethical review, reporting and disclosure, and posttrial benefits where necessary. These morals are strengthened and implemented with ICH GCP (E6), which offers specific procedural guidelines to promote adherence. This necessity helps us to remember that all protocols reflect real lives. A consent form is not just a piece of paperwork; it is a dialog founded on trust. Review by the ethics committee is not bureaucracy; it is a protection that prevents participants from unnecessary exposure to harm. Open publication of findings honors the input of volunteers whose involvement propagates knowledge (Yank and Rennie, 2002).

Contemporary discourses point out that compliance involves continuous monitoring as well. An ethical obligation is not confined to the approval of a study, but extends to monitoring, adverse event reporting, data integrity audits, and public disclosure of results. International harmonization provides participants with the same level of protection regardless of their geographical location. Trials that adhere to the Declaration of Helsinki and other guidelines demonstrate responsibility to both subjects and society. Ethical statements set universal standards, and regulatory frameworks ensure those standards are implemented.

9.3.7 Well-defined trial protocol

An elaborate and precise protocol should be designed, including the study aims, inclusion and exclusion criteria, monitoring, data analysis, and publication policy. The trial protocol is more than a technical document; it is the roadmap or guide that outlines all the steps a clinical study will follow. It is explicit about the study's purpose, who may participate, and how the study will be conducted, including the analysis and discussion of results. In the absence of this clarity, otherwise good research will be inconsistent or ethically flawed. The protocol starts with specific study objectives. These aims define the research question and help to choose endpoints. Evident primary and secondary outcomes will also ensure the study measures what is really important. If the objectives are unclear, the results may be subject to interpretation or bias (Rudolph et al., 2020).

Exclusion and inclusion criteria are also significant. They establish the criteria to whom they are offered and assist in protecting vulnerable persons. Considerate criteria are used to ensure that participants are appropriate candidates for the intervention, in addition to being scientifically valid. To the participants, these criteria are not merely checkboxes that determine whether the study is right and safe for them individually.

The protocol in the monitoring procedures protects participants' well-being and data quality. These are assessment schedules, safety reporting schedules and procedures for adverse events management. Monitoring is planned to avoid confusion and achieve consistency within study sites. An effective protocol also contains an elaborate plan of statistical analysis. The use of prespecified data analysis methods minimizes the chances of selective reporting and manipulation. Publication policy transparency also ensures that positive, negative, or inconclusive findings are truthfully disseminated to the scientific community. This process is a reminder that all the components of the protocol affect actual individuals. Well-defined processes imply small mistakes. Safety measures are defined, which implies greater protection. Open reporting respects time, trust, and participants' contributions (Sookdeo et al., 2020).

9.3.8 Regulatory and guideline adherence

When initiating and conducting a clinical trial, investigators and sponsors must comply with all applicable regulatory provisions, guidelines, and ethical principles. The observance of regulatory guidelines ensures that clinical trials are conducted in a responsible, safe, and transparent manner. Although administrative regulations might seem human, they are more substantive than that: to protect participants and ensure medical knowledge is based on reliable evidence. Whenever initiating and conducting a trial, the investigators and sponsors should comply with all relevant local, national, and international regulations. This will involve compliance with GCP (ICH E6), national drug legislation, ethics committee guidelines, and data protection guidelines. The explicit roles of such frameworks are outlined. Researchers have a responsibility to conduct research in accordance with accepted protocols, protect participants, and document and report adverse events promptly. In their turn, sponsors have to guarantee the adequate design of the trial and its regulatory submissions, its monitoring, quality management, and supervision of the delegated activities (Organization WH, 2005).

Adherence does not occur at a single point in time when a trial is initiated. It is an ongoing commitment. Regulatory bodies can inspect the trial sites and sponsor systems to ensure compliance. Forms (e.g., informed consent forms), logs (e.g., delegation logs), safety reports, and monitoring records are all documents which can be used as evidence that ethical and legal standards are being achieved. This is because regulatory compliance reminds us that all regulations are based on experiences of research failures. Consent requirements were established to protect autonomy. The reason behind safety reporting obligations is that delayed reporting may be detrimental to the participants. Monitoring systems advance responsibility to ensure that scientific zeal does not supersede participants' welfare. Internationalized guidelines, such as ICH GCP, have been used in multinational trials to achieve uniformity across countries. This internationalization ensures that the protections given to participants are similar, regardless of location (Ravinetto et al., 2015).

9.3.9 Data integrity and confidentiality

Clinical trial data shall be properly captured, stored, and processed in a manner that enables reliable reporting, interpretation, and validation, and that assures the confidentiality and privacy of participants. Data integrity and confidentiality can ensure that all information obtained during a clinical trial is credible and appropriately maintained. Data in research are not numbers; they are experiences, health status, and personal stories of different people. It is not only a scientific but also an ethical obligation to protect that information (Van den Broeck et al., 2007).

Data integrity implies that data must be correct, complete, and verifiable from entry into the system through end reporting. The data from the clinical trial should be recorded promptly, with corrections made transparently and audit trails maintained. According to the modern GCP (ICH E6), systems should prevent unauthorized alterations and follow up all entries. Regardless of whether the information is captured on paper or in electronic data capture systems, it should accurately reflect what actually transpired during the study. Analyzing meaningful data requires reliable data. Conclusions about safety and efficacy would be misleading in the presence of incomplete, modified, or inconsistently recorded data. This may influence regulatory decisions and, eventually, patient care. As such, procedures, validation, and quality management systems are used to protect accuracy throughout the trial lifecycle (Tremaine et al., 2005).

Confidentiality complements integrity by ensuring participants' privacy. Identifiable information is usually replaced with coded numbers, and only authorized staff can access it. Sensitive health information is not exposed by means of secure storage systems, password protection, encryption, and adherence to data protection laws. This value reminds us that members of the group provide personal medical information, hoping it will be kept confidential and respected. A confidentiality breach may negatively affect trust and reputation and deter future research participation. Preservation of privacy respects participants by treating them not just as subjects of the research (Gomase et al., 2025).

9.3.10 Risk-benefit assessment

The potential risks and inconveniences that are likely to be encountered must be weighed against the expected benefits to the trial subjects and future patients before the trial commences. A trial must be undertaken if the net benefits outweigh the risks. One of the most significant ethical choices to be made before a clinical trial commences is the risk-benefit assessment. It asks the researchers to consider the potential harms and inconveniences relative to the expected benefits for participants and future patients. A research project must proceed provided the benefits are reasonably outweighing the hazards (Joffe and Miller, 2006).

Some foreseeable risks include physical side effects, psychological distress, time commitments, or social burdens. Prior laboratory research, animal experiments, and clinical results determine these risks. The investigators have to examine this evidence to determine the probability and degree of probable harm. Simultaneously, they consider possible outcomes, such as improved treatment modalities, better understanding of the disease, or direct therapeutic impact on participants. Such assessment, in accordance with international ethical standards such as the Declaration of Helsinki and ICH GCP (E6), must be performed before trial approval and maintained throughout the research. The role of ethics committees is critical in determining whether the risk-benefit balance is acceptable. If new safety information is found, the evaluation should be reconsidered, and the study can be altered or discontinued as needed. This principle is to keep in mind that participants are not abstract points of data. They are people who may be willing to take risks in the hope of contributing to medicine. Adhering to that commitment means they are not subjected to unreasonable or unbalanced risks. Notably, the concept of risk-benefit evaluation is not exclusive to the participants. It also takes into account the value of society. Even a minor risk is unethically justifiable in research that has no scientific merit or purpose. Thus ethics and scientific rigor are closely related (Ulrich et al., 2016).

9.3.11 Medical responsibility

Medical treatment of the trial participants and all medical judgments regarding the trial should be under the supervision of a qualified physician or, in suitable cases, a qualified dentist. In clinical trials, medical responsibility assures that the care of participants will not be severed from a professional clinical judgment. No matter how complicated a study may be or how many group members participate, the health and safety of the subjects involved should always be under the supervision of a properly qualified physician or, in some cases, a qualified dentist (Luce et al., 2004).

As ICH GCP (E6) says, all medical decisions concerning the trial at the study site are in the hands of the investigator. Although some activities may be assigned to nurses, coordinators, or other trained personnel, the physician will remain the final authority for clinical oversight. This involves evaluating eligibility, observing participants' health, managing adverse events, and determining whether a participant should proceed or drop out of the study. The principle implies its practical significance. In case of any side effects that the participant has that are not anticipated, it is not a protocol deviation but an instance that needs professional consideration and decision. The interpretation of clinical signs, adjustment of care, and prioritizing the well-being of the research participant over research goals are only performed well by a qualified physician or dentist who is trained and licensed to carry out this role. Trust is also enhanced through medical responsibility. Respondents tend to join trials when they are in a weak state of ill health. The fact that their care is under the supervision of a licensed healthcare professional is reassuring, as they will be given top priority in terms of safety. Ethics such as the Declaration of Helsinki reinforce the idea that physicians engaged in research should safeguard participants' lives, health, privacy, and dignity. Further, effective medical supervision is used to ensure that the trial process is consistent with prevailing clinical treatment standards. Concomitant treatment choice, emergency surgery or cessation decisions should be well informed by sound medical reasoning rather than research schedules (Chin and Lee, 2008).

9.3.12 Ethics and regulatory approval

A clinical trial can only be initiated with the approval of an IEC and the authority responsible for implementing the relevant regulations, which must establish that the benefits outweigh the risks. This requirement should also remain in compliance during the trial. Neither ethics nor regulatory approval permits a clinical trial to commence unless its safety, scientific value, and ethical acceptability are carefully and independently reviewed. A review and approval of the study protocol by an IEC or IRB and the appropriate regulatory authority should be done before the enrollment of the first participant. These institutions scrutinize the possibility of benefits exceeding anticipated risks and the protection of participants' rights (Bernabe et al., 2019).

It is not merely a procedural process. The scientific rationale, inclusion criteria, exclusion criteria, informed consent forms, monitoring safety plans, and compensation policy of the study are examined by ethics committees. They are supposed to be neutral protectors of participants' welfare. Meanwhile, regulatory authorities evaluate adherence to national laws, the safety of investigational products, and the feasibility of conducting the trial in general. This principle reminds us that behind any protocol are people who can encounter uncertainty and use it to improve their treatment. An independent review will minimize the risk that decisions are made solely by those who carry out or fund the study, within acceptable limits. It establishes a required balance between scientific ambition and morality. They do not stop oversight at approval. Constant observation is needed during the trial. Serious adverse events (SAEs), protocol modifications, and emergent safety data should be reported by the investigators to ethics committees and regulators as soon as possible. These authorities still have the authority to demand changes, halt, or discontinue a study in case the safety of the participants is at stake (Cox et al., 2022).

9.3.13 Protection of participant rights and data

The trial participants must be protected in accordance with the relevant data protection laws regarding their physical and psychological integrity, privacy, and personal data. The assurance of the rights and confidentiality of participants means that clinical studies do not treat individuals as participants in the scientific world, but as human beings deserving dignity, privacy, and security. In all clinical trials, detailed medical histories, laboratory findings, and even genetic information are collected. It is ethically right and legally obligatory to protect this information (Ross et al., 2018).

Physical and mental wholeness should never be violated. They should not subject the participants to unwarranted harm, coercion, or emotional distress. Ethical monitoring, informed consent, and follow-up monitoring are used to ensure that every person feels safe and honored throughout the study. Protection of personal data is also important. European and national regulations, such as the General Data Protection Regulation and other data protection laws, require that personal information be processed in a lawful, fair, and transparent manner. Only relevant data should be gathered, and access to it should be limited to personnel with prior permission. The text is usually coded or pseudonymized to minimize the chances of inadvertent disclosure. This principle reminds us that participants share very personal information with researchers. A violation of confidentiality may lead to embarrassment, discrimination, or even emotional damage. The aspects of responsible research conduct should include secure electronic systems, encryption, audit trails, and cybersecurity protections. Participants' rights to their data are also guaranteed. They should be made aware of how their information will be used, stored, and shared. Where legally allowed, they can demand access to their information or revoke permission for its use in some jurisdictions (Hallinan et al., 2023).

9.3.14 Insurance and indemnity

A clinical trial needs proper provisions to cover liabilities through insurance or indemnity, including potential losses to both investigators and sponsors. Insurance and indemnity clauses are provided to accept responsibility when clinical research is risky, ensuring that the responsibility is clear and managed. Despite the trials being organized to limit harm, unexpected negative effects may arise. Ethical research must ensure that participants, investigators, and institutions are not left unguarded in the event of an injury or liability. According to ICH GCP (E6), the sponsor should ensure that insurance or indemnification is provided to cover claims arising from trial-related injuries. This can include reimbursement for medical treatment, disability, or any other losses that directly resulted from participation. Most regulatory bodies and ethics committees in countries worldwide require evidence of insurance before authorizing a study (Hedgecock, 2005).

A clinical trial needs proper provisions to cover liabilities through insurance or indemnity, including potential losses to both investigators and sponsors. Insurance and indemnity clauses are provided to accept responsibility when clinical research is risky, ensuring that responsibility is clear and managed. Despite the trials being organized to limit harm, unexpected negative effects may arise. Ethical research must ensure that participants, investigators, and institutions are not left unguarded in the event of injury or liability. According to ICH GCP (E6), the sponsor should ensure that insurance or indemnification is provided to cover claims arising from trial-related injuries. This can include reimbursement for medical treatment, disability, or any other losses that directly resulted from participation. Most regulatory bodies and ethics committees in countries worldwide require evidence of insurance before authorizing a study (Kulkayeva et al., 2024).

FIGURE 9.3

Activities in clinical trial conduct. Clinical trial flow chart.

9.4 Overview of the clinical research process

The clinical research process is a controlled and orderly series of events aimed at ensuring that clinical trials are carried out in an ethical, scientific, and in accordance with GCP and regulatory standards. The stages of the process will enhance participant safety, data integrity, and the validity of trial results (Wang et al., 2025).

9.4.1 Activities in clinical trial conduct

The success of a clinical trial involves several interrelated activities that range from planning the study to final reporting. The activities were best represented in Fig. 9.3 and Table 9.3.

9.4.2 Trial protocol development

The programmer will develop the trial protocol in a separate section. The programmer will develop the trial protocol in a separate section. An elaborate guideline is developed to outline the study goals, design, methodology, eligibility criteria, and statistical analysis plan (Turner, 2010).

Table 9.3 The process by which the Food and Drug Administration (*FDA*) implements good clinical practice.

S. No.	21 CFR part	Regulatory focus
1.	Part 11	Standards for the use of electronic records and electronic signatures in clinical trials.
2.	Part 50	Protection of human subjects participating in clinical research, including informed consent requirements.
3.	Part 54	Mandatory financial disclosure by clinical investigators to identify potential conflicts of interest.
4.	Part 56	Oversight and responsibilities of Institutional Review Boards in reviewing and approving clinical studies.
5.	Part 312	Requirements for submitting Investigational New Drug (IND) applications.
6.	Part 312.120	Regulations governing foreign clinical trials not conducted under an IND.
7.	Part 314	Procedures for submitting applications to obtain FDA approval to market a new drug.
8.	Part 320	Guidelines for demonstrating bioavailability and bioequivalence of pharmaceutical products.
9.	Part 601	Submission requirements for FDA approval of biologics licenses.
10.	Part 812	Requirements for Investigational Device Exemptions for medical device trials.
11.	Part 814	Premarket approval process and standards for medical devices.

9.4.3 Development of standard operating procedures

Under the standard operating procedures (SOPs), the following will be created: SOPs are put in place to standardize activities during a trial and ensure uniformity, compliance, and quality across all sites.

9.4.4 Support systems and tools

This involves creating support systems and tools to help organizations manage change. Proper operational tools and systems are developed, including an electronic data capture platform, a safety reporting system, and monitoring tools, to support trial conduct (Malikova, 2016).

9.4.5 Preparation and approval of trial documentation

In this stage, the trial documentation is prepared and approved. Informed consent forms, case report forms (CRFs), investigator brochures, and regulatory submissions are necessary documents which are generated, reviewed, and approved before a study can start.

9.4.6 Trial site and study personnel selection

The trial sites will be carefully selected, and investigators and study staff with relevant qualifications, training, and experience will be appointed.

9.4.7 Review and approval by ethics committee

The study protocol and other pertinent documents are reviewed by IECs to determine whether they are ethically acceptable and do not infringe on participants' rights.

9.4.8 Review of regulatory authorities

The trial is also examined and approved by the relevant regulatory authorities in line with both national and international regulations. Setting up studies involving human subjects requires developing a subject enrollment and informed consent process. Only after voluntary and informed consent is obtained are participants recruited, evaluated for eligibility, and enrolled. Investigational products: this section outlines the procedures for managing investigational products. The manufacturing, storage, handling, dispensing, and accounting of investigational products are conducted in accordance with a set of quality and accountability guidelines. the trial conduct and data collection process: the procedures outlined in the study protocol are carried out, and clinical information is gathered and recorded.

9.4.9 Safety management and reporting

Adverse events and SAEs: adverse events and SAEs are continuously observed, recorded, and reported in accordance with regulatory timelines. Trial monitoring: constant monitoring is done to ascertain the following: protocol adherence, data accuracy, and safety of the participants. Clinical data management: gathered information is revised, authenticated, and safely handled as a measure of reliability and traceability.

9.4.10 Quality assurance activities

Audits and QCs are conducted to ensure compliance with GCP and regulatory standards (Bhatt, 2011).

9.4.11 Trial reporting

Study findings are analyzed and reported in a transparent and scientifically sound manner.

9.5 Common good clinical practice challenges during clinical studies

Certain GCP-related problems are often encountered in clinical trials, even within the scope of accepted procedures, necessitating proactive measures to address and rectify them. Despite thorough planning and execution of the clinical trials in accordance with established GCP requirements, overcoming the challenges remains difficult which are best described in Fig. 9.4. These issues are not necessarily due to negligence; they can also be caused by the complexity of people, processes, and data coordination in realistic healthcare (Organization WH, 2005). Types of clinical studies were given in Table 9.4.

One problem is protocol deviation. The processes described in the studies do not necessarily need to be performed as written due to misinterpretation, timing challenges, or clinical crises. Some may be inconsequential, whereas others may contribute to participant safety or data validity. Such incidents need to be reduced by ensuring clear training and continuous monitoring. Another common obstacle is missing or late documentation. GCP compliance depends on accurate records, but hectic clinical settings may lead to missing signatures, delayed data entry, or a lack of source documentation. Every piece of information that was not provided is not only paperwork, but a possible loophole in transparency (Ablett, 1997).

FIGURE 9.4

Common good clinical practies (*GCP*) Challenges during clinical studies. GCPs challenges.

Table 9.4 Some types of clinical trials.

S. No.	Type of clinical trial	Description
1.	Treatment/ interventional trials	These trials assess new treatments or innovative combinations of existing therapies. They may involve new drugs, surgical techniques, radiation therapies, or improved clinical management strategies to determine their effectiveness in treating a disease.
2.	Preventive/ prophylactic/risk-reduction trials	Prevention trials aim to identify effective ways to reduce the risk of developing a disease in healthy individuals or to prevent disease recurrence. Strategies may include medicines, vaccines, or lifestyle and behavioral changes.
3.	Diagnostic trials	These trials focus on improving methods used to diagnose diseases or medical conditions. They evaluate the accuracy, safety, and usefulness of new diagnostic tests or procedures.
4.	Screening trials	Screening trials are the studies of the most appropriate approaches in identifying the presence of a disease or health condition as early as possible to enable appropriate measures before it's too late.
5.	Quality of life (supportive care) trials	The trials fall under supportive care and palliative care to explore useful interventions that support comfort, the improvement of symptoms, and quality of life for chronic or seriously ill persons.

Another issue is late notification of adverse events. Safety information should be communicated promptly to safeguard participants. Critical safety updates can be delayed when reporting systems are unclear or when personnel are overworked. The credibility of trial results can also be jeopardized by problems with data integrity, such as transcription errors or insufficient auditing trails in electronic systems. Excellent QC tools and trained staff should support the production of trustworthy data. Making these issues more human reminds us that clinical trials involve real patients, healthcare, and complex workflows. Staff turnover, resource constraints, and changes in regulations can cause additional pressure.

Meanwhile, the participants will count on the research teams to maintain the highest standards of care and transparency. These barriers can be overcome through proactive risk management, including risk monitoring, ongoing training, internal audit, and corrective and preventive actions (CAPA). It is also necessary to create a culture of quality in which team members feel accountable and empowered to report the issue (Tritsmans, 1993).

9.5.1 Informed consent-related issues

- Inform that outdated or incorrect consent documents are being used.
- Unreasonably long or complicated consent forms. Failure to give the signed copy of the consent form to the participants.
- Poor monitoring of consent discussions in source documents.
- Lacks of translations or imprecise ones.
- Failure to have revised consent documents ready in time for the ethics committee. Lack of an unbiased witness where necessary.
- Very Poor communication about protocol changes or additional safety data to continuing participants.

9.5.2 Recruitment and study procedure issues

- Screening tests are conducted before informed consent is given.
- Incomplete lab or diagnostic testing as required in the protocol.
- Lack of consulting the medical monitor on the inclusion or exclusion deviation.
- Study visits conducted outside permitted protocol-defined time windows.
- Use of recruitment materials without prior sponsor or ethics committee approval.
- Enrollment challenges without appropriate corrective actions by investigators or sponsors.

9.5.3 Monitoring and documentation issues

- Inaccurate or incomplete source data documentation.
- Poor quality or inconsistentCRF completion.
- Inadequate training of study personnel.
- Missing or incomplete essential trial documentation.
- Insufficient or irregular monitoring activities.

9.5.4 Clinical trial supply management issues

- Errors in subject randomization.
- Inadequate accountability of investigational products.
- Improper storage or handling conditions.
- Poor participant compliance with study medication.
- Unauthorized or premature unblinding of treatment assignments.

9.5.5 Safety reporting issues

- Failure to identify or report SAEs.
- Delays in SAE reporting to sponsors, regulators, or ethics committees.

- Incomplete documentation of adverse events in source records.
- Failure to communicate new safety information to participants.
- Delayed updates to investigator brochures and informed consent addenda (Table 9.1–9.4).

9.6 Conclusion

GCP is the keystone of scientifically sound and ethical clinical research. GCP guarantees the security, safety, and health of human subjects across all research and study categories, whether pharmaceutical research, behavioral interventions, or observational studies, by providing an overarching framework that integrates ethical principles and rigorous scientific criteria. In addition to protecting participants, GCP encourages the collection of high-quality, trustworthy data capable of scientific analysis and regulatory assessment. Its focus on thorough documentation, SOPs, and QC creates transparency, accountability, and integrity throughout all phases of a clinical trial. The development of GCP over time, since the Nuremberg Code and the Declaration of Helsinki, to the ICH E6 (R2/R3) guidelines, demonstrates that the world is determined to balance the ethical obligation with the scientific excellence. Throughout, GCP remains a global resource guiding clinical research and can accommodate novel technologies, including decentralized trials, digital databases, and patient-centered research designs, without compromising ethical, scientific, and socially responsible research conduct. For investigators, sponsors, and research teams, adopting GCP can help ensure that the results of their research are credible, effective, and in accordance with the best standards of clinical research worldwide.

References

Ablett S. Clinical trials: towards good practice. Archives of Disease in Childhood 1997;76(3):283–5. Available from https://doi.org/10.1136/adc.76.3.283, http://adc.bmjjournals.com.

Allin S, Mossialos E, McKee M, Holland W, Organisation WH. Making decisions on public health: a review of eight countries [Internet] 2004 (accessed 19.02.26). Available from. https://iris.who.int/handle/10665/107600.

Berghammer G. Good Clinical Practice (GCP): a universal call for ethics in biomedical research. Medical Writing 2014;23(2):106–12. Available from https://doi.org/10.1179/2047480614z.000000000209.

Bernabe RDLC, van Thiel GJMW, Breekveldt NS, Gispen-de Wied CC, van Delden JJM. Ethics in clinical trial regulation: ethically relevant issues from EMA inspection reports. Current Medical Research and Opinion 2019;35(4):637–45. Available from https://doi.org/10.1080/03007995.2018.1528214, http://www.tandfonline.com/loi/icmo20.

Bhatt A. Quality of clinical trials: a moving target. Perspectives in Clinical Research 2011;2(4):124. Available from https://doi.org/10.4103/2229-3485.86880.

Brown L. Quality control and data monitoring: a guide to Good Clinical Practice. In: Guiloff RJ, editor. Clinical Trials in Neurology [Internet]. London: Springer London; 2001. p. 81–94. Available from http://link.springer.com/10.1007/978-1-4471-3787-0_8, http://doi.org/10.1007/978-1-4471-3787-0_8.

Bunn GP. Good Clinical Practices in pharmaceuticals [Internet]. CRC Press; 2024 [cited 2026 Feb 19]. Available from https://books.google.com/books?hl=en&lr=&id=wIgoEQAAQBAJ&oi=fnd&pg=PT2&dq=GCP+guidelines+are+formulated+for+pharmaceuticals,+biologics,+and+medical+devices,+&ots=K-RSeKF6d4&sig=ZZ8sYBhjujdb7b7CF4tLxVpDqmk.

Chin R, Lee BY. Principles and practice of clinical trial medicine [Internet]. Elsevier; 2008 (accessed 19.02.26). Available from https://books.google.com/books?hl=en&lr=&id=HgmrI2r5eo4C&oi=fnd&pg=PP1&dq=11.%09Medical+Responsibility++Medical+care+provided+to+trial+participants,+as+well+as+all+medical+decisions+related+to+the+study,+must+remain+under+the+responsibility+of+a+suitably+qualified+physician+or,+&ots=TiIBNUAeHi&sig=NBJXegxD2VCc8KAlbbkM-p71U08.

Cox S, Solbakk JH, Bernabe RDLC. Research ethics committees and post-approval activities: a qualitative study on the perspectives of European research ethics committee representatives. Current Medical Research and Opinion 2022;38(11):1897–907. Available from https://doi.org/10.1080/03007995.2022.2115773, http://www.tandfonline.com/loi/icmo20.

DeVito NJ, Bacon S, Goldacre B. Compliance with legal requirement to report clinical trial results on ClinicalTrials.gov: a cohort study. The Lancet 2020;395(10221):361–9. Available from https://doi.org/10.1016/S0140-6736(19)33220-9, http://www.journals.elsevier.com/the-lancet/.

Dixon JR. The international conference on harmonization Good Clinical Practice guideline. Quality Assurance (San Diego, Calif.) 1999;6(2):65–74. Available from https://doi.org/10.1080/105294199277860.

Dorsey ER, Venuto C, Venkataraman V, Harris DA, Kieburtz K. Novel methods and technologies for 21st-century clinical trials a review. JAMA Neurology 2015;72(5):582–8. Available from https://doi.org/10.1001/jamaneurol.2014.4524, http://archneur.jamanetwork.com/article.aspx?articleid=2173272.

Foroughi-Heravani N, Hadavandkhani M, Arjmand B, Larijani B, Baradaran-Rafii A, Nasli-Esfahani E, et al. Principles of Good Clinical Practice. In: Arjmand B, Payab M, Goodarzi P, editors. Biomedical product development: bench to bedside [Internet]. Cham: Springer International Publishing; 2020. p. 99–106. Available from http://link.springer.com/10.1007/978-3-030-35626-2_10, http://doi.org/10.1007/978-3-030-35626-2_10.

Funning S, Anders Grahnén K, Eriksson Å, Kettis-Linblad. Quality assurance within the scope of Good Clinical Practice (GCP)—what is the cost of GCP-related activities? A survey within the Swedish Association of the Pharmaceutical Industry (LIF)'s members. The Quality Assurance Journal 2009;12(1):3–7. Available from https://doi.org/10.1002/qaj.433.

Gomase VS, Ghatule AP, Sharma R, Sardana S. Leveraging artificial intelligence for data integrity, transparency, and security in technology-enabled improvements to clinical trial data management in healthcare. Reviews on Recent Clinical Trials 2025. Available from https://doi.org/10.2174/0115748871371119250818102753, www.eurekaselect.com/640/journal/reviews-recent-clinical-trials.

Grimes DA, Hubacher D, Nanda K, Schulz KF, Moher D, Altman DG. The Good Clinical Practice guideline: a bronze standard for clinical research. Lancet 2005;366(9480):172–4. Available from https://doi.org/10.1016/S0140-6736(05)66875-4, http://www.journals.elsevier.com/the-lancet/.

Hallinan D, Franziska Boehm A, Külpmann M, Elson. Information provision for informed consent procedures in psychological research under the general data protection regulation: a practical guide. Advances in Methods and Practices in Psychological Science 2023;6(1). Available from https://doi.org/10.1177/25152459231151944.

Hedgecock J. Insurance in clinical trials [Internet]. Inst of Clinical Research 2005 cited 2026 Feb 19]. Available from https://books.google.com/books?hl=en&lr=&id=t5wOVpBCgpQC&oi=fnd&pg=PP2&dq=14.%09Insurance+and+Indemnity++Adequate+provisions+must+be+in+place+to+ensure+insurance+or+indemnity+coverage+for+liabilities+arising+from+the+clinical+trial,&ots=zRSrnTqd96&sig=3xKkBRRVanUxymrfBj-vjLUd_9s.

Idänpään-Heikkilä JE. WHO guidelines for Good Clinical Practice (GCP) for trials on pharmaceutical products: responsibilities of the investigator. Annals of Medicine 1994;26(2):89–94. Available from https://doi.org/10.3109/07853899409147334.

Joffe S, Miller FG. Rethinking risk-benefit assessment for phase I cancer trials. Journal of Clinical Oncology 2006;24(19):2987–90. Available from https://doi.org/10.1200/JCO.2005.04.9296.

Kulkayeva G, Graf M, Tarassova V, Tabarov A. Insurance management in clinical trials (international and domestic experience). Astana Medical Journal 2024;122(3):27–35. Available from https://doi.org/10.54500/2790-1203-2024-3-122-27-35.

Luce, Cook, Martin, Angus, Curtis, et al. The ethical conduct of clinical research involving critically ill patients in the United States and Canada: principles and recommendations. American Journal of Respiratory and Critical Care Medicine 2004;170(12).

Malikova MA. Optimization of protocol design: a path to efficient, lower cost clinical trial execution. Future Science OA 2016;2(1). Available from https://doi.org/10.4155/fso.15.89, http://www.future-science.com/loi/fso?open=2015.

Noorzurani MHR, Aziz NA, Abdul Aziz AF, Abd Hamid MZ, Mohamed M, Othman S, et al. The need for 'Good Clinical Practice' in health care research. South African Family Practice 2009;51(3):202–5. Available from https://doi.org/10.1080/20786204.2009.10873848, http://www.safpj.co.za/index.php/safpj/article/view/1174/1543.

Ojha A, Bhargava S. International council for harmonisation (ICH) guidelines. Regulatory affairs in the pharmaceutical industry [Internet]. Elsevier; 2022. p. 47–74. [cited 2026 Feb 19]. Available from https://www.sciencedirect.com/science/article/pii/B9780128222119000083.

Organisation WH. Handbook for good clinical research practice (GCP): guidance for implementation [Internet]. 2005 [cited 2026 Feb 19]. Available from https://iris.who.int/bitstream/handle/10665/43392/924159392X_eng.pdf.

Rana BK, Yadav D, Yadav S. Strategic association for quality excellence: coordination between GMP, GLP, GCP, Metrology, QC, and QA. In: Bhatnagar A, Yadav S, Achanta V, Harmes-Liedtke U, Rab S, editors. Handbook of quality system, accreditation and conformity assessment [Internet]. Singapore: Springer Nature Singapore; 2024. p. 1–21. cited 2026 Feb 19]. Available from https://link.springer.com/10.1007/978-981-99-4637-2_29-1.

Ravinetto R. The revision of the ICH Good Clinical Practice guidelines: a missed opportunity? Indian Journal of Medical Ethics 2017;2(4):255–9. Available from https://doi.org/10.20529/IJME.2017.057.

Ravinetto R, De Nys K, Boelaert M, Diro E, Meintjes G, Adoke Y, et al. Sponsorship in non-commercial clinical trials: definitions, challenges and the role of Good Clinical Practices guidelines. BMC International Health and Human Rights 2015;15(1). Available from https://doi.org/10.1186/s12914-015-0073-8.

Ross MW, Iguchi MY, Panicker S. Ethical aspects of data sharing and research participant protections. American Psychologist 2018;73(2):138–45. Available from https://doi.org/10.1037/amp0000240, http://www.apa.org/pubs/journals/amp/.

Rudolph JE, Naimi AI, Westreich DJ, Kennedy EH, Schisterman EF. Defining and identifying per-protocol effects in randomized trials. Epidemiology (Cambridge, Mass.) 2020;31(5):692–4. Available from https://doi.org/10.1097/EDE.0000000000001234, http://journals.lww.com/epidem/pages/default.aspx.

Saltman RB, Figueras J. Analyzing the Evidence on European Health Care Reforms: experience in Western European health care systems suggests lessons for reform in the United States, according to a major international comparison by the World Health Organisation. Health Affairs (Project Hope) 1998;17(2):85–108. Available from https://doi.org/10.1377/hlthaff.17.2.85.

Shanley TP, Calvin-Naylor NA, Divecha R, Wartak MM, Blackwell K, Davis JM, et al. Enhancing Clinical Research Professionals' Training and Qualifications (ECRPTQ): recommendations for Good Clinical Practice (GCP) training for investigators and study coordinators. Journal of Clinical and Translational Science 2017;1(1):8–15. Available from https://doi.org/10.1017/cts.2016.1.

Sookdeo A, Kromer B, Büntgen U, Friedrich M, Friedrich R, Helle G, et al. Quality dating: a well-defined protocol implemented at ETH for high-precision 14C-dates tested on late glacial wood. Radiocarbon 2020;62(4):891–9. Available from https://doi.org/10.1017/rdc.2019.132.

Towolawi SA, Ethics in clinical research [Internet]. Available from https://www.researchgate.net/profile/Samson-Towolawi/publication/400601185_ShivKesh_Publication_ETHICS_IN_CLINICAL_RESEARCH/links/698a7ce912f837212a18f0e4/ShivKesh-Publication-ETHICS-IN-CLINICAL-RESEARCH.pdf (accessed 19.02.26).

Tremaine WJ, Carlson MR, Isaacs KL, Motil KJ, Robuck PR, Wurzelmann JI. M. Merrick, Ethical issues, safety, and data integrity in clinical trials. Inflammatory Bowel Diseases 2005;11(1):S17. Available from https://doi.org/10.1097/01.MIB.0000184850.46440.ea.

Tritsmans L. Good clinical trials and good clinical practice. Rev Port Cardiol Orgao of Soc Port Cardiol Port J Cardiol Off J Port Soc Cardiol [Internet] 1993 cited 2026 Feb 19]. Available from https://www.academia.edu/download/46999784/Good_clinical_trials_and_good_clinical_p20160704-5039-1153rhu.pdf.

Turner JR. Conducting clinical trials II: operational execution. New drug development [Internet]. New York, NY: Springer New York; 2010. p. 79–97. Available from http://link.springer.com/10.1007/978-1-4419-6418-2_7, http://doi.org/10.1007/978-1-4419-6418-2_7.

Ulrich CM, Ratcliffe SJ, R. Wallen G, Zhou Q, Knafl K, Grady C. Cancer clinical trial participants' assessment of risk and benefit. AJOB Empirical Bioethics 2016;7(1):8–16. Available from https://doi.org/10.1080/23294515.2015.1034381, http://www.tandfonline.com/loi/uabr20.

Van den Broeck J, Melanie. Mackay N, Mpontshane A, Kany Kany Luabeya M, Chhagan ML, Bennish. Maintaining data integrity in a rural clinical trial. Clinical Trials 2007;4(5):572–82. Available from https://doi.org/10.1177/1740774507084106.

Vijayananthan A, Nawawi O. The importance of Good Clinical Practice guidelines and its role in clinical trials. Biomedical Imaging and Intervention Journal 2008;4(1). Available from https://doi.org/10.2349/biij.4.1.e5.

Wang F, Wang X, Liu E, Song F, Chen Y. Impacts of clinical research units on clinical research—a systematic review of empirical studies. Systematic Reviews 2025;14(1). Available from https://doi.org/10.1186/s13643-025-02813-3.

Yank V, Rennie D. Reporting of informed consent and ethics committee approval in clinical trials. Journal of the American Medical Association 2002;287(21):2835–8. Available from https://doi.org/10.1001/jama.287.21.2835.

CHAPTER

10 Regulatory considerations: from farm to shelf

Ramana Gangireddy[1], Ravi Shankar Kunderu[1], Indu Sree Golla[2], Kanaka Durga Devi Nelluri[1] and Tanvitha Vadlamudi[3]

[1]*Department of Pharmaceutics and Biotechnology, KVSR Siddhartha College of Pharmaceutical Sciences, Vijayawada, Andhra Pradesh, India,*

[2]*Department of Chemistry, Siddhartha Academy of Higher Education, Deemed to be University, Vijayawada, Andhra Pradesh, India,*

[3]*Department of Health and Clinical Outcomes Research, Saint Louis University, Saint Louis, MO, United States*

10.1 Introduction

10.1.1 Why regulations are important for herbal and botanical products

Protect public health: regulations establish minimum quality, safety, and efficacy requirements, ensuring that products reaching consumers are not adulterated, contaminated, or falsely labeled. In the absence of regulatory safeguards, harmful exposures to toxins, contaminants, or adulterants, and unexpected adverse events rise. This fits into the conversation that has taken place regarding the need for quality assurance and standardization to ensure the safety of botanical dietary supplements. Assure consistent product quality and reproducibility: plant materials vary with species, organ used, harvest time, geography, and processing. Regulations and standards reduce batch-to-batch variability by requiring identity testing, standardized assays, and manufacturing controls. The intent is to enable informed risk–benefit decisions and responsible use by providing accurate labeling and claims about regulated products so that clinicians and consumers can weigh benefits versus risks, avoiding dangerous drug–herb interactions or inappropriate self-medication. Harmonized standards—pharmacopeial monographs, good manufacturing practices (GMP), good agricultural and collection practices (GACP), testing methods—support international trade and consumer confidence by reducing the risks of exposure to low-quality products. Legal accountability and traceability: regulation establishes a framework for manufacturer accountability, adverse-event reporting, and recalls in the event of safety issues (Kim, 2021).

Healing Botanicals. DOI: https://doi.org/10.1016/B978-0-443-44295-7.00001-X

10.1.2 Problems with unregulated botanicals: adulteration, toxicity, and variability

- Adulteration and botanical misidentification: Many reports cite substitution, contamination, or intentional adulteration—for instance, black cohosh products containing other *Cimicifuga* species; unprocessed botanicals are often misidentified, which alters both efficacy and safety profiles. Chemical and biological contamination: unregulated products can contain pesticide residues, heavy metals, microbial contamination, or undeclared pharmaceutical drugs that impart health risks.
- Variable active constituent and potency profiles: Natural variability and inconsistent processing lead to wide variation in active constituent profiles and potency, undermining predictability in efficacy and safety (Huxtable, 1990).
- Adverse events and interactions due to poor manufacturing or labeling: Consumer self-medication and incomplete labeling may lead to drug–herb interactions, perioperative risks (herb discontinuation before surgery), and polypharmacy complications in elderly patients (Li and Wertheimer, 2023).
- Gaps in regulation and weak enforcement lead to the market entry of low-quality products and the proliferation of unsupported claims, contributing to misinformation and consumer risk.

10.1.3 Role of regulations along the supply chain: cultivation → manufacturing → marketing

- Cultivation/collection: Regulations/GACP establish requirements for proper species identification, selection of seed/plant material, agronomic practices, timing of harvest, and documentation/traceability, necessary to preserve the quality of the raw material and limit contamination.
- Processing and manufacturing: GMP for botanicals and a validated process (extraction, concentration, drying, storage) to control contamination, degradation, and variability, and to require validated analytical methods and batch records. Quality control (QC) and testing: pharmacopeial or other validated monographs on identity, purity, and strength (marker assays, TLC, HPLC, genetic barcoding in some cases) confirm the quality of raw materials and finished products. Safety assessment and pharmacovigilance: regulations determine requirements for safety data, labeling (including contraindications and interactions), adverse event reporting, and recall mechanisms. For products intended as therapeutics, clinical data or evidence of traditional use under specific frameworks may be required.
- Marketing and claims: Regulatory bodies control the type of claims permitted, disease versus health claims, labeling accuracy, and advertising to prevent

misleading statements. The exact classification (drug, supplement, or traditional medicine) dictates the burden of evidence required for claims.

10.1.4 Global regulatory frameworks for herbal and botanical products

The regulation of herbal and botanical products varies widely across regions due to differences in historical use, healthcare systems, and risk assessment philosophies. In the United States, botanicals are primarily regulated by the Dietary Supplement Health and Education Act of 1994. Under this governance structure, the manufacturer bears the burden of ensuring safety and quality before marketing. In contrast, the regulatory body is responsible for monitoring the product after it has reached the market. Claims about the treatment of diseases are banned, and the evidence is not very strong.

In the European Union, there is a more organized and harmonized system. Herbal medicines can be classified as traditional herbal medicinal products or well-established medicinal products in the European Medicines Agency (EMA). This requires evidence of safety, quality, and potential efficacy, with standardized manufacturing processes and monographs established by the Committee on Herbal Medicinal Products (HMPC). Likewise, countries such as China and India have included traditional systems of medicine (Traditional Chinese Medicine [TCM] and Ayurveda, respectively) into their national regulatory systems, thereby integrating the best of the past with the best of the present in terms of quality and safety standards.

International bodies such as the World Health Organization (WHO) have been instrumental in facilitating harmonization by developing guidelines on GACP, quality, and safety. However, the lack of standardization in the definition of evidence and the levels of acceptability is one of the largest obstacles in international trade.

10.1.5 Emerging challenges and future directions in botanical regulation

With the growing demand for herbal and botanical products worldwide, new challenges are emerging for regulatory frameworks. The first of these is the growing complexity of botanical products, such as multiherb formulations and new extraction techniques, which make it difficult to standardize and qualify them. Traditional bioassays using markers may not fully capture the chemical and biological properties of these products, and there is a need to incorporate new technologies, such as metabolomics, DNA barcoding, and systems biology approaches.

Online shopping and cross-border e-commerce also pose new challenges for regulatory systems, as such products can evade national regulatory oversight and pose a danger to consumers through substandard or counterfeit products. There is also a need to incorporate real-world evidence (RWE) and pharmacovigilance data

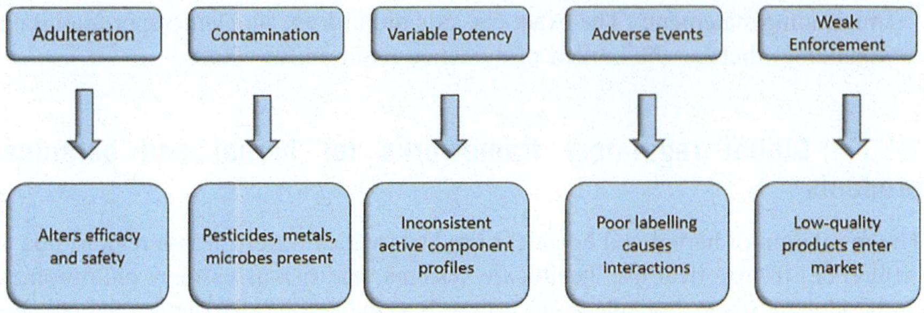

FIGURE 10.1 Common regulatory and safety issues in consumer products.

Safety issues of consumer products.

into regulatory decision-making, especially for products that have been in use for a long time in traditional medicine but lack clinical data. Consumer issues were represented in Fig. 10.1.

However, the future regulatory policies are expected to focus more on risk management and lifecycle approaches. This will include improved postmarket surveillance, enhanced adverse-event reporting systems, and greater clarity on the evidence requirements for efficacy claims. It will also be important to encourage transparency, traceability, and sustainability in the botanical supply chain. Ultimately, it is important to have science-informed regulation that ensures botanicals are used safely and effectively in modern healthcare systems.

10.2 Regulatory frameworks for herbal and botanical products

10.2.1 Overview of global regulatory standards

No single global standard exists; regulatory approaches vary from a strict medicinal product pathway to lighter dietary supplement regimes. Several research papers note the lack of uniformity and call for harmonization to ensure product quality and safe international trade (Poveda OMP, 2015).

10.2.1.1 Examples of regulatory approaches

Full medicine/MA/NDA pathway drugs need to present robust data on quality, safety, and efficacy.

Traditional schemes of registration, such as EU Traditional Herbal Registration (THR), allow simplified routes for long-standing traditional use with defined quality/safety requirements, but with limited claims.

Dietary supplement frameworks (e.g., US DSHEA) allow market entry with less premarket evidence, based on postmarket surveillance; reviewers have criticized this latter aspect as insufficient.

There is a need for global pharmacopeial monographs and harmonized testing standards to manage the complexity of botanical materials.

10.2.2 Differences between pharmaceuticals, herbaceuticals, and supplements

- Pharmaceuticals (botanical drugs): Regulated as medicines, require data on quality, nonclinical toxicology, and clinical efficacy; subjected to GMP and full regulatory review.
- Herbaceuticals/traditional herbal medicines: Intermediate, may be subject to simplified traditional registration, provided there is evidence of historical safe use; quality and safety documentation required; clinical efficacy standards may be adapted in the context of THR-type schemes.
- Dietary supplements/food supplements: Often regulated as food with ingredient and labeling requirements; premarket evidence requirements are limited in some jurisdictions, such as DSHEA, with emphasis on postmarket surveillance and manufacturer responsibility, sometimes allowing variable quality products to circulate (Dou and Weathers, 2022).
- Implications: Classification determines the premarketing data requirements, allowable claims, and the intensity of regulatory oversight; misclassification or inconsistent classification creates quality and public-health risks.

10.2.3 Classification of botanicals by major regulatory bodies (FDA, EMA, AYUSH, TGA, WHO)

- FDA (U.S.): Botanicals commonly enter as dietary supplements under DSHEA or as drugs when therapeutic claims or intended to affect structure/function beyond supplemental uses. The FDA does have botanical drug guidance for the drug pathway (NDA), but the supplement framework is less rigorous and relies on GMP and postmarket monitoring.
- EMA/EU: Two major routes exist, full marketing authorization for herbal medicinal products, as with any other medicine, and THR, for products with plausible evidence of traditional use and sufficient documentation of safety/quality; pharma GMP and pharmacovigilance apply depending on the category.
- AYUSH (India): National frameworks address Ayurveda, Unani, Siddha, and Homeopathy and include traditional knowledge-based registration and pharmacy/quality standards specific to those systems; local regulations govern manufacturing and claims (reviewed in country-specific regulatory chapters).

- TGA (Australia), other national regulators: Each has its own classification rules for listed (lower risk) vs registered (higher risk) products, with varying pre-market evidence and post-market obligations.

10.2.4 Quality, safety, and efficacy standards varying with regulatory status

The quality, safety, and efficacy standards for herbal and botanical products vary considerably depending on their regulatory status. For botanicals that enjoy a status similar to that of pharmaceuticals, a quality dossier is required, which encompasses the identification of starting materials, validated manufacturing procedures, stability studies, and compliance with pharmaceutical GMP. The safety assessment of botanicals is typically conducted through nonclinical toxicology studies, while efficacy is demonstrated through clinical studies, as with synthetic substances.

For traditional herbal medicines, such as those under the THR scheme in the European Union, the regulatory process emphasizes traditional use and safety. While clinical efficacy information may not be required, a justification for quality and pharmacological activity is required. Safety evaluation is based on bibliographic data, postmarketing experience, and, if necessary, specific nonclinical studies. Manufacturing practices and quality standards remain strict, often requiring compliance with pharmaceutical GMP.

The frameworks of dietary supplements are expected to impose the least stringent premarket requirements. The standards of quality may be limited to identity tests and food-grade GMP, and safety evaluations may be highly contingent on the manufacturer's liability. Efficacy information may not be required, and labeling may be permitted only for structure-function or general health claims. These differences may result in variations in product quality and the level of consumer protection. The quality and safety requirements may be standardized across regulatory agencies to minimize public health concerns while ensuring the availability of traditional and botanical dietary supplements.

10.2.5 Gaps in regulation, enforcement challenges, and the need for harmonization

Despite the presence of different regulatory systems for herbal and botanical products, gaps and challenges in enforcement persist. One challenge is the lack of uniformity in definitions and classification criteria for these products across countries, which allows manufacturers to take advantage of less stringent regulations in other countries.

Another issue is the lack of enforcement capabilities within regulatory bodies represented in Table 10.1. The inability to conduct inspections and testing may delay the identification of substandard or unsafe products. Reporting of adverse events associated with herbal products is also low, leading to a delay in the

Table 10.1 Global regulatory approaches.

Section	Key points
Overview of global regulatory standards	• No single global standard; regulations vary from strict medicinal pathways to lighter dietary supplement frameworks. • Research highlights the lack of uniformity and calls for harmonization for quality assurance and safe international trade. • Examples of regulatory approaches: – Full medicine/MA/NDA pathway: requires robust quality, safety, and efficacy data. – Traditional herbal registration (e.g., EU THR): simplified routes based on long-standing use; limited claims; defined quality and safety documentation. – Dietary supplement frameworks (e.g., US DSHEA): minimal premarket evidence; relies on postmarket surveillance (often criticized as insufficient) • Need for global pharmacopeial monographs and harmonized testing standards for complex botanical materials.
Differences between pharmaceuticals, herbaceuticals, and supplements	**Pharmaceuticals (botanical drugs):** regulated as medicines requiring quality, toxicology, and clinical efficacy data; subject to good manufacturing practice (GMP) and full regulatory review. **Herbaceuticals/traditional herbal medicines:** intermediate category; simplified registration based on historical safe use; quality and safety required, efficacy requirements adapted to traditional-use frameworks. **Dietary/food supplements:** often regulated as foods; limited premarket evidence; labeling and ingredient requirements vary; postmarket surveillance emphasized. **Implications:** classification influences data requirements, allowable claims, and regulatory oversight intensity; inconsistent classification may pose quality and public-health risks.
Classification of botanicals by major regulatory bodies	**FDA (U.S.):** botanicals classified as dietary supplements under DSHEA unless therapeutic claims are made; botanical drug pathway (NDA) available for drug claims; supplements rely on GMP + postmarket monitoring. **EMA/EU:** two pathways—full marketing authorization for herbal medicinal products, and THR for long-used traditional medicines; GMP and pharmacovigilance depend on category. **AYUSH (India):** system-specific regulations for Ayurveda, Unani, Siddha, Homeopathy; traditional-use-based registration with defined quality/claims rules. **TGA (Australia):** distinguishes between listed (low-risk) and registered (high-risk) products; varying levels of premarket and postmarket requirements. **TGA (Australia):** distinguishes between listed (low-risk) and registered (high-risk) products; varying levels of premarket and postmarket requirements.

regulatory body's response. Moreover, the complexity of herbal products is not adequately addressed by existing testing systems.

All these issues indicate the need for more international harmonization of strategies. The development of global pharmacopeial monographs, analytical procedures, and minimum quality standards would enhance the comparability and safety of products. More collaboration among regulatory bodies, with the help of organizations such as the WHO, would be beneficial for sharing information, conducting joint inspections, and taking joint actions on safety concerns. A harmonized risk-based regulatory framework would offer a balanced perspective between consumer accessibility, retention of traditional knowledge, and health protection in the rapidly globalizing botanical products market.

10.3 Good agricultural and collection practices

10.3.1 Cultivation, collection, and traceability

GACP guidelines ensure correct botanical identification, sourcing of seed/plant material, cultivation practices, optimal time of harvest (organ and phenological stage), postharvest handling, drying, and storage conditions to maintain active constituents and avoid degradation or contamination. GACP documentation ascertains traceability from the field to the finished product. Traceability mechanisms may include, but are not limited to, batch records, chain-of-custody documentation, GPS/geographic-origin records, lot numbers, and supplier qualification programs that facilitate the identification and isolation of supply-chain problems and support recalls as required. The authentication methods involve verifying species identity and detecting adulterants using morphological, chemical (HPLC, LC–MS), and genetic features. Genetic approaches have been successfully used, for instance, in studies of black cohosh authenticity.

10.3.2 Prevention of contamination (chemical, microbial, heavy metals)

- Preharvest and postharvest controls: Use of quality agricultural inputs, pesticide control, testing of soil and water, and avoiding areas potentially contaminated can minimize chemical/heavy metal risk. Control of drying and storage as a postharvest measure prevents moisture, thus minimizing the potential for microbial growth/mycotoxins. Quality specifications should include testing requirements for raw materials and finished products for microbial limits, mycotoxins, pesticide residues, and heavy metals, as well as the analytical methods used (Van Breemen et al., 2008).
- Regulatory thresholds and action: Maximum allowable levels of contaminants in botanicals have been established by various regulatory agencies. Quality

documentation and certificates of analysis from the manufacturer must reflect compliance with the requirements.

10.3.3 Sustainable sourcing and ethical wildcrafting

- Sustainability concerns: Sustainable cultivation methods, responsible wild harvesting with quotas, sharing, and traceability are encouraged through regulations and industry best practices to avoid depletion and biodiversity loss (Mihai et al., 2025).
- Ethical concerns: The utilization of local/traditional knowledge, fair payment, and adherence to country- and international-conservation regulations (the Convention on International Trade in Endangered Species of Wild Fauna and Flora [CITES], if applicable) are increasingly demanded by regulatory and consumer organizations. Sustainable sourcing also helps in maintaining a consistent quality raw material stream and a stable supply chain.
- Certification and market trends: Organic, fair trade, and sustainability certifications, if obtained, and proof of sustainable wild-harvesting practices can be helpful for market development and consumer acceptance. Industry best practices increasingly support sustainably harvested botanicals.

10.3.4 Role of good agricultural and collection practices in quality standardization and raw material consistency

GACP play a highly significant role in reducing variability and increasing consistency in botanical raw materials, which is a prerequisite for QC. Botanicals are known to be highly variable due to genetic, environmental, and agricultural factors. The GACP guidelines have been developed to reduce the effects of these factors by optimizing agricultural practices such as soil type, irrigation, fertilization, and pest control, which, in turn, optimize phytochemicals.

Raw material standardization can be achieved by optimizing plant species, subspecies, or chemotypes, and by standardizing agricultural practices such as plant part and harvest time. Standardized postharvest processing practices, such as drying temperature, time, and storage conditions, also help ensure that the active pharmaceutical ingredients are not degraded. The GACP facilitates the standardization of differences in agricultural practices and ensures the consistency of assays for marker compounds for pharmacopeial testing.

From a regulatory point of view, GACP ensures that there is a better correlation between the quality of the starting material and the finished product specifications. The documentation process developed by GACP is very helpful for supplier qualification and risk assessment, especially when sourcing botanical materials from different geographic areas. Harmonizing GACP with GMP ensures quality is maintained from the growth stage through to the finished product. Finally, GACP ensures consistency in therapeutic performance and improves confidence in the quality of herbal and botanical products.

10.3.5 Integration of good agricultural and collection practices with risk management and regulatory compliance

GACP is a fundamental risk-management approach for the regulatory control of herbal and botanical products. GACP reduces risks in the earliest stages of the supply chain, thereby lowering the likelihood of quality failures that may affect the safety and regulatory status of the final product. Risks associated with misidentification, agricultural contamination, and harvesting practices can be better managed once GACP is fully incorporated into risk management.

There is a continually growing need for manufacturers to offer risk-managed supplier control, particularly for botanicals that are imported from different parts of the world. GACP documentation is fundamental in risk assessment and supplier audits. In cases of noncompliance, traceability systems make it easy to identify affected product lots, thereby reducing the risk to public health.

In addition, GACP ensures compliance with international regulatory standards, such as pharmacopeial standards, as stated by various organizations, including the WHO. In the context of companies entering markets in various countries, GACP compliance documentation would be a valuable addition to regulatory documentation and inspections, as it would demonstrate the ability to control raw material quality. As the regulatory environment for herbal medicines becomes increasingly stringent, particularly regarding safety and adulteration, the addition of GACP to the overall quality structure will become increasingly imperative.

10.4 Good manufacturing practices for herbal products

10.4.1 Raw material standardization

What "raw material standardisation" means: consistent identity (species and plant part), defined quality attributes—moisture, foreign matter—quantifiable chemical markers or fingerprints (marker compounds or metabolic fingerprints) applied as incoming lot acceptance criteria (Kumar et al., 2025).

Authentication approaches:

Macroscopic/microscopic botanical identification, organoleptic tests.

Chemical profiling includes: HPLC; LC–MS fingerprints; marker assays to confirm the presence/level of known markers.

DNA-based methods for species authentication include ITS2 and DNA barcoding, but these may sometimes be limited by processing/degradation during finished product production. Proper sampling of the raw material becomes very important in this regard (Wang et al., 2023).

10.4.2 Processing, extraction, and purification

- Extraction parameters to be validated and controlled: Nature of solvent, solvent-to-plant ratio, time, temperature; steps involved in purification and concentration, with a view to validating and controlling these parameters, since they would determine the phytochemical profile and the potency of the extract. Research papers emphasize the use of a validated extraction and processing strategy to ensure consistent yields of phytochemicals.
- Selection of extraction technology: The traditional choice is based on the desired marker recovery, safety, and scalability. Papers focused on reviewing various extraction and concentration strategies discuss yield and impurity profiles (Hlatshwayo et al., 2025).
- This includes removing or reducing unwanted components. Explain ways to reduce impurities, residual solvents, and unwanted phytotoxins, such as pyrrolizidine alkaloids, and microbial bioburden. Cleaning and solvent removal steps shall be validated in accordance with GMP. In-process control and sampling include monitoring marker concentrations, moisture, microbial load, and extract density during processing; this helps achieve uniform quality in intermediates. GMP involves proper sample retention and stability testing (Wang et al., 2023).
- Purification and standardization: Application of chromatographic or other separation methodologies to concentrate or standardize marker compound(s) when necessary; the chemical markers would be used as targets for the release assay.

10.4.3 Packaging, labelling, and shelf-life stability

- Packaging materials and shelf life: Packaging should protect the botanical integrity, preventing moisture uptake, oxidation, and exposure to light. It must also be product chemistry-compatible. In studies reported on packaging and shelf-life, material selection and testing were emphasized under accelerated/ stability conditions to establish expiry dates. GMP and regulatory expectations on labeling include proper ingredient listing-both botanical name and part of the plant, batch/lot number, manufacturer, and expiry date, storage instructions, health/usage claims within regulatory allowances, warnings and contraindications, and compliance with regional labeling regulations.
- Stability testing: Conduct accelerated and real-time stability studies to define the shelf life and storage conditions. Study the marker compound content, microbial counts, and physical attributes over time to define shelf life and packaging suitability. Tamper evidence, child-resistant packaging, traceability: include lot coding and barcoding for recalls, and postmarkct survcillance (Bohra, 2025).

10.4.4 Documentation and batch-to-batch consistency

The documentation system provides an account of standard operating procedures (SOPs), batch records, certificates of analysis, deviation reports, change controls, and retention samples. It is the backbone of establishing a consistent process and provides traceability. Batch records and QC release criteria: a full manufacturing and QC record shall be on file for each batch, showing that it meets the in-process and final specification requirements for identity, assay, purity, and microbial limits; each batch shall be released by authorized persons only after review. Stability and retained samples: retain reference samples from each batch for possible future testing if needed due to adverse events or stability failures; retain long-term records for traceability. Continual improvement and corrective action: tracking of deviations; corrective and preventive actions (CAPAs), trending of QC results to maintain and enhance batch-to-batch consistency.

10.4.5 Quality control, in-process testing, and release specifications

QC is a GMP and is essential for producing herbal products that meet preestablished specifications (identity, purity, strength, and safety). Unlike conventional synthetic drugs (e.g., single-molecule drugs), herbal products exhibit greater chemical complexity and variability; therefore a multidimensional QC strategy is necessary for these products.

The release specification is established from the results of botanical identity testing, chemical marker analysis/chromatographic fingerprints, and microbiological/contaminant testing. Additionally, there is extensive in-process quality testing (i.e., monitoring moisture content, extract density, pH, marker compound concentration, and microbial load) throughout the manufacturing process to detect potential issues early and prevent noncompliance at the final release point. All methods used to evaluate in-process and final samples must be validated before use; this includes determining specificity, accuracy, precision, and robustness in accordance with both GMP and compendial requirements.

The CoA is an organized collection of all QC test results for each batch of a product, serving as a record of compliance with the most important specifications before releasing a new batch to qualified personnel. Only batches that meet all acceptance criteria will be released. Additionally, evaluating multiple batches will help verify processing and monitor variations in the quality attributes of batches over time. A good QC system, integrated with GMP and change control, creates uniformity across batches. It helps protect the public by ensuring that herbal products released will meet quality standards.

10.4.6 Good manufacturing practices audits, regulatory inspections, and compliance management

External audits may be conducted by regulatory agencies, certifying bodies, or customers as part of an audit program that assesses supplier compliance with GMP. Once an employee of the regulatory agency is assigned to the inspection, they will assess the overall facilities, equipment, qualified personnel, and quality management systems to assess compliance with GMP regulations.

The inspector will then assess all records, including batch records, deviations, change-control records, and stability studies; they may also take samples for independent laboratory analysis. The results of the inspection can range from a report with minor recommendations to one with serious deficiencies requiring immediate action.

The purpose of compliance management is to effectively manage audit outcomes with CAPAs. To ensure long-term improvements, you will want to use root cause analysis to identify systemic issues rather than determine what went wrong. In addition, proper training, regular SOP reviews, and management reviews are essential to ensuring ongoing compliance with GMP standards. Herbal manufacturers operating in multiple jurisdictions should be aware of the GMP requirements for each market they serve and how those requirements relate to their operations. A well-organized auditing and compliance process will reduce the risk of regulatory violations while improving the overall quality of your herbal manufacturing operations. Quality issues were described in Fig. 10.2.

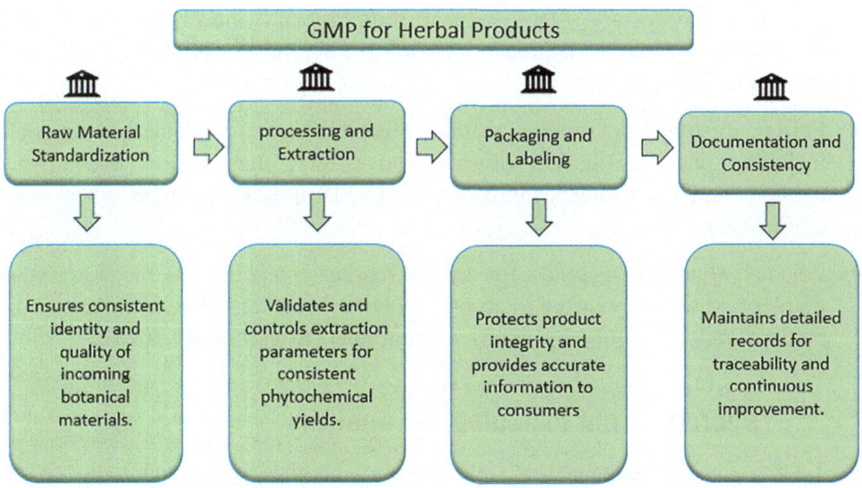

FIGURE 10.2 Key pillars of quality assurance in botanical products.

Quality assurance process.

10.5 Quality control and analytical testing

10.5.1 Phytochemical profiling (HPLC, GC–MS, LC–MS)

10.5.1.1 Tools and roles

- HPLC and UPLC (with UV, PDA, or MS detection): Quantification of marker compounds, fingerprinting, and stability assay development-validated HPLC/UPLC methods are normally applied for release and stability testing.
- GC–MS: Involves those volatile oils and small, usually volatile phytochemicals best suited for essential-oil profiling and residual solvent testing.
- LC–MS/HRMS: Profiling at a high resolution for multiple constituents and impurity identification; used when more comprehensive profiling is needed for complex extracts.
- Fingerprinting and metabolomics: Use chromatographic fingerprints or FTIR profiles as orthogonal methods to marker assays to detect batch-to-batch variation and adulteration (Sahoo and Umashankara, 2022).
- Validation: For release testing under GMP conditions, methods must be validated for specificity, accuracy, precision, LOD/LOQ, and robustness.

10.5.2 Identity, purity, and potency testing

- Identity: Botanical identification combined with chemical marker confirmation and, where necessary, DNA authentication for raw materials. For finished products, DNA may be degraded—chemical fingerprints and marker assays are often relied upon.
- Purity: Testing for foreign matter, related species, undeclared materials, including pharmaceuticals, solvents, and known plant toxins; use targeted methods and wide-screen approaches—LC–MS untargeted scans when suspicious results arise.
- Potency: Quantitative assays for defined marker compounds or bioassays when markers are not fully linked to activity; chemical markers serve as surrogates for potency and are required in many monographs (Kumar et al., 2025).

10.5.3 Microbial and toxicological analysis

- Testing for microorganisms: Total count of aerobes, yeast and mold counts, absence of pathogens (*Salmonella, E. coli*), mycotoxin screening where relevant; GACP/GMP measures to control moisture and drying to limit microbial growth.
- Toxicological screening: Heavy metals (lead, mercury, cadmium, arsenic), pesticide residues, residual solvents, and known phytotoxins (e.g., pyrrolizidine

alkaloids) should be tested in accordance with risk-based specifications and regulatory limits in relevant jurisdictions.

- Method validation and reference limits: Analytical methods used for safety release testing should be validated, and the pharmacopeial or regulatory limits must be adhered to.

10.5.4 Adulteration detection and marker compounds

Adulteration types include substitution with cheaper species, addition of undeclared herbs, addition of pharmaceuticals, and/or dilution with excipients. Research papers document frequent problems with adulteration, indicating a need for routine screening.

- Detection strategies: Targeted assays for known adulterants (pharmaceutical adulterants), and LC–MS untargeted scans for unexpected adulterants. Multimarker strategies and metabolic fingerprinting are used to detect species substitution and quality deviations. DNA barcoding for raw materials; chemical fingerprinting for processed products, where DNA may be degraded. Selection of marker compounds: choose markers that are stable, abundant, and correlated with activity where possible; multiple markers or fingerprinting are preferred for complex botanicals.

10.5.5 Stability testing and shelf-life determination

Stability testing is important to show that botanical raw materials and finished products maintain their identity, potency, purity, and safety throughout their intended shelf life. Due to the chemical complexity of herbal products, stability studies must consider not only the degradation of active or marker compounds and changes in chromatographic fingerprints, but also the formation of degradation products under defined storage conditions. Stability protocols shall follow ICH guidelines, where applicable, and be adapted for botanicals in accordance with the recommendations in pharmacopeial monographs and guidance documents issued by regulatory authorities.

Accelerated and long-term stability studies are generally conducted using validated stability-indicating methods, such as HPLC/UPLC coupled with UV or MS detection. For instance, instrumental approaches can distinguish between markers and degradation products and detect alterations in chemical fingerprints. Physical parameters, such as moisture content, normal water activity, appearance, and odor, also need to be monitored, as they may impact microbial growth and stability chemistry. In products containing volatile constituents, loss may require monitoring and the use of GC-based approaches.

Also, environmental variables such as temperature, humidity, and light must be documented and monitored, as botanicals tend to absorb moisture, increasing the risk of hydrolysis, oxidation, and/or microbial proliferation. Stability data can be used to mathematically calculate the time to retest a material and the time to finish an item's shelf life, as well as through trend analysis of stability data. If you do not have sufficient stability data, you should use conservative estimates of a product's shelf life and conduct continual stability studies in accordance with GMP.

10.5.6 Reference standards, documentation, and data integrity

To ensure accurate QC of botanical products, it is necessary to use an appropriate selection and control of reference standards along with accurate documentation and protection of data integrity. Reference standards are considered either pharmacopeial standards, qualified secondary standards, or well-characterized in-house reference materials. If no marker is commercially available, in-house reference materials must be purified and characterized structurally (e.g., by NMR or MS), and assigned a purity value using a scientifically accepted method.

Chromatographic fingerprint reference standards should be established from authenticated batches stored under controlled conditions for use as reference points in routine testing. The acceptance criteria for fingerprints should be scientifically defined to accommodate the natural variability of plant materials while remaining sensitive to potential adulteration or changes in quality. Any changes made to the reference standards or fingerprint libraries must be subject to proper change-control procedures.

GMP requires significant documentation to support methods, including method validation reports, system suitability, sample traceability, and analysis results for each production run, so that the reviewer can assess whether all GMP requirements were met. Electronic data systems used for chromatographic and spectroscopic methods must comply with principles of data integrity, specifically the ALCOA+ principles: attributable, legible, contemporaneous, original, accurate, complete, consistent, durable, and available. Companies must provide audit trails and access controls, and perform frequent data reviews to justify the trustworthiness of their analytical output.

An appropriately maintained reference standard, combined with good data integrity practices, is the foundation upon which quality decisions that can be defended are made, compliance with regulations is achieved, and a product's characteristics remain consistent from one lot to another over an extended period of time. QC and analytical testing for botanical products were mentioned in Table 10.2.

10.6 Safety evaluation of herbal products

10.6.1 Preclinical toxicity

- Types of studies: Acute, sub-acute/sub-chronic toxicity, genotoxicity, and, if indicated based on specific findings, reproductive toxicity or specific organ-toxicity studies; these are generally required for botanical drugs and higher-risk products and are recommended if there are data gaps.

Table 10.2 Summary of quality control and analytical testing for botanical products.

Section	Key elements
Phytochemical profiling (HPLC, GC–MS, LC–MS)	• HPLC/UPLC (UV, PDA, MS): quantification of marker compounds, fingerprinting, stability testing; validated methods used for release and stability studies. • GC–MS: ideal for volatile oils and low-molecular-weight compounds; essential-oil profiling and residual solvent testing. • LC–MS/HRMS: high-resolution profiling for multiple constituents and impurity identification; essential for complex extracts. • Fingerprinting/metabolomics: chromatographic fingerprints or FTIR profiles used to detect batch variation and adulteration. • Validation requirements: specificity, accuracy, precision, LOD/LOQ, and robustness for good manufacturing practice (GMP)-compliant release testing.
Identity, purity, and potency testing	• Identity: botanical ID plus chemical marker confirmation; DNA barcoding for raw materials (DNA often degraded in finished products, so chemical fingerprints preferred). • Purity: testing for foreign matter, related species, undeclared materials, pharmaceuticals, solvents, and plant toxins; targeted methods plus wide-screen LC–MS for suspicious results. • Potency: quantitative assays for marker compounds or bioassays when markers are not fully linked to activity; markers serve as potency surrogates and are required in many monographs.
Microbial and toxicological analysis	• Microbial testing: total aerobic count, yeast and mold counts, absence of pathogens (*Salmonella*, *Escherichia coli*), mycotoxin screening; good agricultural and collection practice/GMP measures used to control moisture and drying. • Toxicology: heavy metals (lead, mercury, cadmium, arsenic), pesticides, residual solvents, phytotoxins (e.g., pyrrolizidine alkaloids); risk-based specifications aligned with regulatory limits. • Method validation and limits: all analytical methods for safety testing must be validated; pharmacopeial or regulatory limits must be followed.
Adulteration detection and marker compounds	• Types of adulteration: species substitution, addition of undeclared herbs or pharmaceuticals, dilution with excipients; widespread issues reported in research. • Detection strategies: targeted assays for known adulterants; untargeted LC–MS scans for unexpected contaminants; multimarker and metabolomic fingerprinting for species verification; DNA barcoding for raw materials; chemical fingerprinting for processed products. • Marker compound selection: choose stable, abundant, activity-linked markers; use multiple markers or full fingerprinting for complex botanicals.

- Dose selection and characterization: Standard toxicology approaches are to be used, preferably with well-characterized extracts defined by the marker content; relate toxicology findings to chemical constituents to identify potential toxicants. Role of traditional use: historical/traditional use can inform safety expectations, but does not replace targeted toxicology when the exposure or formulation differs significantly from traditional preparations (Hlatshwayo et al., 2025).

10.6.2 Herb-drug interactions

Mechanisms include pharmacokinetic interactions, such as induction/inhibition of Cytochrome P450 (CYP) enzymes, transporter effects, and pharmacodynamic interactions, including additive or antagonistic effects with conventional drugs. Research papers warn that inadequate labeling and variable potency increase the risk of interactions in clinical use.

- Approaches to evaluation: In vitro CYP induction/inhibition assays, transporter studies, and, when appropriate, clinical interaction studies for common or high-risk botanicals; provide evidence-based labels with warnings about known interactions.
- Risk mitigation: Clear labeling of potential interactions, communication with healthcare providers, and pharmacovigilance to detect interaction signals post-market (Wang et al., 2023).

10.6.3 Allergens and contraindications

- Allergen identification: Determine whether botanical excipients or pollen residues are an allergenic hazard; identify common allergens and excipients (e.g., milk and gluten) where they are used in dosage forms (Bohra, 2025).
- Contraindications: List known contraindications, including pregnancy, lactation, and pediatric or geriatric cautions, if documented based on constituent safety profiles and histories; labeling is to include these according to regulatory expectations.

10.6.4 Post-marketing surveillance and pharmacovigilance

- Need: Because many herbal/supplement products enter the marketplace with limited premarket safety data, postmarket surveillance is necessary to detect adverse events, product quality failures, or adulteration-based harms. Research papers support the calls for stronger pharmacovigilance systems and adverse event reporting to protect public health and guide regulatory action (Ram et al., 2023).
- Systems and responsibilities: Manufacturer responsibility for collecting adverse events and reporting to regulatory authorities as required; maintain complaint logs and product investigations. Regulatory databases and signal detection: central AE databases and collaboration of regulators and industry on the

detection of safety signals and execution of recalls or labeling changes. Consumer and clinician reporting: facilitate accessible reporting mechanisms, and educate clinicians to report suspected herb-related adverse events.

10.6.5 Special populations and vulnerable groups

Safety assessment of herbal products must consider vulnerable populations who may have altered pharmacokinetics or pharmacodynamics, or heightened sensitivity to botanical constituents. Specifically, these include pregnant and lactating women, infants and children, older adults, and persons suffering from chronic diseases such as hepatic, renal, or cardiovascular conditions. Differences in metabolism, organ function, and body composition play an important role in efficacy and toxicity, leading to significant potential for harmful consequences even when products are considered safe for the general adult population.

Special attention is required during pregnancy and lactation because of the potential for embryotoxicity, teratogenicity, or transferring constituents into breast milk. For most botanicals, reproductive toxicity data are lacking, and a precautionary approach is generally recommended unless safety is well documented. Pediatric populations are also particularly challenging due to immature detoxification pathways and uncertainties about appropriate dosing. On the other hand, geriatric populations may be exposed to polypharmacy and age-related declines in drug clearance.

Risk assessment should include preclinical information from well-established human observational studies for which constituent toxicology is well understood. If data are missing, cautious label use with clear warnings or use restrictions is appropriate. In addition, formulations, including excipients and alcohol content, are likewise reviewed for appropriateness for use in the populations of concern.

Clearly, labeling, education of healthcare providers, and education of consumers are critical risk mitigation strategies. There is a regulatory requirement in some countries that mandates statements about special populations, and integrating such factors into safety assessments helps ensure appropriate marketing, preventable injury avoidance, and alignment of herbal products with public health values.

10.6.6 Benefit-risk assessment and regulatory

A structured assessment of the benefit-risk ratio is integral to evaluating the safety of herbal medicines, especially when scientific evidence is varied or scarce. A logically based evaluation involves balancing the health benefit against the risk of adverse effects associated with the proposed use. There are many ways to demonstrate the benefits of traditional herbal products, including their historical use and the experiences of those who have used them.

In addition, the product risk assessment includes a review of inherent product risks. Examples of inherent product risk include the potential for ingredient(s) to introduce toxicity and for ingredient interactions. When considering the use of herbal medicines as a means of self-care for long-term use, or by an at-risk group, the conservative safety factor will be the key consideration. If the benefit of a herbal product is considered low, the assessment must include an evaluation of the level of risk relative to the likelihood of occurrence.

The way that herbal products are regulated varies greatly around the world, with classifications as food, supplement, traditional medicine, or botanical drug, but there are commonalities regarding the principles of how to make decisions about all the products (e.g., providing adequate justification of the safety conclusion, clearly documented assumptions, and providing sufficient risk-management measures important for the classification(s)). Such measures may include: restrictive limits on the product (e.g., dosing, time of use, contraindications, or postmarket monitoring obligations), and the reasonable assumption of potential risks associated with its use.

10.7 Clinical trials for botanicals

10.7.1 Study design considerations for herbal products

10.7.1.1 Key design elements

When appropriate, choose randomized, controlled (placebo or active comparator), double-masked designs to reduce bias.

Clearly define the intervention in terms of the standardized extract, dose, and form of administration. Ensure batch identity and stability throughout the trial period.

- Outcome measures: Use validated clinical endpoints with appropriate safety monitoring; tailor endpoints to the intended claim, symptom relief versus disease modification.
- Population and inclusion/exclusion: Define narrowly to restrict confounders (such as concomitant medications, comorbidities) and capture potential interaction risks.
- Evidence: A recent descriptive analysis of herbal medicine clinical trials found that randomized, double-masked, parallel designs were commonly used and that placebo controls were widely utilized, reflecting methodological norms and the need for rigorous study design for herbal products (Koonrungsesomboon et al., 2024).

10.7.2 Challenges in standardising herbal extracts

10.7.2.1 Key challenges

Natural variability in constituents due to genetics, environment, time of harvest, and processing; reproducible extracts with defined marker content are established only through upstream controls such as GACP/GMP and validated manufacturing processes (Rajini et al., 2024).

- Choosing markers: Single markers may not accurately reflect activity in complex multiconstituent extracts. Multimarker panels or bioassays/bio-fingerprints are often preferable (Brendler et al., 2022).
- Analytical limitations: DNA methods may be unsuitable for finished extracts, and chemical fingerprints and LC–MS profiling are important for documenting consistency across trials (Wu and Shaw, 2022).

10.7.3 Evidence-based validation of safety and efficacy

- Approach: Combine preclinical toxicology with well-designed clinical trials and postmarket surveillance to build evidence of safety and efficacy. In areas where the trials are few or conflicting, regulators and researchers should focus on rigorous randomized trials and scrupulous reporting of adverse events (Kaggwa et al., 2022).
- Practical priority: Emphasize trials utilizing well-characterized materials, adequate blinding, and placebo controls that report efficacy and safety outcomes in formats that make data useful for regulatory review and meta-analysis (Koonrungsesomboon et al., 2024).

10.7.4 Ethical and regulated aspects of botanical clinical trials

The ethical issues in clinical trials of botanical products are similar to those of any other investigational agent. Still, there is a specific requirement to take into account the botanical nature of the intervention, evidence of its traditional use, recognition of its risks, doubts regarding its therapeutic efficacy, and the possibility of herb-drug interactions, while obtaining informed consent. Ethical clearance for the clinical trial protocol is mandatory from institutional review boards/scientific ethics committees, and the study should be conducted in accordance with the Declaration of Helsinki and the Good Clinical Practice guidelines.

The requirements differ depending on the classification of the BP, which may include dietary supplements, traditional medicine, and botanical drugs. It is important to determine whether an investigational new drug application is required before the trial is initiated, as this may vary depending on the type of BP used as an investigational agent. Product manufacturing must adhere to GMP to ensure the safety and stability of the agent used for clinical purposes.

Ethical issues that are particularly pertinent to botanical trials:

The main challenge when considering botanicals is managing trial participants' expectations, given that these products are rooted in tradition. It is particularly important that, when conducting trials in which the region owns the herb, therapeutic misconception be avoided. Transparency of trial registration, publication of trial results, whether positive, negative, or neutral, along with addressing any issue of conflict of interest, is of vital importance to maintain the scientific integrity of the trial, rendering it more easily acceptable.

10.7.5 Placebo design, blinding, and participant compliance

Blinding and placebo use in clinical trials of herbal products have proven difficult. Each type of herbal medicine produces a specific taste, color, smell, and gastrointestinal effect, which is dependent on the herbs and preparation used, and this can potentially reveal characteristics of the treatment(s) to participants or investigators.

Placebo development presents difficulties when matching a placebo to the herbal medicine, because the herbal medicine's characteristic sensory properties must be incorporated into the placebo formulation. Poorly matched placebos can undermine the internal validity of the study data.

In addition, to the study participants, investigators, outcome assessors, and data analysts must also be blinded. The "double dummy" method should also be used to assess studies that compare herbal medicines with allopathic medicines or use different dosage forms. Randomization should be clearly described, as should the methods of blinding used in clinical trials.

Ensuring participants' compliance in longer clinical research trials (with multiple daily doses) can pose serious challenges. The general idea that herbal products (medicinal plants) are "safe" may lead to noncompliance with dosing regimens/directions when used with other supplements. Some methods of participant compliance will include: counting of capsules, use of diaries/journals, electronic monitoring and potentially through analysis of markers/metabolites present in various biological fluids.

Increased education and simplified dosing schedules for participants, in conjunction with follow-up visits, should help to improve participant compliance. If you design your placebo, blind your study, and monitor the participant's compliance with these aspects of your study, it will help to ensure that the efficacy and safety analyses will be valid. In addition, it will maximize the potential contribution of clinical studies using botanical products to evidence-based decision-making. Critical factors were represented in Fig. 10.3.

10.8 Regulatory approval pathways

10.8.1 United States—FDA (DSHEA, botanical drug development pathway)

10.8.1.1 Categories

- Dietary supplement (under DSHEA): Marketed without prior FDA preapproval if ingredients are dietary and meet DSHEA definitions; manufacturers are

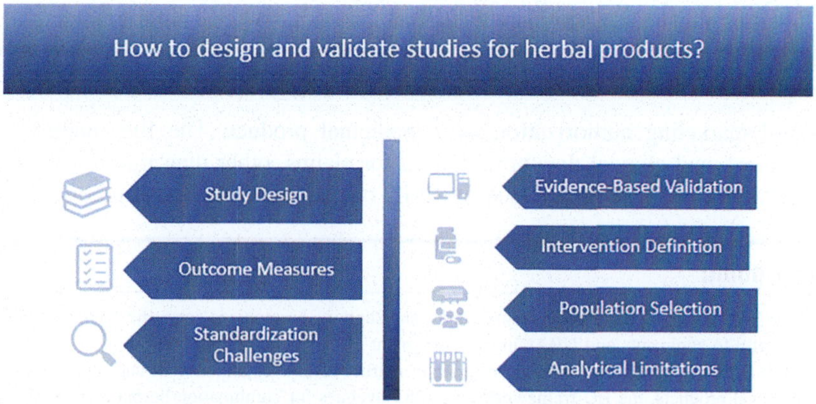

FIGURE 10.3 Critical factors influencing botanical research and validation.

Validation of botanical.

responsible for safety and truthful labeling; structure-function claims allowed with disclaimer; disease claims require a drug approved.

Botanical drug-as-drug/biologic, if it is promoted for diagnosis, cure, mitigation, treatment, or prevention of disease, has to follow FDA approval pathways for drugs, either through NDA or a biologics license. The FDA has guidance on the development of botanical drugs; complex mixtures may follow NDA 505(b)(1) or, in some cases, rely upon other pathways 505(b)(2) when some reliance on literature or reference products is appropriate.

Key point:

DSHEA permits many botanical supplements to be marketed with less premarket evidence than drugs, which affects their claims and labeling.

Sponsors would need to provide strong safety and efficacy data to support therapeutic claims, market a drug, and pursue FDA drug approval. The FDA issued specific guidance on botanical drug development covering quality, manufacturing, and clinical trials (Nordberg et al., 2022).

10.8.2 Europe—European Medicines Agency regulations for herbal medicinal products

10.8.2.1 Categories and routes

- Traditional herbal medicinal product: A simplified route is available for products with long traditional use and plausible safety/effectiveness for minor conditions, relying on bibliographic evidence and safety data without the need for full clinical trial dossiers.

- Well-established use/human medicinal product: Where bibliographic/clinical evidence exists to demonstrate efficacy/safety comparable to conventional medicinal products.
- Full marketing authorization as a medicinal product: The full quality, non-clinical, and clinical data are required for claims, other than traditional use, or for a new indication, assessed by EMA or national competent authorities.

Key point:

The HMPC of the EMA develops monographs and supports harmonized scientific opinions, which are useful to applicants and NRAs in the assessment of claims and registrations.

While particularly problematic in the case of market access for multiherbal and non-European traditional products, the EU framework does offer avenues for combination herbal medicinal products (HMPs) where appropriate documentation is supplied.

10.8.3 India—AYUSH and FSSAI guidelines

Regulatory landscape: India has a parallel framework: AYUSH prescribes the norms for the traditional medicine systems Ayurveda, Siddha, Unani, Yoga and Naturopathy, Sowa-Rigpa, Homeopathy, CDSCO, and the Drugs and Cosmetics Act for modern pharmaceuticals; FSSAI regulates foods and nutraceuticals/dietary supplements.

Key point:

Thousands of traditional formulations are recognized under AYUSH in India, with approvals and claims determined by the category, if any, the product falls into: traditional medicine, Ayurvedic proprietary medicine, nutraceutical/food, or pharmaceutical requiring oversight from CDSCO. They must go through various regulators covering manufacturing standards, quality, labeling, and claims. Reviews generally emphasize Indian compliance with GMPs and applicable monographs, where available (Rajini et al., 2024).

10.8.4 China and East Asia—traditional medicine and modern regulatory integration

China and other East Asian countries have developed regulatory systems that combine traditional medicine systems with modern drug regulations. In China, herbal medicines can be classified into TCMs, health foods, and drugs depending on their intended use. The National Medical Products Administration is responsible for registering drugs, including TCMs, under schemes that take into account classical prescriptions, modified traditional compounds, and new botanical drugs. Classical TCM prescriptions recorded in official pharmacopoeias may qualify for simplified registration procedures, while new compounds or new indications require non-clinical and clinical data.

The QC standards are very high, focusing on raw material identification, Chinese pharmacopoeia standards, and validated analytical procedures for markers and impurities. For new botanical drugs, applicants are expected to meet modern pharmaceutical development standards, including compliance with GMP and conducting clinical studies.

In Japan, Kampo medicines are controlled as prescription or over-the-counter medications and are covered by the national health insurance system. Kampo preparations are standardized and produced in accordance with pharmaceutical GMP. South Korea also controls herbal medicines through formal drug approval procedures, integrating traditional concepts with contemporary safety and quality standards.

The above East Asian models show that traditional herbal medicine can be made compatible with strict regulatory control. However, variations in pharmacopeial standards, formulation concepts, and clinical approaches might make global harmonization and global market entry of botanicals developed according to these models challenging.

10.8.5 International harmonization and global market access challenges

Large differences exist in the regulation, evidence base, and claims associated with botanicals, posing challenges to their global commercialization. The same product (i.e., a botanical dietary supplement approved as a dietary supplement in the United States) may be considered a medication in Europe or a traditional medicine in Asia. These divergent perspectives can impede a manufacturer's ability to gain global market access and properly position their product.

Harmonization is limited; however, there are efforts to improve the situation. The WHO has developed recommendations on the quality, safety, and efficacy of herbal medicines, while pharmacopeial harmonization projects are developing standards for analytical requirements across geographic regions. Unfortunately, there remains limited consistency in the acceptance of evidence from foreign clinical trials, traditional use, or quality standards. Moreover, there is heightened scrutiny of multiherbal products and nonlocal traditional medicines due to a lack of familiarity with their traditional uses or the individual herbs that comprise these products.

For companies to succeed in product development, they must adopt regulatory methodologies as early as practicable, including comprehensive quality documentation, conservative claim practices, and labeling by region of sale. Including, in clinical research protocols, the requirements of the various regulatory authorities, for example, the use of globally accepted endpoint criteria and well-characterized products, will assist in maximizing the potential utility of clinical data across multiple jurisdictions. Another significant piece of the overall strategy for navigating this complexity will be the use of regulatory intelligence, regulatory engagement, and regulatory strategies relating to postmarket compliance. While complete harmonization of all regulatory requirements is not imminent, harmonization of the requirements related to the three critical areas of quality, safety, and transparency will facilitate comprehensive integration and greater acceptance of botanicals, enabling their appropriate use in healthcare systems worldwide.

10.9 Labeling, claims, and marketing regulations

- Permitted and prohibited claims:

Claims permitted for foods/dietary supplements—for example, under the DSHEA—are typically structure-function claims describing an effect on normal structure or function, for example, "supports relaxation." These require truthful, nonmisleading statements and a disclaimer that the FDA has not evaluated the claim.

- Prohibited claims: These are the disease claims-diagnose, mitigate, cure, treat, or prevent disease-which are reserved for approved drugs and, without appropriate approvals, are prohibited on foods/supplements.
- Health claims, structure-function claims, and risk reduction claims:

Health claims are strictly regulated because they link a substance to a reduction in disease risk and often require preapproval or authorization, depending on the jurisdiction, with substantial supporting evidence. Supplements may make structure-function claims, but must have adequate substantiation and include required disclaimers (U.S.).

- Transparency, consumer protection, and misbranding:

Regulators emphasize appropriate labeling, ingredient identity, potency, and contamination control—misbranded products with incorrect ingredient listing, false claims, or adulteration face enforcement actions. Harmonization of indications can be supported by EMA monographs and national registers, which also inform consumers/providers about approved uses (Etaware et al., 2025).

10.9.1 Requirements for label content and mandatory disclosure

Regulatory compliance and consumer protection for herbal supplements depend on accurate and complete labeling of product(s). Generally, a label includes: the name of the product, the genus and species of the herb used, the part of the plant used, the dosage form, the net quantity of the product, and instructions for use. If using standardized extracts, the extract ratio, solvent system, and/or the legally permitted amount of specific marker compounds must be listed on the label.

Botanical nomenclature regarding the potential for substitute botanicals and the associated ambiguity is extremely critical to ensuring consumer safety.

All products should include the name and address of the manufacturer or supplier; batch/lot number; and expiration/best-by date to assist in tracing and/or recalling products. Other safety-related information, such as warnings, contra-indications, possible allergens, and age/pregnancy restrictions, should be included upon availability of applicable scientific data and/or regulatory requirements. United States dietary supplement labeling requires a supplement facts panel, and European Union Herbal medicinal products follow the requirements set by their template registration process.

Another key point to bear in mind is claims-related disclaimers. In the U.S., for example, structure/function claims must contain a disclaimer explaining that the FDA has not reviewed the claim and that the product should not be used to treat, diagnose, cure, or prevent any disease. If required disclaimers are omitted from labeling materials, this could lead to litigation for alleged misbranding or misleading information. An effective label meeting all applicable rules and regulations will not only reduce the likelihood of litigation but also improve consumer confidence in your claims.

10.9.2 Advertising, digital marketing, and enforcement trends

Regulatory authorities carefully examine advertising and marketing methods to make sure promotional statements correspond to a product's formal classification in the same way they do product labeling. Press releases, websites, social media posts, influencer endorsements, and testimonials can all be considered extensions of product labeling, so claims in these forms must adhere to the same claim restrictions as product labels. Disease claims made through indirect or digital means can also result in enforcement, even if the product label itself complies with regulations.

Regulatory agencies are increasingly monitoring online marketplace advertising and digital media where herbal supplements are promoted with inflated or inferred statements about their benefit to people with chronic or serious health issues, and/or based on usage practices not substantiated by relevant clinical studies. Promotional statements that imply the product will cure, treat, or help reduce the effects of a chronic or serious disease, implied statements regarding clinical effectiveness based solely on a long history of use, or references that do not correlate to relevant clinical study data, may be considered misleading or deceptive unless supported by adequate evidence. Promotional testimonials that imply a health outcome from using an herbal supplement, without any error disclaimers and/or without other legally required balanced information, are often the most problematic.

Warning letters, product seizure, fines, mandatory recall, and public notice are examples of enforcement mechanisms. The FDA and FTC share jurisdiction over advertising in the United States. The FTC regulates deceptive advertising and ensures that marketing claims are substantiated, while the FDA regulates the products themselves. Each member country of the EU has a competent authority responsible for ensuring regulatory compliance by advertisers. There are also self-regulatory bodies that help to regulate the advertising industry.

To reduce regulatory risk, advertisers and companies should establish a system for reviewing all marketing materials; establish a review system that ensures approved claims are used in compliance; and implement a training program to educate employees on regulatory boundaries. Maintaining market access and consumer confidence in the advertised products depends on proactive compliance with applicable federal advertising and digital communication regulations.

10.10 Future trends in botanical regulation

The use of digital traceability and blockchain in the supply chain:

Regulatory and industry stakeholders seek reliable, collaborative information-sharing frameworks to improve the efficiency of reviews and ensure the continuous supply of food. Digital traceability and blockchain have been identified as enablers of physical provenance and quality assurance throughout the different countries/regions that make up the international supply chain (Barceló et al., 2023).

10.10.1 The use of AI systems for quality control and smart farming processes

Reviews have documented the potential for AI and ML to increase in the fields of analytics, QC, and sustainability in food sourcing. For example, these technologies would improve product extraction and standardization, and enhance the detection of adulteration in botanical mixtures, thereby allowing regulators and companies to ensure the consistent high quality of complex botanical mixtures (Girman et al., 2019).

10.10.2 Harmonization of global regulations

The majority of regulatory authorities (e.g., EMA and WHO) are looking to share assessments with other NRAs through reliance efforts, such as pilot projects or WHO-led reliance initiatives, to save time and eliminate redundancy when making postmarketing approval changes, thereby expediting global access to plant-based medicines (Motrinchuk and Belousov, 2023).

10.10.3 Evolving evidence frameworks and real-world data integration

In the future, there will be a shift in how regulations for botanical products are developed. There will likely be greater focus on flexible evidence frameworks that incorporate RWE alongside traditional forms of evidence, such as preclinical and clinical trials. Many botanicals have been used for many years, and, as such, it is often not feasible to perform large randomized trials of all botanical products. As a result, regulators have become increasingly interested in how to utilize multiple sources of RWD (observational studies, electronic health records, pharmacovigilance databases, and patient-reported outcomes) to assess the safety and effectiveness of botanical products.

RWE could also be very useful in identifying rare adverse events associated with botanical products, determining long-term patterns of use, and assessing effectiveness among populations that are often underrepresented in clinical trials. The use of RWD, in combination with a robust system of quality and traceability, can help to establish links between the safety of specific products, lots, and/or formulations—this is particularly important when evaluating complex botanical mixtures.

The use of RWE also raises several methodological challenges, including confounding, quality of data used as a basis for understanding outcomes attributed to the use of herbal products when used in conjunction with other medicines (polypharmacy), as well as an expectation that regulatory agencies will provide additional clarity regarding acceptable study designs, data sources, and analytical approaches related to RWE used in support of botanicals. The development of standardized report frameworks and enhanced interoperability among national adverse event databases will also help regulatory agencies build confidence in the use of RWE.

As evidence models develop over time, RWE can be expected to complement, but not replace, controlled clinical trials in determining the ultimate acceptance of botanical products as safe and effective therapeutic entities across their lifetimes (i.e., usefulness and effectiveness), thereby facilitating adaptive regulation and providing timelier risk-management decisions.

10.10.4 Sustainability, biodiversity protection, and ethical sourcing

Sustainability and biodiversity preservation are expected to be key components of future regulations for botanical products. As demand for herbal medicines and supplements grows worldwide, consumers will drive a greater need for wild plants; this will create pressure on wild plant populations, along with increased concerns about harvesting, habitat loss, and the stability of supply. Without a doubt, an increasing number of organizations—regulatory, international, and industry—are realizing that product safety and quality cannot be sustained without implementing responsible harvesting and sourcing practices.

New regulations will require companies to demonstrate sustainable collection/sourcing, cultivation, and fair-trade practices. To this end, reconciling with existing agreements such as CITES with respect to endangered species will likely continue; developing and using traceability systems to ensure legal/ethical collection and sourcing will be required; and encouraging growing cultivated alternatives to wild-harvested materials will be promoted. In addition, the expansion of GACP is likely to require specific consideration regarding environmental and social impact.

Another important aspect of ethical sourcing is the protection of Indigenous knowledge regarding the use of medicinal plants and equitable benefit-sharing with communities that have traditionally used those plants. Sourcing transparency and respect for access and benefit-sharing agreements are gaining regulatory and societal awareness through international biodiversity frameworks. Integrating sustainability metrics into regulatory decisions will enable authorities to help ensure the continuity of supply, protect ecosystems, and enhance public confidence in botanical products. These trends indicate a movement toward a more comprehensive regulatory model that balances innovation, public health, and environmental stewardship.

10.11 Conclusion

In other words, all parts of the whole supply chain, at all levels—which starts with a farm and ends up at a table—must operate under a set of rules that are metered about the food they process. These rules become the basis for assurance of safety in the supply chain, traceability of food products, and properly marked labels. The role of regulators in promoting safety for the general public through an open process of providing consumers with public assurance is through transparency. Each regulator provides an environment for safe and reliable food product to consumers by having stringent requirements for farmers/processors; including compliance with agricultural practices; food processing; accurate labeling and proper packaging and delivery; therefore providing consumers with an environment that is safe/secure and minimizing the risk (to some extent) of food product not being safe; innovative; or uncaring about sustainability or fair-trade practices. The success of providing a strong regulatory framework will produce a high-quality food production system that ultimately supports the overall trustworthiness of the food supply chain from production to retail.

References

Barceló E, Dimić-Mišić K, Imani M, Spasojević Brkić V, Hummel M, Gane P. Regulatory paradigm and challenge for blockchain integration of decentralized systems: example—renewable energy grids. Sustainability 2023;15(3):2571. Available from https://doi.org/10.3390/su15032571.

Bohra P. Study on the role of vati kalpana (tablet formulation) in enhancing the shelf life of herbal drugs. Indian Journal of Applied Research 2025:1–2. Available from https://doi.org/10.36106/ijar/9000123.

Van Breemen RB, Fong HHS, Farnsworth NR. Ensuring the safety of botanical dietary supplements. American Journal of Clinical Nutrition 2008;87(2). Available from https://doi.org/10.1093/ajcn/87.2.509s, http://www.ajcn.org/cgi/reprint/87/2/509S.

Brendler T, Brinckmann JA, Daoust M, He H, Masé G, Steffan K, et al. Suitability of botanical extracts as components of complex mixtures used in herbal tea infusions—challenges and opportunities. Frontiers in Pharmacology 2022;13. Available from https://doi.org/10.3389/fphar.2022.1013340.

Dou J, Weathers P. Specialty molecules from plants and in vitro cultures as new drugs: regulatory considerations from flask to patient. Plant Cell, Tissue and Organ Culture (PCTOC 2022;149(1-2):105–11. Available from https://doi.org/10.1007/s11240-022-02287-4.

Etaware P, America O, Egara O, Ekun V. Herbal medicine: scientific validation and future prospects. International Journal of Pharmacy and Chemistry 2025;11(3):67–75. Available from https://doi.org/10.11648/j.ijpc.20251103.12.

Girman CJ, Ritchey ME, Zhou W, Dreyer NA. Considerations in characterizing real-world data relevance and quality for regulatory purposes: a commentary. John Wiley and Sons Ltd, United States Pharmacoepidemiology and Drug Safety 2019;28(4):439–42. Available from https://doi.org/10.1002/pds.4697, http://onlinelibrary.wiley.com/journal/10.1002/(ISSN)1099-1557.

Hlatshwayo S, Thembane N, Krishna SBN, Gqaleni N, Ngcobo M. Extraction and processing of bioactive phytoconstituents from widely used South African medicinal plants for the preparation of effective traditional herbal medicine products: a narrative review. Plants 2025;14(2). Available from https://doi.org/10.3390/plants14020206, www.mdpi.com/journal/plants.

Huxtable RJ. The harmful potential of herbal and other plant products. Drug Safety 1990;5(Supplement 1):126–36. Available from https://doi.org/10.2165/00002018-199000051-00020.

Kaggwa B, Kyeyune H, Munanura EI, Anywar G, Lutoti S, Aber J, et al. Safety and efficacy of medicinal plants used to manufacture herbal products with regulatory approval in Uganda: a cross-sectional study. Evidence-based Complementary and Alternative Medicine 2022;2022. Available from https://doi.org/10.1155/2022/1304839, http://www.hindawi.com/journals/ecam/contents.html.

Kim N-C. Need for pharmacopeial quality standards for botanical dietary supplements and herbal medicines. Food Supplements and Biomaterials for Health 2021;1(1). Available from https://doi.org/10.52361/fsbh.2021.1.e10.

Koonrungsesomboon N, Sakuludomkan C, Na Takuathung M, Klinjan P, Sawong S, Perera PK. Study design of herbal medicine clinical trials: a descriptive analysis of published studies investigating the effects of herbal medicinal products on human participants. BMC Complementary Medicine and Therapies 2024;24(1). Available from https://doi.org/10.1186/s12906-024-04697-7, https://bmccomplementmedtherapies.biomedcentral.com/.

Kumar G, Kumar R, Kumar A, Kumar A, Bharti G. Role of chemical marker in quality control of herbal drugs. Journal of Biomedical and Pharmaceutical Research 2025;14(4):1–14. Available from https://doi.org/10.32553/jbpr.v14i4.1355.

Li WQ, Wertheimer A. Narrative review: The FDA's perfunctory approach of dietary supplement regulations giving rise to copious reports of adverse events. INNOVATIONS in pharmacy 2023;14(1):7. Available from https://doi.org/10.24926/iip.v14i1.4989.

Mihai M, Ciont C, Pop OL, Dumitras DE, Mihai VC, Morariu ID, et al. European consumer and regulatory trends in medicinal plant food supplements and their functional properties: the road from farm to fork. Multidisciplinary Digital Publishing Institute (MDPI), Romania Applied Sciences (Switzerland) 2025;15(21). Available from https://doi.org/10.3390/app152111605 Available from: https://www.mdpi.com/journal/applsci/.

Motrinchuk AS, Belousov DY. Considerations for the design and conduct of externally controlled trials for drug and biological products. Real-World Data & Evidence 2023;3(4):29–40. Available from https://doi.org/10.37489/2782-3784-myrwd-45.

Nordberg RC, Otarola GA, Wang D, Hu JC, Athanasiou KA. Navigating regulatory pathways for translation of biologic cartilage repair products. Science Translational Medicine 2022;14(659). Available from https://doi.org/10.1126/scitranslmed.abp8163, https://www.science.org/doi/10.1126/scitranslmed.abp8163.

Poveda OMP R. Regulation of herbal (Traditional) medicinal products in the European Union. Pharmaceutical Regulatory Affairs: Open Access 2015;04(02). Available from https://doi.org/10.4172/2167-7689.1000142.

Rajini A, Ranjitha N S, Audi Narayana N, Jothieswari D. Unravelling framework for herbal product manufacturing and approval in south India. Future Journal of Pharmaceuticals and Health Sciences 2024;4(1):46–50. Available from https://doi.org/10.26452/fjphs.v4i1.552.

Ram KH, Kothari LP, Mundada AS. Overview of good manufacturing practices requirements for herbal medicines in India and Europe. International Journal of Drug Regulatory Affairs 2023;11(4):52–66. Available from https://doi.org/10.22270/ijdra.v11i4.632.

Sahoo MR, Umashankara MS. FTIR based metabolomics profiling and fingerprinting of some medicinal plants: an attempt to develop an approach for quality control and standardization of herbal materials. Pharmacognosy Research 2022;15(1):163–7. Available from https://doi.org/10.5530/097484900288.

Wang H, Chen Y, Wang L, Liu Q, Yang S, Wang C. Advancing herbal medicine: enhancing product quality and safety through robust quality control practices. Frontiers in Pharmacology 2023;14. Available from https://doi.org/10.3389/fphar.2023.1265178.

Wu H-Y, Shaw P-C. Strategies for molecular authentication of herbal products: from experimental design to data analysis. Chinese Medicine 2022;17(1). Available from https://doi.org/10.1186/s13020-022-00590-y.

PART

5

Future directions and conclusion

CHAPTER

11 The future of healing botanicals: emerging trends and technologies

CH.K.V.L.S.N. Anjana Male[1], S.T.V. Raghavamma[2], Kanaka Durga Devi Nelluri[3], Vidya Perli[3] and Bantupalli Suranjan[4]

[1]*School of Pharmacy, ITM University, Gwalior, Madhya Pradesh, India,*
[2]*Department of Pharmaceutics, Chalapathi Institute of Pharmaceutical Sciences, Guntur, Andhra Pradesh, India,*
[3]*Department of Pharmaceutics and Biotechnology, KVSR Siddhartha College of Pharmaceutical Sciences, Vijayawada, Andhra Pradesh, India,*
[4]*University of Michigan, Ann Arbour, MI, United States*

11.1 Introduction to healing botanicals

11.1.1 Definition and scope

Healing botanicals are plants and plant-based materials used to prevent and treat various diseases. Healing botanicals are made from a wide range of plant materials, including leaves, roots, rhizomes, bark, flowers, seeds, and fruits. These plant materials contain phytochemicals that confer the medicinal properties of the healing botanicals. Some of the major phytochemicals responsible for the pharmacological properties of healing botanicals include alkaloids, flavonoids, terpenoids, phenolics, and glycosides.

The scope of healing botanicals is vast, encompassing applications across various domains of medicine, including wound healing, pain management, inflammation, infection, metabolic disorders, and immunomodulation. Healing botanicals are available in various forms, including crude extracts, powdered extracts, teas, tinctures, and standardized extracts. With the advent of phytochemistry, biotechnology, and drug delivery systems, the scope of healing botanicals has increased from traditional medicine to modern medicine. This shows the importance of healing botanicals in modern medicine and traditional medicine (El-Sherbeni and Negm, 2023).

11.1.2 Importance of healing botanicals in modern healthcare

The importance of healing botanicals in modern medicine has been elevated amid growing concern about their limitations. Issues of resistance to antibiotics, side effects of medications, the cost of treatment, and the long-term consequences of chemical medications have elevated the levels of interest in alternative medicine.

Healing Botanicals. DOI: https://doi.org/10.1016/B978-0-443-44295-7.00003-3

Moreover, the healing botanicals are important in complementary and alternative medicine, where they are used in combination with conventional medicine to achieve better outcomes. Many scientific studies have proven the effectiveness of botanicals in the healing of wounds, reducing inflammation, fighting infections, and generally maintaining health (Barman et al., 2023). As traditional claims are supported by modern science, healing botanicals are increasingly accepted into mainstream healthcare systems worldwide (Colantonio, Rivers, 2017b).

11.1.3 Comparison with conventional pharmaceutical medicine

Generally, conventional pharmaceutical medicine uses synthetic drugs manufactured to target specific biological pathways or disease conditions. Despite being highly efficient, conventional medicine using synthetic drugs, which follow a single active ingredient path, might limit its effectiveness, leading to side effects. Healing botanicals, however, have a complex composition of phytochemicals that enables them to exert multiple effects simultaneously.

The medicinal herb has a comprehensive range of properties, including antiinflammatory, antioxidant, antimicrobial, and immunomodulatory activities, resulting from the synergistic effects of various bioactive compounds. This has enabled many healing plants to address complex, multifaceted health challenges more holistically. The healing herbs offer a marked advantage over other chemical-based medications, which, in many instances, require polypharmacy to address all the health challenges at hand effectively (Colantonio, Rivers, 2017a). This therefore provides a clear account of the unique advantage of the healing herbs compared to other medications, including their effects, especially upon prolonged use (Sun et al., 2024).

11.1.4 Emerging trends in healing botanicals

The future of healing botanicals is being driven by emerging trends resulting from research advances, consumer demand, and healthcare needs. The emerging trend in the future of healing botanicals is their increasing inclusion in evidence-based medicine. There have been developments in the analysis of phytochemicals and in research that have enabled better identification of the active principles of medicinal plants.

The next significant emerging trend in the market is the rising demand for personalized and preventive healthcare solutions. Healing botanicals are being researched for their potential in providing solutions for preventive medicine, wellness, and chronic disease management. There is significant consumer demand for plant-based, sustainable, and responsible medicinal products, thereby driving greater innovation in the botanical market. This factor is also contributing to innovation in the botanicals domain. In addition, there is a significant revival of interest in traditional medicine, also known as indigenous medicine, which uses modern scientific techniques.

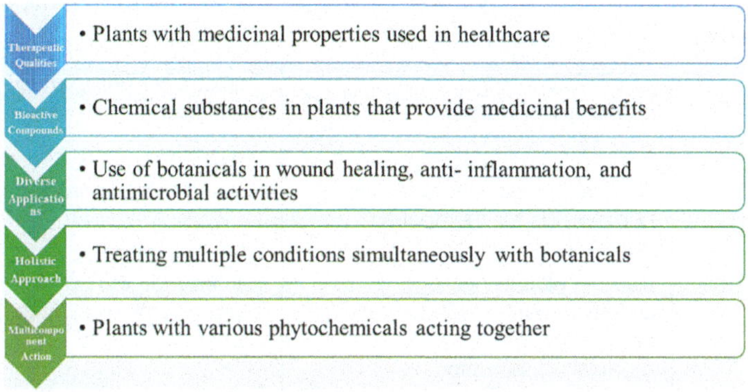

FIGURE 11.1 Significance of healing botanicals.

Botanicals importance.

11.1.5 Technological innovations shaping the future of healing botanicals

Significantly, technological innovations have improved the efficacy, safety, and accessibility of healing botanicals. For example, advanced technologies such as supercritical fluid extraction (SCFE), nanotechnology, and green extraction have greatly improved the bioavailability of these healing botanicals. The use of nanotechnology has improved the delivery of botanical compounds, which were previously limited by solubility and absorption issues.

In addition, advances in biotechnology, including tissue culture, genetic engineering, and metabolomics, are enabling the mass production and standardization of medicinal plants. Artificial intelligence (AI) and machine learning are also being employed to predict interactions between compounds and plants and to accelerate drug development from plant sources (Fig. 11.1). In conclusion, the aforementioned technologies are helping transform healing botanicals into an effective, scientifically validated therapeutic modality, making them relevant in the coming years. Table 11.1 outlines the basics of healing botanicals, their uses, and how they differ from conventional medicine.

11.2 Historical and traditional perspectives on healing botanicals

11.2.1 History and development of plant-based medicine

The use of plant substances as part of a therapeutic (disease-centered) system is one of the earliest therapeutic methods in human culture. Ancient civilizations had a

Table 11.1 Introduction to healing botanicals: definitions, significance, and comparisons.

Aspect	Healing botanicals	Traditional (allopathic) medicine
Definition and scope	Derived from plants (leaves, roots, flowers, seeds) with bioactive compounds for therapeutic purposes	Relies on synthetic drugs and isolated active ingredients
Applications	Used in wound healing, antiinflammatory, antimicrobial, antioxidant, and immune-modulating therapies	Primarily targets specific diseases or symptoms with single-purpose drugs
Healthcare integration	Integrated into both traditional and modern practices, rising popularity due to fewer side effects and resistance	Standard practice in modern clinical settings, especially in acute and emergency care
Mechanism of action	Multicomponent and holistic; acts on multiple physiological pathways simultaneously	Single-target mechanisms; focused therapeutic effects
Side effects & safety	Generally, fewer side effects due to natural composition (though not always risk-free)	Potential for significant side effects, drug interactions, and long-term toxicity
Resistance and sustainability	Less prone to resistance development; often sustainable when ethically sourced	Increasing antibiotic resistance; reliance on synthetic production can be ecologically taxing
Therapeutic philosophy	Holistic and preventive; supports the body's healing mechanisms	Disease-targeted and reactive; often suppresses symptoms rather than addressing root causes

history of using the healing (therapeutic) properties of plants (e.g., Egyptians, Chinese, Indians, Mesopotamians, and Greeks).

Documents are available that depict the use of plants as medicines in ancient Egyptian papyri, Chinese medical texts, and Indian religious texts. Ayurveda, traditional Chinese medicine (TCM), Unani, and Siddha systems of medicine have all incorporated medicinal plants into their practices and established the foundational components of ethnomedicine.

These systems used a whole-person approach, focusing on the body's functioning to promote health rather than on a single area of dysfunction (symptom-based treatment). Over time, practitioners of traditional botanical medicine gained understanding through practical experience, leading to improvements in plant selection, processing, and dosing. Modern science has enabled the pharmacological and clinical evaluation of many traditional botanical medicines and increased the ability to combine traditional knowledge with modern science to support their use today (Momot, 2024).

11.2.2 Global ethnomedicinal practices

Ethnomedicinal practices are quite diverse from one region of the world to another. This is due to the diversity of plant species in different regions. Indigenous people have long relied on plants in their regions to treat diseases. This is not only limited to the choice of plants but also involves the preparation methods and administration of these plants (Kustiyati et al., 2024).

In Africa, South America, Southeast Asia, and Oceania, ethnomedicine is the primary source of healthcare for many people. Diversity in ethnomedicinal practices indicates the vast potential of biodiversity present in the plant kingdom and the need to preserve the knowledge. The documentation and scientific validation of ethnomedicinal practices can help discover new drugs and develop new therapeutic agents. The incorporation of ethnomedicinal knowledge into existing healthcare systems can help develop appropriate, accessible systems (Podder and Anand, 2022).

11.2.3 Cultural significance and indigenous knowledge

Medicinal plants are of immense cultural significance, as they are not limited to being a medicine. In many traditional societies, medicinal plants are inextricably linked to the culture and traditions of those people. In many traditional and indigenous cultures, medicinal plants are intricately linked to the culture, spiritual values, and traditions of those people (Ding, 2024). These plants are also used for rituals and ceremonies, as well as traditional medicine, which shows a holistic approach to medicine.

Indigenous knowledge systems are a rich source of information regarding the sustainable harvesting and use of natural resources. Indigenous knowledge systems also promote a sense of respect for nature and the use of plant resources in ways that sustain these resources for future use. It is very important to preserve indigenous knowledge systems not only to safeguard cultural heritage but also to promote biodiversity and sustainable health practices. There is a tremendous potential to promote the development of effective, ethical, and environmentally responsible botanical medicines through the use of indigenous knowledge systems.

11.2.4 Transition from traditional knowledge to modern applications

The knowledge of healing botanicals has shifted from traditional knowledge to a more documented and even scientifically analyzed approach. Modern pharmaceuticals are derived from traditional botanicals, underscoring the importance of ethnomedicine in the discovery of new medicines. Traditional knowledge has been a major source of information for identifying medicinal plants and their uses.

This has provided an opportunity to verify the botanical knowledge of the past through pharmacological and clinical research, leading to improvements in safety, standardized dosing, and efficacy. While traditional healers relied on empirical knowledge, modern science has isolated the principles of action of healing botanicals. The union of traditional knowledge and modern science has given healing botanicals greater validity and relevance in modern healthcare systems.

11.2.5 Preservation and challenges of traditional botanical knowledge

The history of healing botanicals has been rich, but today, traditional knowledge faces many challenges in the modern world. Urbanization, biodiversity loss, and the decline of indigenous populations are among the major reasons for the loss of ethnobotanical knowledge. The majority of this knowledge is still undocumented and is at high risk of being lost as the older generation of traditional knowledge holders dies out.

In addition, the ethical and legal issues related to intellectual property rights, biopiracy, and the failure to recognize the contribution of indigenous knowledge are among the challenges. To conserve indigenous botanical knowledge, it is necessary to document it and formulate policies to share its benefits. The inclusion of traditional knowledge in modern medicine, with a focus on cultural values and sustainability, is essential to preserving the historical heritage of healing botanicals for posterity.

11.2.6 Movement from traditional medicine to a science-based system

Developing a scientific basis for the use of herbal therapies represents one of the significant advances in medicine today, with the transition of herbal medicine into pharmacological and toxicological sciences validating and allowing the establishment of standards of care and protocols for creating and utilizing herbal products. Establishing a scientific basis supports the long-held belief that herbal remedies are a valid form of medicine and are included in the development of medicine. In addition to demonstrating the historical roots of herbal therapy. Table 11.2 illustrates the diversity of cultures that practice herbal medicine and how traditional herbal knowledge has been preserved within them.

11.3 Phytochemistry and bioactive compounds

11.3.1 Classification of phytoconstituents

Phytoconstituents, which are naturally occurring chemical compounds found in plants, are believed to be the reason for the medicinal properties of these plants. These compounds have been classified into several major groups, including alkaloids, flavonoids, terpenoids, phenolic compounds, glycosides, saponins, and tannins. Terpenoids have many medicinal properties, including antimicrobial, antiinflammatory, and anticancer effects, whereas phenolic compounds have antioxidant and antimicrobial effects. Since a single plant contains multiple phytoconstituents, these compounds can exhibit synergistic effects, making them more effective in treating various health complications.

Table 11.2 Historical and traditional perspectives on healing botanicals

Aspect	Details
2.1 History and development	• Healing botanicals date back to ancient civilizations such as Egypt, China, and India. • Used extensively in Ayurveda and traditional Chinese medicine (TCM). • Ethnomedicine developed over centuries through empirical use. • Modern science is increasingly validating traditional plant-based remedies.
2.2 Global ethnomedicinal practices	• Practices vary globally, reflecting cultural diversity and local flora. • Indigenous communities rely on endemic plants and passed-down knowledge. • Emphasizes the need to document and conserve regional ethnobotanical traditions. • Integration of traditional practices enhances global healthcare innovation.
2.3 Cultural significance and indigenous knowledge	• Healing botanicals hold spiritual and cultural meaning in many communities. • Seen as a part of heritage and identity, not just therapeutic agents. • Indigenous knowledge promotes sustainability and respectful use. • Valuable for guiding conservation and ethical harvesting practices.

Each class of phytoconstituents has its own chemical properties and biological activities that render the plant medicinally important. Alkaloids are nitrogen-containing compounds that are highly valued for their outstanding pharmacological properties. Flavonoids are highly valued for their antioxidant, antiinflammatory, and cardioprotective properties, which greatly contribute to reducing oxidative stress (Vickers and Zollman, 1999).

11.3.2 How they work

The health effects of phytochemicals result from multiple biochemical mechanisms; this means their effects may not be the same for everyone or at all times. Some actions include changing how cells communicate with each other (cell signaling), altering how certain enzymes function, interacting with receptors on the outside of cells (membrane-bound receptors) or inside cells (intracellular receptors), and altering gene expression. As phytochemicals affect many different molecular targets, they can potentially affect many complex diseases or disorders (i.e., inflammation, oxidative stress, bacterial and/or viral infections).

Examples include flavonoids, which can reduce the production of inflammatory substances by blocking cyclooxygenase and lipoxygenase enzymes, and alkaloids, which can bind to receptors in the brain and thereby relieve pain. Many phytochemicals also provide some degree of immune stimulation and/or induce abnormal cell death (apoptosis); therefore they have cancer-fighting properties. Understanding all the molecular mechanisms of action will be important for developing and providing effective botanical medicines within the broader context of evidence-based practice (Grossman, 2021).

11.3.3 Techniques of analysis used in phytochemical

The advances in analytical techniques over the past few years have greatly advanced the field of phytochemistry. High-performance liquid chromatography is useful for quantifying phytoconstituents in plant material. In contrast, gas chromatography-mass spectrometry has been found very useful for the analysis of volatile and semivolatile compounds.

On the other hand, nuclear magnetic resonance spectroscopy is used to obtain precise structural information about phytochemicals. In addition, the use of new techniques, such as liquid chromatography-mass spectrometry and metabolomics, has been reported to provide an exhaustive analysis of herbal metabolites (Tao et al., 2021). Analytical techniques are instrumental in ensuring the quality, safety, and reproducibility of herbal products, advocating the standardization of herbal products and the scientific authentication of herbal medicines (Pearn, 2005).

11.3.4 Synergistic effects of bioactive compounds

Understanding the synergistic effects of bioactive compounds is a defining characteristic of medicinal plants. While pharmaceutical medicines are created primarily by combining a single active ingredient, medicinal plants contain complex mixtures of bioactive compounds, known as phytochemicals, which interact with one another to create a much stronger and more balanced source of bioactive compounds than would be expected based on their individual activities.

An example would be the ability of flavonoids to increase the solubility and stability of alkaloids, and the effects of phenolic compounds to provide additional antioxidant effects that work in conjunction with both antiinflammatory and antibacterial effects. To develop a successful botanical medicine, it is important to understand how these phytochemicals interact with one another and how they are assimilated in traditional medicine practice. Continued research into phytochemical synergy is helping bridge the gap between traditional knowledge and modern scientific validation.

11.3.5 Role of phytochemicals in drug discovery and development

There is significant importance of phytochemicals (i.e., bioactive molecules from plants) in today's drug discovery and development processes. They can serve as lead compounds or templates for the design of new drugs. The vast majority of currently marketed pharmaceuticals (medications) are derived (in part or whole) from plant-based bioactive molecules. Therefore botanicals provide a massive amount of chemical diversity for pharmacological use. Phytochemists can now isolate, modify, and optimize phytochemical compounds to improve their efficacy and safety thanks to advancements in the field.

Furthermore, modern drug discovery techniques, such as computer modeling, molecular docking, and high-throughput screening, have enabled researchers to screen potential phytochemicals for medicinal applications and to predict their biological activities and development timelines. Thus phytochemicals continue to play important roles in the discovery and development of new and innovative pharmaceuticals. This reinforces the continued importance of medicinal plants for future pharmaceutical research.

11.3.6 Importance of phytochemical synergy and complexity

One key attribute of healing plants is the presence of multiple bioactive compounds that work synergistically to promote healing. Unlike conventional synthetic drugs that target a single molecule, botanical drugs can target multiple molecules owing to the complexity of the phytochemicals they contain. The importance of understanding the complexity of phytochemical synergy cannot be overstated, as it is critical to the development of botanical drugs and their holistic healing properties. Further research in this area is expected to improve the predictability of healing botanicals. Table 11.3 summarizes the main types of plant compounds, their actions, and the methods used to analyze them.

Table 11.3 Phytochemistry and bioactive compounds of healing botanicals.

Aspect	Details
3.1 Phytoconstituent classification	- Phytochemicals are classified into: • Alkaloids – analgesic, neuroactive properties • Flavonoids – antioxidant, antiinflammatory Terpenoids – antimicrobial, antitumor • Phenolic compounds – antioxidant, cardioprotective - Each class contributes uniquely to healing mechanisms.
3.2 Modes of action	- Phytochemicals act via: • Modulation of signaling pathways (e.g., inflammation, apoptosis) • Enzyme inhibition (e.g., cyclooxygenase, acetylcholinesterase) • Interaction with cellular receptors (e.g., hormone receptors) - Understanding action mechanisms supports drug discovery and formulation.
3.3 Analytical techniques	- Key methods include: • High-performance liquid chromatography – separation and quantification of compounds • Gas chromatography-mass spectrometry – volatile compound analysis • Nuclear magnetic resonance – structure elucidation - These techniques ensure quality control, standardization, and safety of herbal products.

11.4 Technological advancements in botanical science

11.4.1 Plant biotechnology and genetic engineering

Plant biotechnology has had a major impact on the research, growth, and use of medicinal plants by facilitating the development of modern tools that enhance their therapeutic value and productivity (Brody et al., 2009). The use of genetic engineering techniques has enabled the isolation of genes involved in the synthesis of certain phytochemicals, thereby improving the quality and quantity of these compounds. This technology has enabled the production of highly productive plants that are resistant to pests and environmental factors and have a standardized phytochemical composition.

Genetic engineering technology has enabled the standardization of medicinal plants, which is one of the major problems associated with botanical drugs—their variation in active principles. This technology is useful for developing plants that are genetically optimized for the production of medicinal compounds, a major challenge for medicinal plants—their reliance on natural resources. Therefore plant biotechnology is useful for developing medicinal plants in ways that are not harmful to the environment and that support sustainable production of medicinal compounds (Shahpesandy et al., 2022).

11.4.2 CRISPR and synthetic biology in medicinal plants

There has been a major shift in medicinal plant research with the advent of CRISPR-Cas gene-editing technology. This technology can make precise, efficient changes to plant genetic makeup. This technology differs from other genetic engineering technologies because it does not require inserting foreign genetic material into plants. This technology is very appealing for the use of medicinal plants. This technology can be used to improve the biosynthetic pathways of desired phytochemicals, enhance stress tolerance, and eliminate undesirable traits.

Synthetic biology is an advancement of this technology, which can be used to design or construct new pathways for the production of medicinal compounds. This technology can be used to produce bioactive phytochemicals in a controlled environment, ensuring the availability of the compound without depleting natural plant resources (Mishra and Srivastava, 2024).

11.4.3 Plant tissue culture and clonal propagation

The tissue culture technology has immense applications in the large-scale propagation of medicinal plants. It has been found that micropropagation, callus culture, and somatic embryogenesis are highly effective methods of tissue culture that can be used to produce quality plant material on a large scale. The importance of tissue culture technology lies in its contribution to the large-scale propagation of medicinal plants required for research, industrial processing, and pharmaceutical manufacturing. It is also important for the conservation of endangered and slow-growing medicinal plants.

Tissue culture technology helps conserve valuable germplasm without further exploiting natural habitats. Tissue culture technology is also important for the production of secondary metabolites by cell and suspension culture. It is an important tool to produce medicines.

11.4.4 Advances in technology: extraction/formulation and delivery of bioactive compounds from healing plants

Many technological innovations have resulted in more efficient methods for extracting, formulating, and delivering bioactive compounds from healing plants. Examples of advanced extraction technologies include SCFE, microwave-assisted extraction (MAE), and ultrasound-assisted extraction (UAE). Compared with conventional extraction methods, SCFE, MAE, and UAE provide higher yield, purity, and retention of heat-sensitive phytochemicals. They have improved the efficiency of botanical processing while reducing the volume of solvents required, thereby making it more environmentally friendly and sustainable (Goktas et al., 2020).

Furthermore, advances in nanotechnology have produced new delivery systems that enhance the solubility, bioavailability, and stability of phytochemicals. Several new delivery systems (e.g., nanoemulsions and liposome delivery) created using nanotechnology are available for delivering botanicals into the body, thereby increasing their therapeutic effectiveness and improving patient compliance with prescribed treatments.

11.4.5 Digital technologies, quality control, and standardisation

Digital technologies have fundamentally changed how we view quality control and standardization in scientific research that uses plants as samples for analysis. Examples of these techniques include metabolomics, chemometrics, and bioinformatics—each of which is helping to develop comprehensive profiles (or "fingerprints") of plant metabolites to support the consistent and authentic production of botanicals. All of these techniques are essential for developing quality, safe, and standardized products derived from herbs.

In addition to the aforementioned three technologies, researchers are also utilizing AI and machine learning to predict phytochemical profiles, optimize environmental conditions for plant growth, and determine therapeutic uses for botanicals. Using digital technologies has greatly reduced the time spent conducting research and improved the quality of decision-making, leading to the development of botanical drugs for use in today's healthcare delivery system and facilitating the successful commercialization of botanicals for healing purposes.

11.4.6 Sustainable biotechnological innovations in medicinal plant production

Biotechnological innovations have the potential to offer sustainable solutions for the production of medicinal plants and their phytochemicals. The production of phytochemicals through biotechnological innovations such as metabolic engineering, in vitro culture, and bioreactor production will reduce reliance on wild plants. This will, in turn, reduce the problems associated with plant overexploitation, thereby addressing biodiversity loss (Fig. 11.2). In addition, biotechnology offers sustainable solutions for the production of high-quality plant materials, thereby ensuring therapeutic quality. Sustainable biotechnological innovations are expected to make a vital contribution to the increasing demand for botanical medicines. Table 11.4 highlights modern biotechnological advances that improve the quality and production of medicinal plants.

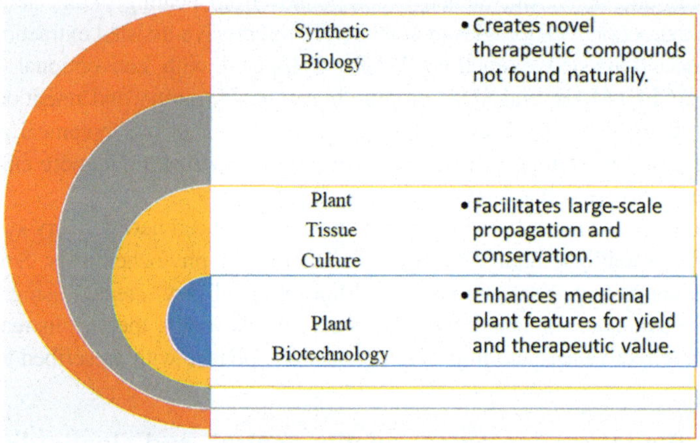

FIGURE 11.2 Advancements in botanical science.

Updates in the botanicals.

Table 11.4 Technological advancements in botanical science.

Aspect	Details
4.1 Plant biotechnology and genetic engineering	- Enhances yield, potency, and consistency of medicinal plants. - Genetic modification allows for amplification of desirable traits (e.g., higher alkaloid content). - Supports the development of sustainable and efficient sources of phytochemicals.
4.2 CRISPR and synthetic biology	- CRISPR-Cas9 enables precise genetic editing to boost targeted phytochemical production. - Synthetic biology can design and build novel bioactive compounds not found in nature. - Expands the scope of therapeutic applications beyond natural limits.
4.3 Plant tissue culture and clonalpropagation	- Allows mass propagation of medicinal plants under controlled conditions. - Ensures uniform quality and purity of botanical materials. - Aids in the conservation of endangered medicinal plant species and the preservation of genetic traits.

11.5 Innovative drug delivery systems

11.5.1 Nanotechnology and targeted deliverys

Nanotechnology has been recognized as a revolutionary tool in drug delivery, providing immense benefits for the delivery of medicinal botanicals. Plant-derived compounds have been found to have limitations, such as low solubility, permeability, and stability, which reduce their efficacy. Nanoparticle-based delivery systems have been found to overcome these problems by increasing the stability and bioavailability of the botanicals.

The targeted delivery of nanocarriers can be engineered to accumulate in target tissues or cells via passive or active targeting strategies, thereby reducing the risk of systemic toxicity. Liposomes, polymeric nanoparticles, solid lipid nanoparticles, and nanoemulsions are some of the most used nanoparticles in carrying out botanical substances. They play a crucial role in delivering phytotherapeutic agents across biological membranes, including the gastrointestinal tract and the blood–brain barrier. This increases their efficacy, leading to better outcomes, for which nanotechnology seems to be a promising tool for future botanical drug therapy.

11.5.2 Encapsulation and bioavailability enhancement

Generally, the encapsulation technologies play an important role in overcoming all the inherent limitations of phytochemicals, including poor stability and low bioavailability. Enclosing within a protective matrix, usually made up of polymers, lipids, or biopolymers, in various micro and nanoencapsulation techniques protects the bioactive plant compounds from environmental degradation and metabolic breakdown. These encapsulation systems allow phytochemicals to be released in a controlled, sustained manner, thereby maintaining therapeutic levels for a longer period. In addition to the above, drug encapsulation would enhance its solubility and bioavailability, especially for hydrophobic drugs. An example of such a drug is curcumin, an antiinflammatory derived from the plant turmeric. This drug can enhance solubility, bioavailability, and efficacy when encapsulated. This has expanded the therapeutic uses of many phytochemicals that were previously not considered due to poor bioavailability (Vidhale et al., 2024).

11.5.3 Transdermal and controlled-release formulations

Transdermal formulations of botanical drugs have been recognized as an efficient mode of noninvasive botanical drug delivery. The transdermal drug delivery system offers sustained-release formulations that improve patient compliance. The transdermal drug delivery system delivers the active drug components directly through the skin, thereby avoiding the first-pass effect in the liver. This leads to steady plasma drug levels with less frequent dosing regimens.

Botanical extracts can be formulated into gels, patches, creams, and ointments using penetration enhancers and polymer matrices. The formulation helps in the increased absorption of the botanical extracts through the skin. Formulating botanical extracts as sustained-release formulations further enhances the efficacy of botanical drugs. The formulation helps increase the drug's stability in the body, thus reducing side effects. Formulating botanical drugs as sustained-release formulations helps increase patient compliance. The formulation of botanical drugs as sustained-release formulations is a significant innovation in the application of healing botanicals.

11.5.4 Smart and stimuli-responsive delivery platforms

One of the most promising areas of botanical therapy is smart drug delivery. Smart drug delivery platforms can respond to specific physiological stimuli, including pH, temperature, enzymes, and oxidative stress, enabling site-specific, on-demand delivery of phytochemicals (Ma et al., 2019). For example, a pH-responsive delivery platform can be used to deliver botanicals to inflamed tissues or tumor sites, thereby improving the therapeutic efficacy of the compounds and minimizing side effects.

In addition, the use of biodegradable polymers and biosensors has enabled real-time monitoring and control of the delivery of plant-derived therapeutic agents. This has not only improved the efficiency of therapeutic agent delivery but also provided the opportunity to utilize the approach of personalized medicine. As the field continues to advance, the role of smart, stimuli-responsive delivery platforms will be invaluable for utilizing the therapeutic potential of healing botanicals.

11.5.5 Smart and patient-centric delivery systems

The future of botanical medicine depends on the development of smart drug delivery systems that overcome the limitations of phytochemicals, such as solubility, stability,

Table 11.5 Innovative drug delivery systems for botanical therapeutics.

Aspect	Details
5.1 Nanotechnology and targeted delivery	- Nanoparticles (e.g., liposomes, polymeric nanoparticles) enhance bioavailability and target specificity. - Enable delivery of phytochemicals to specific tissues or cells. - Reduce systemic side effects by localized therapy.
5.2 Encapsulation & bioavailability	- Micro- and nanoencapsulation protect unstable phytochemicals from degradation. - Improve solubility, absorption, and controlled release. - Example: Encapsulated curcumin shows improved GI uptake and antiinflammatory effects.
5.3 Transdermal & controlled-release	- Transdermal systems (patches, gels, creams) bypass the digestive system and first-pass metabolism. - Allow for sustained release and enhanced patient compliance. - Controlled-release formulations reduce dosing frequency and improve efficacy.

and absorption. Smart drug delivery systems are being developed to overcome the limitations of phytochemicals. The development of smart drug delivery systems would enable targeted, sustained delivery of botanical compounds. The integration of materials science and pharmaceutical technology would enhance the clinical utility of healing botanicals. Table 11.5 presents new drug delivery methods that enhance the effectiveness and stability of botanical medicines.

11.6 Artificial intelligence and data-driven discovery

11.6.1 Machine learning in predicting compounds

Machine learning and AI technologies have now been recognized and established as important tools in the discovery of novel therapeutic compounds of plant origin. Conventional methods for identifying bioactive phytochemicals are often laborious and insufficient due to the enormous chemical complexity of plants. Machine learning can process massive datasets containing information on the structures, biological activities, and targets of different phytochemicals to predict their therapeutic potential.

Therefore by recognizing patterns or relating concepts to complex datasets, the algorithms used in AI Mechanisms identify the compounds most likely to affect biological systems in nature. In this regard, these algorithms significantly streamline the drug discovery process by identifying drugs with a high probability of activity in biological systems. In this way, machine learning technologies significantly accelerate the discovery of new botanical therapeutics.

11.6.2 Big data and phytochemical databases

The combination of big data analytics and extensive phytochemical databases has revolutionized the evaluation and comparison of plant medicinal compounds. Phytochemistry databases contain detailed information on plant species, plant chemical constituents, pharmacological and toxicological data, and medicinal uses of plants. Phytochemistry databases are useful reference tools for scientists interested in the medicinal value of plants.

Big data analytics helps process and interpret these data in an efficient and meaningful manner by uncovering hidden correlations in the datasets. Big data analytics encompasses various types of information, including chemical, biological, and ethnomedicinal data. Therefore the scope of use of botanicals can be better established, their safety profile improved, and the certainty of botanical medicines enhanced.

11.6.3 Artificial intelligence in personalised botanical therapy

AI plays a critical role in advancing individualized botanical therapy. This is because, at a fundamental level, AI can provide unique treatments tailored to an individual's needs. When discussing data, this would include, but not be limited to, genetics, metabolism, lifestyle, and environmental influences, among other factors that can be used to develop a botanical therapy formula that leads to optimal results for specific individuals.

Personalized botanical therapies eliminate the need for trial-and-error treatment, reducing the risk of adverse reactions and accounting for the variability in drug responses among individuals. This is particularly because of the broader movement to ensure precision medicine that considers treatment and healthcare on a case-by-case basis, rather than in a standardized fashion. This means that the use of AI for such a purpose would not only be effective but also useful for formulating specific healthcare schemes using healing botanicals.

11.6.4 Artificial intelligence-assisted drug design and virtual screening

The application of AI has greatly improved the drug design process using virtual screening and in silico modeling of phytochemicals. Using AI, thousands of phytochemicals can be scanned to determine their binding affinities, pharmacokinetics, and toxicity levels without conducting any laboratory experiments. This helps efficiently and accurately discover botanical drugs by saving time, money, and resources.

With virtual screening technologies and molecular docking simulations, scientists can study the interactions between phytochemicals and biological targets, thereby selecting lead compounds at an early stage. This AI technology not only accelerates the discovery of botanical drugs but also increases the likelihood of successful development of effective botanical drugs by prioritizing compounds with high therapeutic value.

11.6.5 Ethical issues, data integration, and future challenges

Although AI has introduced robust tools for botanical research and drug development, several important issues arise from its application. The most important of these issues concerns the quality of the information used to train AI systems. Phytochemical, ethnobotanical, and related information used to train AI systems is in a state of complete disarray, which may compromise the quality of the information these systems generate. Information on the training of AI systems may also be insufficient, leading to the complete neglect of new therapeutic agents or to unreliable inferences.

The second most important issue in AI is the ethical use of indigenous knowledge. Although a large amount of information on plant healing agents is of indigenous origin, it has been developed by indigenous cultures over many generations. Although information related to indigenous knowledge can be digitized, there is concern about intellectual property rights.

Table 11.6 Artificial intelligence (AI) and data-driven discovery in botanical research.

Aspect	Details
6.1 Machine learning in compound prediction	- Uses AI algorithms to predict the therapeutic potential of phytochemicals. - Analyzes large datasets of compound properties and activities. - Accelerates drug discovery by prioritizing promising candidates.
6.2 Big data and phytochemical databases	- Integrates big data analytics with vast phytochemical information. - Compares chemical profiles, biological effects, and ethnobotanical uses. - Uncovers new insights on the efficacy and safety of botanicals.
6.3 AI in personalized botanical therapy	- Utilizes patient-specific data (genetics, environment, lifestyle) to tailor treatments. - AI suggests customized botanical regimens for improved efficacy and fewer side effects. - Supports personalized medicine trends in healthcare.

Data integration is also a challenge, especially because botanical research spans multiple fields, including pharmacology, genomics, chemistry, and clinical science. The integration of these data sources into a standardized, compatible format is a challenge, but an important step towards maximizing the full potential and benefits of AI applications. Finally, the transparency and interpretability of AI models are also important considerations, especially in the healthcare sector, where they are critical for gaining approval and acceptance.

The future will require the collaboration of several actors, including scientists, ethicists, policymakers, and indigenous populations, towards addressing the aforementioned challenges and concerns. The development of sound data governance policies and the improvement of data quality are critical steps towards the responsible use and integration of AI in botanical research. By considering the aforementioned factors, it is possible to successfully and responsibly integrate AI into botanical research to promote the sustainable discovery and development of healing botanicals in the future healthcare sector. Table 11.6 explains how AI and big data help discover new plant compounds and personalize treatments.

11.7 Integration of traditional knowledge with modern science

11.7.1 Ethnobotany and evidence

Ethnobotany is an important area that bridges traditional healing practices and scientific research. Ethnobotany is the documentation of the medicinal use of plants by different societies. Traditional systems of medicine have been developed through years of experience. They provide vital information about the healing potential of plants. These ethnobotanical documents, therefore provide a basis for scientific research, providing a direct guide for researchers to follow in identifying plants and formulations for pharmacological and clinical research.

Traditional knowledge is thus preserved, and scientific methods such as phytochemical screening, in vitro and in vivo studies, and clinical trials support traditional medicine. The combination of ethnobotanical studies and scientific methods, therefore encourages evidence-based approaches to evaluating traditional or alternative medicine. This combined approach ensures that traditional knowledge or alternative medicine with a scientific basis is incorporated into medical practice.

11.7.2 Preservation of indigenous practices

The preservation of indigenous knowledge and traditional healing practices plays a critical role in safeguarding indigenous cultures and biodiversity, and in developing sustainable healthcare systems. Indigenous people have a rich culture of knowledge about herbal medicine and medicinal plants, encompassing the identification, harvesting, and use of various plants for healing. However, all this knowledge and information is on the verge of extinction.

It requires the collective efforts of researchers, indigenous practitioners, and policymakers to preserve it. Community-based documentation initiatives, participatory research, and policy frameworks could ensure the respect and preservation of indigenous practices. At its essence, integrating traditional healing practices into modern healthcare systems will enable societies to harness substantial benefits from sustainable, indigenous healing practices, thereby preserving indigenous intellectual and cultural contributions.

11.7.3 Challenges in harmonisation

However, despite the potential advantages of integrating traditional knowledge and modern science, several challenges hinder this integration. The first challenge concerns the epistemological differences between traditional and scientific systems or approaches. For instance, traditional medicine takes a comparatively holistic approach, while modern scientific research takes a reductionist approach.

The other challenges are linked to Intellectual property rights and ethics. The issue of ownership of traditional knowledge is a greater challenge. The other challenges include differences in standards, documentation processes, and clinical validation issues. The challenges associated with traditional knowledge and modern scientific research must be addressed to ensure a balanced relationship between the two.

11.7.4 Models for collaborative research and community engagement

In integrating traditional knowledge and modern science, there is a need for models of collaborative research and community engagement. This will ensure that traditional knowledge and modern science are well integrated. Participatory research models will ensure that the owners of the knowledge are actively involved in the research, including the selection of plants and the interpretation of results. This will ensure that the traditional knowledge is well captured. This will also ensure that trust levels between the two parties improve and that traditional healing practices are not misrepresented or oversimplified.

Community engagement will also improve the applicability of scientific research, as traditional healers know how to prepare plants and determine appropriate dosages.

11.7.5 Policy frameworks, knowledge protection, and future directions

Policy frameworks play an important role in ensuring support for the sustainable integration of traditional botanical knowledge into modern research and healthcare systems. Protection of Indigenous communities' interests through intellectual property rights and equitable benefit-sharing provisions in policy frameworks is important for ensuring that Indigenous communities are protected from biopiracy and have access to the benefits of their traditional knowledge through international and/or national laws.

The future direction of integrated botanical medicine will focus on developing standardized protocols that support cultural diversity and the principles of science, and on educational and training investments that facilitate the integration of botanical medicine into the healthcare system. Ethical policies aligned with scientific innovation will support the coexistence of traditional and scientific knowledge synergistically, helping build more sustainable healthcare systems.

11.7.6 Ethical collaboration and models of knowledge integration

There is a need for ethical collaboration to integrate traditional knowledge and modern science effectively. This will ensure the participation of indigenous people, traditional healers, scientists, and policymakers in documenting and validating botanical knowledge. There is a need for collaboration to ensure the fair sharing of benefits and the preservation of culture. This will ensure the development of botanical medicine (Fig. 11.3). There is a need for collaboration to ensure the integration of traditional knowledge into accepted practices. Table 11.7 describes the blending of traditional knowledge with scientific research, including challenges and benefits.

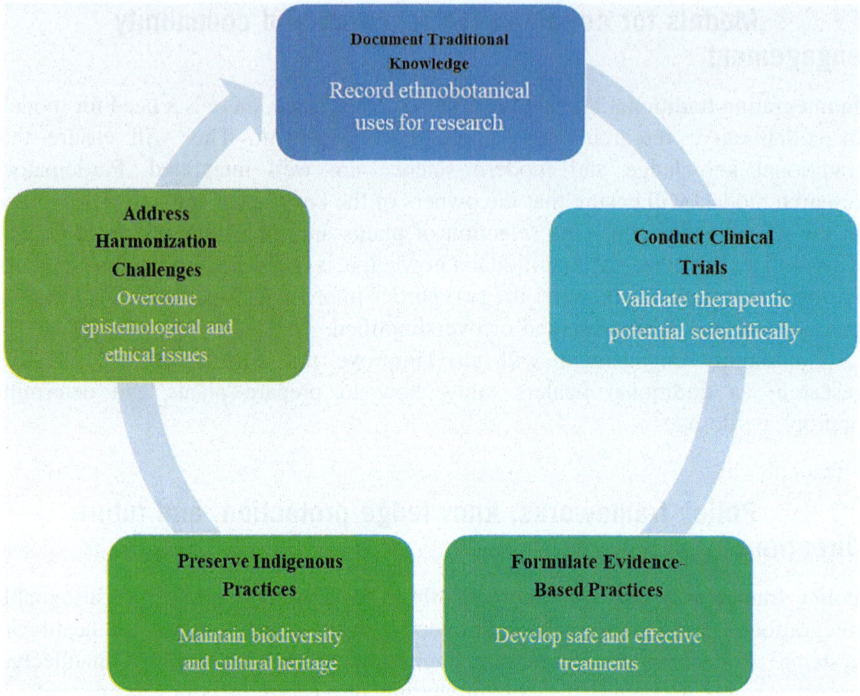

FIGURE 11.3 Cycle of harmonising traditional and modern practices.

Cyclic view of traditional practices.

Table 11.7 Integration of traditional knowledge with modern science.

Aspect	Details
7.1 Ethnobotany and evidence-based validation	- Combines traditional plant use knowledge with modern scientific validation. - Ethnobotanical records guide clinical and pharmacological research. - Supports development of evidence-based, safe, and effective botanical therapies.
7.2 Preservation of indigenous practices	- Emphasizes conservation of indigenous knowledge, biodiversity, and cultural heritage. - Encourages collaboration among researchers, indigenous communities, and policymakers. - Aims to document, respect, and integrate traditional practices in healthcare.
7.3 Challenges in harmonization	- Differences in knowledge systems and methodologies between traditional and modern medicine. - Intellectual property rights and benefit-sharing issues. - Necessitates mutual respect, understanding, and ethical frameworks for collaboration.

11.8 Regulatory and quality aspects

11.8.1 Global regulatory frameworks

The nature of regulatory oversight for the development of healing botanicals varies worldwide, which may significantly affect their development, marketing, and eventual use in healthcare planning. It is important to note that while some countries regulate botanicals as dietary supplements, others require them to undergo rigorous clinical trials as medications. Thus it is important to develop global regulatory frameworks for botanical medications that can ensure their safety, quality, and efficacy.

International and regulatory bodies may be highly effective in achieving harmonization through the development of joint guidelines and quality standards. This process of harmonization may not only ensure the global availability of genuine botanicals but also help develop confidence and innovation in botanical medicine.

11.8.2 Standardization and safety testing

Standardization is a highly important factor that contributes to the uniformity, reliability, and therapeutic efficacy of botanicals. Differences in plant species, growing conditions, harvesting practices, and extraction methods result in marked variations in the concentrations of bioactive principles. Standardization in the selection, processing, and preparation of raw materials may also help minimize differences and ensure the reproducibility of clinical responses.

Alongside standardization, there will be a need for appropriate toxicological testing for the acceptance of botanicals in today's healthcare system. This includes toxicological and pharmacokinetic studies, as well as properly designed clinical trials, which are highly essential for assessing potential toxicities and establishing therapeutic efficacy.

The development of a validated analytical procedure and the standardization of testing procedures will certainly give botanicals additional marks of authenticity and help integrate them into evidence-based medicine.

11.8.3 Intellectual property and ethical issues

One ethical issue associated with intellectual property rights in botanical research is the use of traditional knowledge in the development of new medicines. Some botanical medicines have been used traditionally, and the exploitation of this knowledge without proper recognition and compensation has raised several ethical issues (See, Baum 2023).

Ethical research, collaboration, and the sharing of benefits with local and indigenous communities worldwide are among the practices that should be followed in the development of botanical medicines. This will ensure that the development of botanical medicines is carried out in a socially just and scientifically progressive manner.

11.8.4 Quality assurance, good manufacturing practices, and traceability

Assurance of quality is a significant component of regulating botanicals, as their quality and therapeutic efficacy can be greatly affected by differences in species, cultivation, harvesting, and processing. Good manufacturing practices (GMPs) should be adhered to during the production and control of botanicals to ensure consistency with predetermined quality standards. GMP regulations cover all aspects of manufacturing, including the procurement, processing, packaging, storage, and distribution of raw materials; these regulations ultimately help reduce the risk of contamination, adulteration, and variability in the final product.

Traceability is another important aspect of quality control; it enables complete documentation of the botanicals' supply chain, from farming to the finished product. Correct identification of the botanical species, verification of raw materials, and evaluation of the growing conditions of the botanicals will help assure the quality of the botanicals and the safety of the end-consumer.

11.8.5 Future regulatory challenges and harmonisation strategies

Despite progress in regulating botanicals, challenges remain worldwide. The categorization of botanicals across countries remains a challenge, creating barriers to trade and acceptance. In some countries, botanicals are recognized as supplements, while in others they are recognized as medicines.

To address the challenges, there is a need for greater collaboration and harmonization of botanical regulations worldwide. The development of guidelines for evaluating safety, efficacy, and quality will be significant for the acceptance of botanicals worldwide. There is a need to strengthen regulations on botanicals and ensure they meet modern healthcare standards.

11.8.6 Enhancement of regulatory compliance and quality assurance systems

Regulatory systems and quality assurance systems are important for the development of healing botanicals in the healthcare industry. Standardization of growth, processing, and testing of botanicals will enhance consumer acceptance and regulatory approval. Harmonization of regulatory requirements across international markets may facilitate the entry of botanicals. Surveillance and regulatory compliance with quality standards will be important in maintaining the purity of

Table 11.8 Regulatory and quality aspects of botanical medicines.

Aspect	Details
8.1 Global regulatory frameworks	- Regulatory standards for plant products vary widely between countries. - Harmonized international frameworks promote safety, efficacy, and quality. - Collaboration between regulatory agencies facilitates global trade and consumer confidence.
8.2 Standardization and safety testing	- Standardized production ensures consistent quality and therapeutic effects. - Safety testing (toxicity, clinical trials) is crucial for validation. - Development of protocols improves acceptance in modern healthcare systems.
8.3 Intellectual property and ethical concerns	- Protecting intellectual property, especially traditional knowledge, raises ethical challenges. - Ensuring indigenous communities benefit fairly from commercialization. - Benefit-sharing agreements and ethical frameworks are essential.

botanical medicines. Table 11.8 discusses the global regulations, quality standards, and ethical issues associated with botanical products.

11.9 Sustainability and environmental impact

11.9.1 Ethical sourcing and conservation of biodiversity

The long-term survival of medicinal herbs has been linked to ethical sourcing. By ethical sourcing, it is meant sourcing plants in a manner that does not lead to overexploitation, habitat destruction, or human-induced imbalance in plant populations. This has been a concern for medicinal plants, which are often exploited or considered endangered. If we do not adopt sustainable harvesting methods, we might lose such plants or even disturb their role in our ecosystem.

Biodiversity conservation aims to maintain the genetic diversity of medicinal plants, which are of greatest importance for maintaining ecosystem resilience and providing a future source of medicinal resources.

Community-based conservation, ethical use guidelines, or certifications are also important in promoting ethical sourcing (Ma et al., 2019). These not only help conserve natural resources but also empower communities by ensuring sustainable livelihoods.

11.9.2 Eco-friendly cult

The adoption of sustainable cultivation practices is crucial for reducing the environmental impact of large-scale medicinal herb cultivation. Eco-friendly cultivation practices include organic, permaculture, and agroforestry. The practices improve soil fertility and water conservation, while avoiding the use of inorganic fertilizers and pesticides. The practices also support nature's balance by improving soil biotic processes and conserving useful insects.

Aside from that, sustainable cultivation can help develop high-quality, potent medicinal plants. This is achieved through a combination of factors, including the cultivation of medicinal plants using sustainable systems that produce higher concentrations of active compounds under lower chemical stress, along with increased soil nutrient availability. Moreover, agroforestry systems exist, which are beneficial due to their combination of biodiversity and efficient land utilization (Shree, Patra, and Sahoo, 2024).

11.9.3 Green chemistry strategies

Green chemistry strategies are primarily used to reduce the environmental impacts of processes such as the selection, processing, and formulation of botanicals. The traditional processing of botanical extracts involves the use of enormous quantities of toxic solvents, which require extensive processing, resulting in hazardous waste and pollution. The green chemistry strategy proposes using less toxic solvents, avoiding waste generation, or employing beneficial technologies to overcome such challenges.

Other innovative approaches, such as SCFE, UAE, and MAE, are also efficient methods for extracting compound constituents and are environmentally friendly. These approaches may also improve the extraction of herbal medicine compound constituents, thereby reducing solvent use and saving time. Green chemistry strategies are important in the herbal medicine industry to minimize environmental impacts without undermining the safety, efficacy, and quality of herbal medicine, as the relationship between sustainability and technological advancement is a major milestone in the development of healing botanicals (Jahan, 2022).

11.9.4 Impact of climate change on medicinal plant resources

Climate change has been identified as a major threat to the sustainability of medicinal plant resources, driven by changes in temperature patterns, rainfall, and ecosystem stability. Climate change can affect the geographical distribution, growth, and phytochemical composition of medicinal plants, potentially affecting their availability and efficacy. Most medicinal plants are highly sensitive to environmental changes, and climate change can lead to habitat destruction or reduced yields, particularly in biodiversity hotspots.

In addition, climate change stress can alter the concentration of bioactive compounds in plants, potentially affecting their medicinal properties. The above points highlight the importance of developing strategies to mitigate the effects of climate change, including climate-resilient agricultural practices, natural resource conservation, and research on the effects of environmental stressors on phytochemical composition. It is important to address the challenge posed by climate change to ensure the sustainability of healing botanicals.

11.9.5 Sustainable supply chains and future environmental responsibility

Establishing sustainable supply chains is crucial to ensuring the future of the botanical medicine industry is environmentally responsible. Sustainable supply chains involve responsible procurement, fair trade, transportation, and waste management. The botanical medicine industry should strive to ensure that environmental responsibility is incorporated into the development of healing botanicals.

Environmental responsibility will be a major determinant of the future of healing botanicals. The use of sustainability certification programs, renewable energy, and the circular economy model will help diminish negative environmental effects. The botanical medicine industry can ensure sustainability while protecting natural resources.

11.9.6 Addressing environmental challenges and climate resilience

Environmental sustainability is a critical issue for the future of healing botanicals. This is especially the case in relation to climate change. Climate change can influence the growth patterns, phytochemicals, and geographical distribution of

Table 11.9 Sustainability and environmental impact in healing botanicals.

Aspect	Details
9.1 Ethical sourcing and biodiversity conservation	- Ensures harvesting does not deplete natural resources or harm ecosystems. - Protects threatened species and genetic diversity. - Promotes community conservation programs and sustainable harvesting certifications.
9.2 Eco-friendly cultivation techniques	- Uses organic farming, permaculture, and agroforestry to reduce chemical use and improve soil health. - Enhances biodiversity and medicinal plant quality. - Supports sustainable agricultural practices benefiting the environment and crop health.
9.3 Green chemistry strategies	- Applies nontoxic solvents and waste minimization in extraction/formulation. - Employs energy-efficient and solvent-free methods (e.g., supercritical fluid extraction, microwave-assisted extraction). - Balances environmental sustainability with product safety and efficacy.

medicinal plants. This can pose a challenge to the sustainability of medicinal plants. Environmental sustainability will ensure the future availability and efficacy of healing botanicals. Table 11.9 emphasizes sustainable sourcing, eco-friendly farming, and green extraction techniques for botanicals.

11.10 Future directions and emerging trends

11.10.1 Personalized and precision herbal medicine

The future path of healing botanicals is moving increasingly towards personalized medicine. Instead of prescribing generic herbal medicines to patients in general, there is a shift towards personalized herbal medications tailored to each individual's needs. Genetic factors, metabolic factors, dietary habits, and external influences are among the important factors known to influence an individual's response to a particular herbal remedy. New advancements in pharmacogenomics are helping scientists develop methods to create personalized markers that can determine which herbal remedy is best suited to an individual.

The field of precision herbal medicine aims to ensure the efficacy of drug treatment while limiting side effects by providing the appropriate herbal ingredients to the appropriate patient. This would involve analyzing the genetic mechanisms underlying an individual's drug metabolism to determine the levels of phytochemicals present and simultaneously reassure the patient about the treatment provided (Mohanta et al., 2022). The emerging diagnostics technology would definitely involve herbal medicine as an important part of individualized healthcare systems.

11.10.2 The addition of medicinal plants to preventive healthcare strategies

The incorporation of medicinal plants into a preventative healthcare strategy is a new, growing trend. Many healthcare systems around the world have been changing over the last few decades to place greater emphasis on preventing disease and maintaining a healthy lifestyle than on treating disease after it occurs. Culturally relevant medicinal plants can help us achieve long-term health (i.e., disease-free status) through their potent antioxidant, antiinflammatory, and immunomodulatory effects.

Regular use of plant-based supplements, functional foods, and herbal teas will allow our bodies to develop more generalized forms of resistance, reduce oxidative stress, and maintain overall physiological balance. The use of botanical medicine as part of various public health initiatives will likely yield additional benefits, such as greater use of traditional preventive approaches (e.g., nutrition and behavioral change education). As a result of using more traditional preventive approaches through these public health initiatives, overall population health is expected to be increased, and the overall cost of delivering healthcare services will be lower.

11.10.3 Innovations for global health equity

Similarly, advancements in healing botanicals offer tremendous potential to promote health equity on a global scale. Around the world, in many low- and middle-income countries, access to conventional pharmacological therapies for healthcare management is limited owing to financial, geographical, and/or infrastructural constraints. The application of herbal medicines may be an effective and cost-effective approach.

For example, collaborative work among scientists, healthcare experts, and communities across a region can help ensure that a form of botanical therapy is used to help address certain health issues within that region. Furthermore, modern technologies such as digital platforms, telemedicine, and advanced supply chain management systems may help make herbal medicine more accessible worldwide (Yıldırım et al., 2023). Therefore innovations in healing botanicals may help minimize healthcare inequalities.

11.10.4 Digital health technologies and smart therapies will play a role in shaping future use of botanical healing

There will be continued growth of the role of digital health technologies in determining the future of botanical healing. For example, mobile health (mHealth) may be utilized through mHealth applications, wearable devices, and electronic health records to monitor a patient's response to a given botanical therapy. These systems enable ongoing measurement of efficacy, safety, and compliance, thereby enhancing the accuracy of botanical therapy.

The combined use of AI with digital platforms may assist in making real-time decisions regarding herbal medicine by utilizing patient information and predicting future outcomes based on herbal treatment. Additionally, the combined use of smart therapeutic systems that incorporate botanicals, along with digital monitoring platforms, can increase patient involvement, improve treatment engagement, and enhance long-term health outcomes. Such technological developments will further position botanical healing within the ongoing evolution of modern healthcare systems.

11.10.5 Translational research, development, and future potential of botanicals for healing

Translational research will help develop commercially viable, effective products based on laboratory research into the healing properties of botanicals. A coordinated effort among scientists, clinicians, and commercial representatives will enable the effective transfer of results from basic research, clinical trials, and commercial products. Regulatory agencies will work together to develop consistent regulatory standards and measurements, further expediting the development of safe, efficacious botanical-based therapies (Pradhan, Mohanty, and Kumar, 2025).

The future of commercializing botanicals depends on balancing ethics and innovation to ensure that the appropriate and respectful commercialization of these plants retains the traditional cultural and ecological components that brought them to market (Mărgăoan et al., 2021). With the proper balance of all of these components, botanicals will be a major part of healthcare in the future.

11.10.6 Technological convergence shaping the future of botanical medicine

The fusion of new technologies and botanical medicine marks a major shift in the future of healthcare. New technologies, such as AI, biotech, precision medicine, and digital health platforms, will help bring botanical medicine to patients faster than ever through the discovery, validation, and personalization of botanical medicines (Gumisiriza et al., 2023). New technologies will enable us to detect bioactive compounds faster, accelerate formulation development, and provide customized therapeutic solutions for patients. As new technologies converge with traditional knowledge of botanicals, botanical medicines can serve as the foundation of future healthcare (Sharma et al., 2023). For a summary of future trends in botanical medicine, please see Table 11.10 for key areas related to personalized herbal medicine, preventing illness before it occurs, and achieving health equity globally.

Table 11.10 Future directions and emerging trends in healing botanicals.

Aspect	Details
10.1 Personalized and precision herbal medicine	- Tailors herbal therapies based on individual genetics, lifestyle, and health status. - Utilizes pharmacogenomics and biomarkers for optimized efficacy and reduced side effects. - Aligns with the broader shift towards personalized medicine.
10.2 Incorporation into preventative healthcare	- Uses botanicals proactively to prevent diseases and promote wellness. - Leverages immunomodulatory, antiinflammatory, and antioxidant properties. - Supports public health programs integrating herbal medicine into community health.
10.3 Innovations for global health equity	- Focuses on affordable, accessible herbal medicines for low-resource settings. - Encourages collaborations with local communities to develop culturally relevant interventions. - Employs technology to enhance distribution and maximize equity in healthcare access.

11.11 Conclusion

Healing botanicals are being rediscovered in a whole new light due to advances in science/technology, the return to traditional knowledge, and the importance of whole-body approaches to addressing the ever-increasing complexities of today's global health challenges. Chronic disease burden and environmental degradation are only two examples of how global healthcare systems are adapting to new ways to find contemporary and holistic solutions through the use of botanical medicine.

The exciting new development is the incorporation of drug delivery systems that utilize a variety of methods, including nanotechnology-based formulas, encapsulated phytochemicals, and transdermal delivery methods. By doing this, the bioavailability, effectiveness, and target specificity of herbal chemicals have increased dramatically, allowing botanicals to meet clinical standards and be increasingly used as evidence-based medicine.

Applying AI and computer software to research in herbal medicine is greatly enhancing the speed and accuracy of identifying and verifying plant-based medicines (i.e., phytomedicines) by leveraging sophisticated chemistry databases. AI systems can discern the complexities of interactions between the herb and the human body, including potential interactions that have yet to be identified, while also developing new formulations. The combination of data science and ethnobotany will enable rapid, consistent, and customized use of plants in the development of interventions.

At the same time, there is a resurgence of traditional healthcare practices worldwide, including Ayurvedic medicine, TCM, and Indigenous healing practices. With their historical context and strong ties to the environment, these practices are gaining acceptance within the scientific community and are being integrated into contemporary research as legitimate treatment options. The integration of these treatment modalities into contemporary healthcare provides patients with increased treatment options and greater opportunities for inclusion and preservation of cultural heritage.

There has been a significant development in how we source, grow, and sell medicinal plants, creating eco-friendly solutions. Many of these plants can be harvested in ways that help to keep biodiversity intact while also providing benefits to local communities. The following approaches will be part of the growth strategy for the botanical industry and align with the growing movement toward environmental and social justice: ethical wildcrafting, regenerative agriculture, and fair trade.

Alternatives: As we move forward, botanical medicines will become the foundation for integrative, patient-centered healthcare. They are an alternative or complementary solution to synthetic medications, and many may be used to prevent or treat chronic conditions, promote healthy immune function, enhance mental well-being, and support the use of lifestyle medicines. Botanicals represent a more eco-friendly approach to drug discovery and public health, in line with a broader movement to develop green pharmaceuticals.

Recent advancements in biotechnological areas such as metabolomics, genomic editing, and synthetic biology will allow increased access to rare and endangered plant phytochemical compounds while improving their standardization and safety. These new technologies must also have supportive, enforceable ethical and regulatory frameworks in place to prevent the misappropriation of indigenous knowledge and to ensure that both parties benefit fairly from any collaboration.

The ongoing success of this emerging field will depend on multi-disciplinary collaboration across multiple sectors, a healthy regulatory environment, and integrated/interconnected education. Healthcare professionals need to be informed on how to incorporate botanical medicine into their practices appropriately, and patients deserve simple information about botanical medicine that respects the cultural differences between the two parties. Additionally, all parties involved in research, traditional medicine practice, and the business community working together will be necessary to develop best practices aimed at promoting efficacy, access, and sustainability through codevelopment efforts.

Overall, the future of medicinal plants will not necessarily be the preservation of what has gone before, but rather a combination of ancient teachings used for today's new medicines. Plant-derived medicines will continue to offer hope in improving world health, as they bridge the gap between ancient plant medicine's teachings and today's scientific evidence. They will also help restore ecological balance. They are a powerful reminder that health can never be separate from the environment in which we live, the culture that shaped us, and the community from which we come.

Working together, tradition and technology will be essential as we look to the future. We need to find ways to ensure that the healing plants of the Earth will continue to exist as living, genuine, and ethical elements of the world's healthcare systems through the appropriate use of wisdom, stewardship, science, and a common vision for the future. In this way, plant medicine can continue to promote people's health and heal the Earth itself.

References

Barman T, Samant SS, Jyoti, Dey A, Nandy S, Maitra R, et al. Sustainable employment of folkloric botanicals and conservation practices adopted by the inhabitants of Parbati Valley of North Western Himalaya, India in healing substantial corporeal disorders. Advances in Traditional Medicine 2023;23(2):443–82. Available from https://doi.org/10.1007/s13596-021-00605-3, https://www.springer.com/journal/13596.

Brody H, Meghani Z, Greenwald K. Michael Ryan's writings on medical ethics. Philosophy and Medicine. Springer; 2009. Available from https://doi.org/10.1007/978-90-481-3049-8.

Colantonio S, Rivers JK. Botanicals with dermatologic properties derived from first nations healing: part 1—trees. Journal of Cutaneous Medicine and Surgery 2017;21(4):288–98. Available from https://doi.org/10.1177/1203475417690306, http://cms.sagepub.com/content/by/year.

Colantonio S, Rivers JK. Botanicals with dermatologic properties derived from first nations healing: part 2—plants and algae. Journal of Cutaneous Medicine and Surgery 2017;21(4):299–307. Available from https://doi.org/10.1177/1203475416683390, http://cms.sagepub.com/content/by/year.

Ding R. 3D printing with self-healing materials and applications. Science and Technology of Engineering Chemistry and Environmental Protection 2024;1(9). Available from https://doi.org/10.61173/a520hc60.

El-Sherbeni SA, Negm WA. The wound healing effect of botanicals and pure natural substances used in in vivo models. Inflammopharmacology 2023;31(2):755–72. Available from https://doi.org/10.1007/s10787-023-01157-5.

Goktas Z, Zu Y, Abbasi M, Galyean S, Wu D, Fan Z, et al. Recent advances in nanoencapsulation of phytochemicals to combat obesity and its comorbidities. Journal of Agricultural and Food Chemistry 2020;68(31):8119–31. Available from https://doi.org/10.1021/acs.jafc.0c00131.

Grossman LA. Choose Your Medicine. Oxford University Press; 2021. Available from https://doi.org/10.1093/oso/9780190612757.001.0001.

Gumisiriza H, Olet EA, Mukasa P, Lejju JB, Omara T. Ethnomedicinal plants used for malaria treatment in Rukungiri District, Western Uganda. Tropical Medicine and Health 2023;51(1). Available from https://doi.org/10.1186/s41182-023-00541-9, https://tropmedhealth.biomedcentral.com.

Jahan I. Nanotechnology for Drug Delivery and Cancer Therapy. IGI Global; 2022. p. 338–62. Available from 10.4018/978-1-7998-8936-6.ch015.

Kustiyati S, Fajri IK, Masithoh RF, Dewi RDC, Nurseha N. Towards holistic healthcare: integrating traditional and modern medicine for improved patient outcomes. Global International Journal of Innovative Research 2024;2(5):942–51. Available from https://doi.org/10.59613/global.v2i5.144.

Ma Z, Fan Y, Wu Y, Kebebe D, Zhang B, Lu P, et al. Traditional chinese medicine-combination therapies utilizing nanotechnology-based targeted delivery systems: a new strategy for antitumor treatment. International journal of nanomedicine 2019;14:2029–53. Available from https://doi.org/10.2147/IJN.S197889, https://www.dovepress.com/getfile.php?fileID=48693.

Mishra M, Srivastava D. An ethnomedicinal study of a few indigenous plants used by the folklore of Gorakhpur district, Uttar Pradesh. Acta Botanica Plantae 2024;3(1):21–35. Available from https://doi.org/10.51470/abp.2024.03.01.21.

Mohanta YK, Chakrabartty I, Mishra AK, Chopra H, Mahanta S, Avula SK, et al. Nanotechnology in combating biofilm: a smart and promising therapeutic strategy. Frontiers in Microbiology 2022;13. Available from https://doi.org/10.3389/fmicb.2022.1028086, https://www.frontiersin.org/journals/microbiology#.

Momot SL. Promoting comprehensive health services: harnessing the synergy of traditional and modern healing approaches. Global International Journal of Innovative Research 2024;2(5):905–14. Available from https://doi.org/10.59613/global.v2i5.152.

Mărgăoan R, Topal E, Balkanska R, Yücel B, Oravecz T, Cornea-Cipcigan M, et al. Monofloral honeys as a potential source of natural antioxidants, minerals and medicine. Antioxidants 2021;10(7):1023. Available from https://doi.org/10.3390/antiox10071023.

Pearn J. The world's longest surviving paediatric practices: Some themes of Aboriginal medical ethnobotany in Australia. Journal of Paediatrics and Child Health 2005;41(5-6):284–290. Available from https://doi.org/10.1111/j.1440-1754.2005.00612.x.

Podder V, Anand A. Integrative healthcare in india: a paradigm shift in modern medical education and practice of 21st century. Integrative Medicine Case Reports 2022;3(1):2. Available from https://doi.org/10.38205/imcr.030102.

Pradhan S, Mohanty JN, Kumar S. Ethnomedicinal plants used by santhal community of India. Annals of Agri Bio Research 2025;30(1):86–97. Available from https://doi.org/10.53941/agrbio.2025.1000011, https://agribiop.com/.

See PC, Baum N. Use of botanicals in modern healthcare. Health Administration, Leadership, and Management Journal 2023;1(1):27–30. Available from https://doi.org/10.55834/halmj.5949966089.

Shahpesandy H, Al-Kubaisy T, Mohammed-Ali R, Oladosu A, Middleton R, Saleh N. A concise history of islamic medicine: an introduction to the origins of medicine in islamic civilization, its impact on the evolution of global medicine, and its place in the medical world today. International Journal of Clinical Medicine 2022;13(04):180–97. Available from https://doi.org/10.4236/ijcm.2022.134015.

Sharma N, Vasisht K, Kaur J, Sandhu SK, Dey K, Hameed BA, et al. Blending ethnomedicine with modern technology—from conventional to tailored products: modulating biopharmaceutical properties of berberis extract by solid lipid nanoparticles for wound healing. Journal of Functional Biomaterials 2023;14(8):418. Available from https://doi.org/10.3390/jfb14080418.

Shree D, Patra CN, Sahoo BM. A review on nanotechnology mediated-herbal drug delivery for the treatment of obesity. Natural Products Journal 2024;14(1):18–34. Available from https://doi.org/10.2174/2210315513666230412111914, https://www.eurekaselect.com/article/130885.

Sun S, Lin W, Yang L, Zhang C, Kan H, Xu C, et al. Near-infrared light-actuated on-demand botanicals release and hyperthermia by an antibiotic-free polysaccharide-based hydrogel dressing for the synergistic treatment of wound infections. Journal of Materials Chemistry B 2024;12(5):1307–16. Available from https://doi.org/10.1039/d3tb02714g.

Tao HX, Zhao GD, Linghu KG, Xiong W, Sang W, Peng Y, et al. Botany, traditional use, phytochemistry, pharmacology and toxicology of Sigesbeckiae Herba (Xixiancao): a review. Phytochemistry Reviews 2021;20(3):569–87. Available from https://doi.org/10.1007/s11101-020-09714-4, www.springer.com/journal/11101.

Vickers A, Zollman C. ABC of complementary medicine: herbal medicine. British Medical Journal 1999;319(7216):1050–3. Available from https://doi.org/10.1136/bmj.319.7216.1050.

Vidhale DK, Mora VR, Pasam S, Kotha A, Konda U, Boggula N. Critical review on pongamia pinnata: a versatile medicinal plant. Asian Journal of Pharmaceutical Research and Development 2024;12(4):138–46. Available from https://doi.org/10.22270/ajprd.v12i4.1455.

Yıldırım M, Sessevmez M, Poyraz S, Düzgüneş N. Recent strategies for cancer therapy: polymer nanoparticles carrying medicinally important phytochemicals and their cellular targets. Pharmaceutics 2023;15(11):2566. Available from https://doi.org/10.3390/pharmaceutics15112566.

CHAPTER

12 Conclusion: harnessing the power of healing botanicals for a healthier future

Awanish Kumar[1], Grandhi Surendra[2], Arifa Begum Shaik[3], Mohana Naga Rajeswari Dasaraju[4] and Kanaka Durga Devi Nelluri[3]

[1]Department of Biotechnology, National Institute of Technology, Raipur, Chhattisgarh, India,
[2]School of Pharmacy, ITM University, Gwalior, Madhya Pradesh, India,
[3]Department of Pharmaceutics and Biotechnology, KVSR Siddhartha College of Pharmaceutical Sciences, Vijayawada, Andhra Pradesh, India,
[4]Department of Chemistry, Siddhartha Academy of Higher Education, Deemed To Be University, Vijayawada, Andhra Pradesh, India

12.1 An overview of key information regarding healing herbs

12.1.1 Bridging traditional wisdom and modern scientific evidence

Traditional systems of medicine, such as Ayurvedic, Unani, Siddha, and indigenous systems, have extensive empirical knowledge of medicinal plants, their dosage forms, and multiingredient formulations. Ethnobotanical studies, such as the one conducted in Uttar Pradesh, document several plant species with medicinal value, including commonly used parts such as leaves, thereby establishing their relevance to the community (Sharma et al., 2024). Modern pharmacology has tools such as phytochemical profiling, mechanism-of-action studies, toxicity studies, and clinical trials, in addition to helping validate claims of traditional medicine. Reviews of such attempts to bridge the gap between traditional and modern medicine emphasize the possibility of integration, contingent on the quality of data, research methods, and community participation (Sharma et al., 2025).

Soymida febrifuga, in addition to *Dendrobium nobile*, represents a successful example of how classical literature, ethnobotany, and chemical/pharmacological studies contribute to the exploration of potential therapeutic options. The aforementioned literature reviews integrate traditional uses, phytochemistry, and preclinical results to identify therapeutic leads worthy of further clinical investigation. The ethnobotanical literature, including the identification of frequently used families such as Fabaceae and Acanthaceae, provides a link to the traditional knowledge base and to species worthy of further phytochemical and pharmacological studies to validate traditional uses and connect the traditional knowledge base to laboratory results.

Healing Botanicals. DOI: https://doi.org/10.1016/B978-0-443-44295-7.00006-9

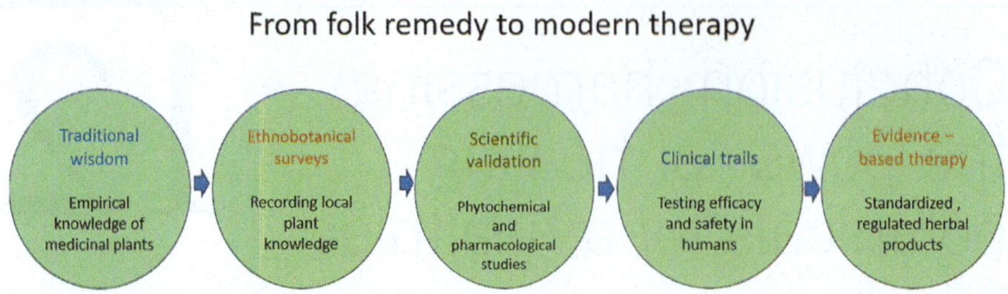

FIGURE 12.1

Translational pathway of medicinal plants. Phyto compounds pathway.

To effectively integrate the traditional and the modern, there is a need to respect both fields: scientists should respect the traditional knowledge base, and traditional systems should be receptive to scrutiny aimed at validating traditional uses with respect to safety, dosing, and mechanism-of-action (Yang et al., 2024).

The alignment between native medicine traditions, medicinal native wisdom, and modern science is another strong rationale for developing standardized, evidence-based herbal therapies. The advanced technologies available today, such as metabolomics, molecular docking, and systems biology, provide a synergistic approach to multicomponent herbal formulations and confirm the validity of the polyherbal concept in traditional systems. The integration of clinical validation, quality control, and regulatory frameworks as the three primary pillars that underpin the conversion of traditional herbal medicine into the next generation of recognized, worldwide pharmaceutical products requires drug discovery to provide solutions for preserving native knowledge. By doing this, drug discovery protects the ethical and sustainable use of native knowledge in the development of contemporary healthcare products (Fig. 12.1). It illustrates the systematic progression of medicinal plant knowledge from traditional wisdom to evidence-based therapy.

12.1.2 Important discoveries and therapeutic consequences of medicinal herbs

Medicinal plants have a wide variety of phytochemicals that produce a broad array of pharmacological effects. Among others, these effects include antiinfective, antiinflammatory, metabolic, and neuroprotective effects. The chapters in this book summarize examples such as *D. nobile*, supporting the development of translational research by mapping chemical constituents and their biological activities. Several consistent scientific themes have been identified throughout this book: the majority of traditional uses of medicinal plants provide a reasonable biological basis for their therapeutic effects; preclinical studies provide evidence of therapeutic potential; and one of the greatest limitations of widespread clinical use of many botanicals is their safety, dosing, and standardization (Halder and Jha, 2023).

A very good example of this is *Crotalaria multibracteata*, which contains some promising bioactive constituents. Still, additional research is needed to define the mechanisms by which they confer therapeutic benefits, assess their potential for toxicity, and develop strategies for mass cultivation before they are introduced into clinical practice. Likewise, numerous studies in the literature have identified the generally low number of high-quality randomized clinical trials supporting many botanicals. By developing improved methods of standardization, following established reporting guidelines (e.g., STROBE), this will improve the quality and comparability of supporting evidence (Fan et al., 2023). In summary, the progression of scientific advancement in botanical research follows a systematic approach: starting with the documentation of traditional knowledge, then proceeding to phytochemical profiling and pharmacological and toxicological assessment, and finally culminating in controlled clinical trials. Transparent methods and standardized reporting procedures will be the primary pathways for converting promising plant-derived products into validated therapeutic agents (Englund et al., 2006).

The use of research on medicinal plants to develop new drugs is crucial to meet the needs of public health, sustainability, and integrated medicine. Herbal components provided many of the pharmacotherapeutics we have today, including treatments for chronic inflammatory diseases, metabolic disorders, cancer, and Alzheimer's disease. The therapeutic properties of a given herbal plant are often determined by factors such as the species from which it is derived, its source, the method of harvesting, and its preparation before use. Therefore, to ensure that patient receives safe and effective treatment, good agricultural and collection practices (GACP), quality control and assurance measures, and pharmacovigilance systems must be implemented for each of these methods. It would also greatly facilitate the scientific and responsible use of herbal medicines as a modern therapeutic modality by enabling communication among botanists, pharmacologists, physicians, and regulators at all levels. This would aid in restoring the medicinal properties of plant-derived medicines and increasing the overall recognition/credibility of using herbal products within the current healthcare delivery system (Fabricant and Farnsworth, 2001).

12.2 Advances in evidence-based botanical medicine

12.2.1 Enhancing research, trials, and documentation

For a plant-based remedy to be accepted as legitimate by current health systems, it must have been adequately supported by strong scientific evidence. This requires well-designed preclinical studies, standardized extracts, appropriately designed clinical trials, and transparent reporting, as set out in existing guidelines, for example, STROBE for reporting observational studies (von Elm et al., 2007).

Recent literature reviews highlight the need for a set of standard methodologies (standardized chemical fingerprinting), compliance with GACP, reproducible extraction methods, and clinically relevant endpoints for patients/decision-makers (Setzer, 2018).

Through ethnobotanical surveys, one way to identify high-value species is to create a priority list of species planned for study, given the limited research resources, using species lists from documented sources such as Uttar Pradesh (Heinrich et al., 2009). Standardizing the analytical methodology will enable comparability of results across laboratories and clinical studies by defining chromatographic profiles, identifying specific marker compounds, and issuing batch-specific certificates of analysis (Burchard, 2009).

When designing a clinical trial, the randomized controlled trial (RCT) design should be utilized whenever feasible. If randomization is not possible, well-documented observational studies in accordance with established reporting frameworks, such as STROBE, should be used to ensure results are clear and useful for health technology assessments and policies. Also, toxicity evaluations should be conducted before testing humans with species that contain known toxins, such as *Crotalaria* species that contain pyrrolizidine alkaloids; therefore comprehensive safety testing must be conducted before subjecting humans to clinical trials involving these species (Englund et al., 2006).

Moreover, technological advancements in digital health devices and data-driven approaches are continuously bolstering the evidence base for botanical medicine. The use of electronic medical records, real-world evidence (RWE) platforms, and bioinformatics enables the collection and analysis of large datasets for multiple purposes, such as herbal use, safety signals, and treatment outcomes. Systems pharmacology and network analysis methodologies are powerful tools for phytomedicine, in part because they enable a better understanding of multitarget and synergistic effects that would be difficult to depict with standard models. If these methodologies are accompanied by high-quality documentation standards and open data sharing, the advancements made through these methods will enhance transparency, facilitate independent validation, and accelerate the transformation of well-characterized botanicals into clinically relevant therapies (Fu et al., 2004).

12.2.2 Challenges in standardisation and consistency

It is difficult to standardize botanicals due to the inherent variability in chemical compounds and their types found in plants. The levels of phytochemicals will fluctuate according to multiple factors, such as genotype, plant part, harvest timing, soil conditions, climate, and processing methods; therefore reproducibility, dosage accuracy, and safety may be affected. Recent research reviews indicate that controlled cultivation methods and manufacturing practices lead to consistent clinical outcomes (Li et al., 2008).

To reduce the amount of variation that occurs in products produced from plants, GACP requires the documentation of the source of the plants that were used, the verification of the plants' genetic identity through voucher specimens, as well as an accurate recording of the conditions under which the plants were harvested and the conditions following harvest. Ethnobotanical studies that support the collection of voucher specimens enable the accurate attribution of traditional uses to specific plant species (Company WAB, Collection HGGN and STC, 1922). Analytical standardization of the final product is also important. It can be accomplished by utilizing a multimarker chromatographic fingerprinting technique rather than traditional methods used to establish the potency of raw materials, or by using established range data for commonly available botanicals to obtain meaningful quality measurements for multiconstituent products. The use of multimarker chromatographic fingerprints, stability testing, and validated analytical methods is also sufficient to meet the regulatory and clinical requirements outlined below. Multiherb combinations pose an additional challenge to establishing the therapeutic value of individual herbs, as multiple herbs will interact with each other and be present in the final product formulation. Network pharmacology and various "omic" approaches (e.g., genomics, proteomics) may provide insight into the nature of interactions that occur when herbs are combined for clinical research, thereby enabling improved study design to demonstrate the efficacy of multibotanical products for disease treatment (Gröber, 2011).

Because of differences in regulatory rules across countries, standardizing botanical medicines can be very difficult. When botanicals are categorized differently (e.g., as foods, dietary supplements, or medicinal products), this results in differing quality standards, varying labeling requirements, and different post-marketing surveillance obligations. The regulatory split between countries has the potential to limit worldwide recognition of botanicals and hinder the ability to compare safety and efficacy data across jurisdictions. Harmonization efforts initiated by international organizations will be essential to ensuring consistent quality of botanicals, thereby maintaining patient safety and confidence in the evidence supporting the use of botanical therapies. Equally importantly, efforts to establish robust pharmacovigilance systems and risk/benefit assessment frameworks for botanicals should ensure the maximum benefit is achieved through harmonization (Hopkins, 2008).

12.2.3 Regulatory, ethical, and translational considerations

Botanical medicines that have been proven to work depend on standards for how they are researched or tested, developed, and used in clinics. Depending on the company and the type of herbal product being produced (herbal supplements or products sold as prescriptions), multiple standards must be followed for each product in each of these locations: US FDA, WHO, and other jurisdictions around the world. As a result, there is a lack of consistency in the level of risk presented to consumers and the amount of evidence required to demonstrate product efficacy, safety, and manufacturing quality for each of these product types globally. Thus improving regulatory clarity and implementing herbal-specific risk assessment tools to ensure the safe use of botanicals in modern healthcare systems is essential to their successful incorporation (Wu et al., 2020).

Ethical considerations are equally important, especially regarding the use of traditional knowledge and community-derived resources. Ethnobotanical studies must be conducted in ways that ensure prior informed consent, the sharing of benefits, and proper credit to indigenous knowledge holders. International treaties, such as the Convention on Biological Diversity (CBD) and the Nagoya Protocol, emphasize the fair and equitable sharing of genetic resources and related traditional knowledge. Ethical conduct safeguards community rights and, at the same time, enhances the trustworthiness and long-term viability of botanical research by fostering trust between researchers and knowledge holders (Bhadra et al., 2025).

A major challenge from a translational point of view is how to connect the findings of the laboratory with clinical and public health applications. Several botanicals have shown promising preclinical activity but have not progressed further due to poor formulation approaches, insufficient pharmacokinetic data, or insufficient funding for large-scale trials. Collaboration among academia, industry partners, clinicians, and regulatory authorities is essential to overcome these barriers. Through the harmonization of ethics, regulatory standards, and translational research, botanical medicine can move from being simply a traditional use and experimentally validated to an evidence-based therapeutic adoption (Thamizhoviya, 2025). Fig. 12.2 highlights the major challenges in advancing evidence-based botanical medicine.

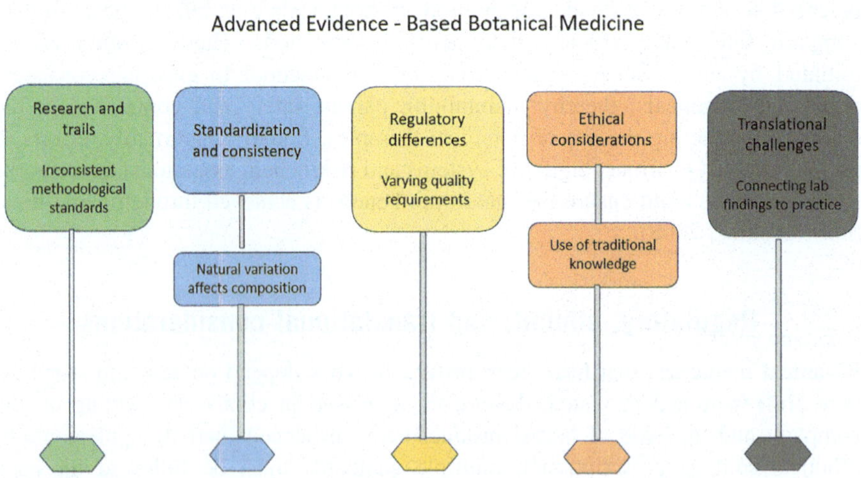

FIGURE 12.2

Challenges in evidence-based botanical medicine. Evidence of botanicals.

12.3 Innovation and technology for the future of herbal medicine

12.3.1 Digital tools and technologies using artificial intelligence for phytochemical analysis

Artificial intelligence (AI) tools and digital technologies are quickly becoming the tools of choice for Pharmaceutical botany and safety assessments, and for personalized herbal medicine. The AI strengths in prediction, pattern recognition, and clinical decision support will directly benefit the development of pharmaceutical botanical drugs (Zhang et al., 2021). AI can perform phytochemical data analysis by classifying chromatographic profiles, predicting bioactivity from chemical fingerprints, and prioritizing compounds to study in the laboratory—thereby decreasing the laboratory workload and expediting the identification of lead compounds. Digital ethnobotany tools, including mobile apps and structured databases, enable the accurate documentation of traditional knowledge with GPS data and voucher images. Ethnobotanical surveys provide raw species lists that may be digitized and algorithmically analyzed for cross-cultural patterns (Louie et al., 2020).

AI models integrating pharmacological and safety data may help predict herb-drug interactions in clinical contexts and identify patients likely to benefit or experience risk, by extending principles demonstrated in AI-supported medical management (Elhaddad and Hamam, 2024). The main limitation is data quality: biased or poorly annotated datasets lead to unreliable predictions. AI tools, therefore require accurate, complete, and verifiable data collection, provenance tracking, consent, and clear reporting to ensure their safe and valid use (Gebru et al., 2021).

Phytochemical analysis is only a part of it, as AI-driven technologies are being expanded to support formulation development, quality control, and postmarketing surveillance of herbal medicines. Machine learning models assist in processes such as extraction parameter optimization, stability profile prediction, and batch-to-batch consistency verification by correlating processing variables with chemical and biological results. In pharmacovigilance, natural language processing tools enable systematic mining of adverse event reports, electronic health records, and published literature to identify safety signals potentially related to herbal products. Moreover, by coupling these technologies with regulatory databases and clinical decision support systems, they not only enhance risk assessment capabilities but also pave the way for the safe, effective, and personalized use of herbal medicines in the next healthcare models (Saslis-Lagoudakis et al., 2012).

12.3.2 New biotech methods

The use of biotechnology improves the uniformity and safety of plant-derived materials and increases the number of plants available for medicinal use. Examples of biotechnological methods include plant tissue culture, metabolic engineering, synthetic biology, and microbial biosynthesis, which are used to grow compound(s) shown to have pharmacological activity, with consistent quality and less impact on wild-harvested materials (Phillips and Garda, 2019). Tissue culture and micropropagation are particularly useful for preserving the existing populations of endangered medicinal plant species by enabling cultivation outside their native environments. Metabolic engineering enables the cultivation of a specific target metabolite by microbes under controlled conditions, thereby reducing both the variability and the risk of contamination that result from producing plant-derived materials using agricultural practices. In addition to using technology to produce a healthy and safe product, the introduction of bioprocessing technology (e.g., bioreactors and good manufacturing practice [GMP] extraction) eliminates the uncertainty inherent in producing a product through agricultural means. In addition, the regulation of biotechnological products falls into different categories, each requiring appropriate quality and safety controls and an evaluation of the potential environmental risks associated with their use (Liu et al., 2017).

Genomics, transcriptomics, and proteomics greatly improve our ability to isolate genes and precisely define the pathways leading to the production of phytochemicals with medicinal value, facilitating advancement in the complete biotechnological process. The complete genomes of medicinal plants and their associated organisms (such as microorganisms) can be used to identify genes encoding enzymes and regulators that control the synthesis of medicinal compounds, which can then be used for gene editing and pathway engineering. These techniques not only increase the production and quality of medicinal compounds but also help track their products and protect related intellectual property rights. In concert with environmentally sound manufacturing processes and regulatory systems, these modern biotechnological tools provide a significant opportunity to upgrade traditional herbal materials into high-quality, safe, and easily scalable medicinal products that can become part of today's healthcare systems (Zeng et al., 2022).

12.3.3 Sustainability, precision medicine, and future directions

Future trajectories for herbal medicines will hinge on integrating sustainable development standards across technology development and precision medicine frameworks. As medicinal plants are now distributed globally, they are putting immense pressure on wild populations; responsible sourcing, cultivation, and conservation measures must be developed and implemented. Examples of such technologies that will assist in reducing reliance on harvesting plants from the wild whilst ensuring consistent quality (as well as continuity of supply) include controlled cultivation systems, tissue culture production, and microbial biosynthesis. Integrating sustainability metrics into the research and development pipelines of herbal medicines will enable the achievement of both environmental and public health goals, as well as assist with global compliance with sustainability standards (Chen et al., 2016).

In addition, precision medicine techniques open up new possibilities to ensure the effective application of herbal medicine in clinical practice by evaluating genetic, metabolic, and lifestyle differences among patients. The application of advanced pharmacogenomics and metabolomics techniques helps to identify unique patient subgroups in clinical practice who are positively or negatively influenced by a specific herbal medicine. The application of clinical decision support systems, aided by AI, helps personalize the dosage of these herbal medicines, thereby increasing their safety and efficacy in achieving therapeutic goals. Therefore these techniques can take herbal medicine from traditional practice to the domain of evidence-based, data-driven, and personalized practice (Li and Zhang, 2013). Looking ahead, the integration of digital technologies, biotechnology, and regulatory science will be the main features of the next generation of evidence-based herbal medicines. Working together across disciplines, supporting clinical research over a long period, and standardizing internationally will be key to the successful transition to innovation and practice. Herbal medicine has an opportunity to maintain its position as a reliable and effective element of today's health care systems if it can balance the scientific approach, sustainability, and patient-centered care with traditional knowledge (Ekor, 2014). As illustrated in Fig. 12.3, technological innovations and sustainability-focused strategies are shaping the future of herbal medicine.

12.3.4 Conservation, climate impact, and sustainable cultivation

A review of ethnobotanical data and medicinal plants indicates significant conservation issues, and many important medicinal plants are at risk due to limited ranges, habitat degradation, and local population threats. Therefore the cultivation

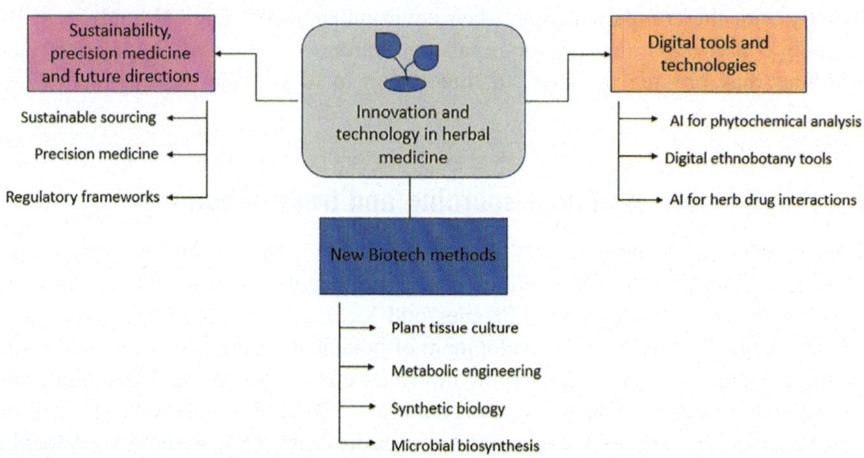

FIGURE 12.3

Innovation and technology in herbal medicine. Innovations in herbals.

of plants using sustainable methods as well as conservation measures will be essential for the responsible development of botanically based therapeutics (Porter-Bolland et al., 2012). Through conservation-oriented cultivation (agroforestry), creating local seed banks, and promoting community nurseries, we can not only cultivate and conserve medicinal plants but also support their recovery and provide local economic benefits to communities. The use of plant leaves for medicinal purposes is indicated by ethnobotanical research and would encourage a move towards nondestructive harvesting to alleviate existing environmental pressure (Rahman et al., 2017).

To promote nurseries' resilience, nurseries should increasingly use climate-resilient plants. This would involve monitoring plant populations and using adaptive cultivation techniques to ensure the availability of raw materials for the medicinal plant sector (Rodhouse et al., 2015). Further, ex situ conservation methods, such as tissue culture and seed banking, could be employed to conserve threatened or endangered species by reducing reliance on wild plant materials and thereby ensuring sufficient raw materials for the medicinal plant sector (Kovalchuk et al., 2014).

Collaboration among different job types and industries can lead to the successful development of sustainability methods through actions taken by researchers, policymakers, business members, and community members to conserve nature by enforcing laws, offering incentives for sustainable agriculture, and regulating the commerce in medicinal plant materials. It will help preserve ecosystems in the herbal medicine industry, develop new drugs through biodiversity conservation, and secure the future of the Earth. By utilizing these methods to verify the sustainability of their material or product source, many companies could succeed by utilizing an established system for traceability/certification to guarantee that products are sourced back to the company's or the manufacturer's production location and comply with the company's established environmental/sociological guidelines. By aligning conservation through the regulatory framework, the herbal medicine trade activities may provide the basis for innovation in future drug development (Tsai et al., 2018).

12.3.5 Benefits of ethical sourcing and trade practices

Ethical sourcing is both a duty to society and a requirement of international agreements, such as the CBD, which establishes regulations regarding access and benefit-sharing with Indigenous Peoples and local communities. The responsible use of botanicals requires the development of botanical products through active and genuine engagement (with free, prior informed consent) with the community and the equitable sharing of benefits (Morgera et al., 2014). To make ethical sourcing operational, three key elements need to be addressed: (1) Community protocols/legal agreements specify the terms for the equitable sharing of benefits and also assist in capacity building and, when/as appropriate, coownership and/or royalties to holders of traditional knowledge. (2) Traceability systems, by which certificates of origin and chain of custody documents will be sent to the end user, will help ensure that procurement is transparent and fair. (3) Sustainable local development initiatives can bring considerable socioeconomic benefits to the community by generating income, providing agronomic training, and including community members in the quality control and production of the products. These initiatives will empower knowledge holders to become active partners in both conservation and the development of botanical products (Jagger et al., 2014).

The scientific validity and market viability of herbal products can all be affected by the ethical and legal implications of responsible sourcing and fair trade. The importance of responsible sourcing of herb materials is that it can greatly lower the capitalization, adulteration, and misidentification risks of herbal products through a transparent onboarding process. Building consumer confidence and supporting regulatory compliance will also enhance the ability of ethical herb products to meet rigorous national and international quality standards and thus be included in global healthcare systems (Thorsen et al., 2006).

12.3.6 Policy integration, certification, and long-term ecological stewardship

Taking into account ecological principles when developing local and global policies for research and development, production, and trade in medicinal plants will ultimately promote the long-term availability of herbal medicines. By implementing policies that promote biodiversity preservation, such as sustainable harvesting quotas and environmental impact assessments, the long-term sustainability of herbal medicines will be enhanced. Additionally, the claim that the international statutory frameworks, including the CBD and the Sustainable Development Goals (SDGs), support this sentiment by promoting equitable use and sharing of the benefits from the ecosystem supports the aforementioned claims (Howes et al., 2020).

Furthermore, certification schemes and sustainability standards are of great importance, as they serve as bridges between policies and the real-world. Such schemes include organic certification, fair trade,GACP certification, among others. These schemes have set measurable targets for achieving environmental and social sustainability. These targets help improve the supply chain by making sourcing more spot-on; thus enforcing environmental protection laws becomes easier. At the same time, a reward awaits those at the forefront of sustainable production methods. From the perspective of end users and medical practitioners, certified herbal products are more trustworthy, as they are invariably tested and found to be of high quality, safe for use, and produced in accordance with the appropriate ethical standards (Robinson et al., 2012). However, long-term ecological stewardship following regulation and certification will also require ongoing monitoring, management, and education. By integrating remote sensing, geographic information systems, and ecological modeling, it is possible to monitor plant populations continuously and assess habitat health. Educating plant growers, harvesters, and industry professionals through various initiatives is also a crucial step in integrating sustainability at all levels of the supply chain. By themselves, policy integration and certification cannot ensure the necessary ethical, scientific, and environmental underpinnings of the herbal medicine industry. It is through their integration with stewardship approaches that a robust framework is established to support the entire chain of herbal medicine development. In addition, training and capacity building in regulatory science are necessary for successful global harmonization (Wang et al., 2024). Table 12.1

Table 12.1 Sustainability, ethical sourcing, and policy frameworks in herbal medicine development.

Thematic area	Key issues identified	Strategic approaches	Outcomes and relevance
Conservation and sustainable cultivation	Habitat loss, overharvesting, climate-induced variability, decline of wild medicinal plant populations	Agroforestry systems, community nurseries, seed banks, nondestructive harvesting (leaf-based use), ex situ conservation (tissue culture, seed storage)	Biodiversity conservation, recovery of endangered species, stable raw material supply, livelihood support
Climate impact and adaptive sourcing	Altered plant distribution, changes in phytochemical composition, reduced availability	Climate-resilient sourcing, adaptive cultivation, continuous ecological monitoring	Long-term stability of medicinal plant quality and availability under climate change
Ethical sourcing and benefit sharing	Risk of exploitation of indigenous knowledge, lack of transparency in trade	Free prior informed consent, community agreements, equitable benefit sharing, traceability systems	Social justice, protection of traditional knowledge, improved product authenticity, and acceptance
Policy integration and certification	Weak enforcement, unsustainable trade practices, regulatory gaps	Alignment with CBD and SDGs, GACP, organic and fair-trade certification, sustainability standards	Regulatory compliance, improved product quality, global market access
Long-term ecological stewardship	Lack of monitoring, fragmented governance	Remote sensing, GIS-based monitoring, education programs, regulatory science capacity building	Resilient herbal medicine supply chains and ecosystem protection

below summarizes the most important sustainability, ethical sourcing, and policy strategies for the responsible development of herbal medicines.

12.4 Regulatory evolution and global harmonisation

12.4.1 Strengthening international standards and cooperation

However, there are significant differences in the regulatory systems of various countries for botanicals, which hinder international research, development, and marketing. Policy reviews have consistently emphasized the importance of harmonized standards in quality, safety, and efficacy as a facilitator of responsible international exchange, with a focus on enhancing the evidence base for botanical medicines (Tilburt and Kaptchuk, 2008). Harmonization efforts should focus on several important elements: first, establishing common quality standards for analytical procedures, reference materials, and acceptable concentrations of marker compounds would help avoid redundant testing and facilitate regulatory approval (Li et al., 2008). Second, common safety evaluation standards, such as standardized toxicology studies and genotoxicity testing, along with common pharmacovigilance practices, would enable the uniform evaluation of botanical risks. Third, international collaborative research platforms would enable the compilation of ethnobotanical information, the development of common clinical trial approaches, and the conduct of multicentre trials (Pelkonen et al., 2014). Such collaborative mechanisms have been employed successfully in other domains of therapeutic research and are encouraged by guidelines on evidence synthesis and reporting (Liu et al., 2025).

The regions that contain most of the medicinal plants, rich areas, are not well-equipped with the infrastructure of analysis, the capacity for clinical research, and also with the capacity for regulatory knowledge. International cooperation programs that include technical training in laboratory and regulatory knowledge can help fill these gaps and ensure that the same standards are applied to all countries. Global cooperation on regulations, through enhanced institutional capacity and policy alignment, can be a catalyst for safe innovation and the development of botanically derived therapeutics (Moher et al., 2018).

12.4.2 Addressing regulatory gaps and future needs

Although there have been some advances in the regulation of botanical products, regulatory gaps remain that are perceived to inhibit the safe and efficacious incorporation of botanical products into contemporary health care systems. These gaps include inconsistent naming of botanicals; insufficient pharmacovigilance for herbal medicines; and a lack of clearly defined regulatory pathways to support the development of complex, multiconstituent preparations. To mitigate these gaps, specific innovations will need to be developed following the best available evidence and policy analysis. Key to improving nomenclature accuracy will be the authentication of botanical samples submitted to regulatory authorities and, when possible, molecular characterization, to prevent misidentification and adulteration of plants submitted as products (Pringsheim et al., 2011). There is also an urgent need to develop framework-based, tiered adaptive regulatory pathways for botanical natural health products that allow low-risk products to use the tiered adaptive regulatory framework, while ensuring that products with therapeutic/disease-modifying claims continue to go through full drug approval processes. There is a need to improve postmarketing surveillance through active pharmacovigilance, including standard adverse event reporting and digital signal detection systems. Ultimately, evidence generation should utilize more flexible approaches. Where traditional RCTs are not feasible, well-designed observational studies and registry-based pragmatic trials that adhere to the STROBE criteria can generate reliable evidence to inform policy (Glienke et al., 2015).

Filling regulatory gaps will involve adopting more flexible, scientifically based governance systems that balance innovation with public health requirements. Also, by incorporating RWE, digital technologies, and cross-border data sharing into the regulation of botanicals, the regulatory community can help to achieve more appropriate and innovative regulatory decisions. Additionally, ongoing collaborations between regulators, researchers, industry stakeholders, healthcare providers (HCPs), and holders of traditional knowledge will help ensure that regulatory frameworks evolve based on evidence and inclusiveness. By proactively adapting regulatory approaches to the unique nature of botanicals, global healthcare systems can ensure safe, effective, and equitable access to herbal medicines (da Silva et al., 2019).

12.4.3 Toward a unified global framework for botanical medicines

The process of integrating herbal remedies into global health systems is still in progress. It depends on establishing uniform regulatory systems that include accepted uses based on historical data and modern scientific evidence. These systems will need to reflect an appropriate level of risk to include: (1) the long- time established traditional products versus; and (2) newly developed botanical formulations with a specific therapeutic claim. Working with a coordinated approach to regulatory structures will provide a clearly defined pathway for the herbal industry; eliminate duplicate work across jurisdictions; and facilitate cross-border research collaboration and product/ingredient sharing (Asahina et al., 2013).

An international harmonization process should provide the foundation for information interoperability and transparency in data sharing across multiple disciplines. An example of this would be through the establishment of a unified database for all botanical medicines, including standardized databases for identity, clinical evidence, safety profiles, and phytochemical/botanical components attributes. These global databases could facilitate regulatory harmonization by synthesizing all relevant clinical and research data on botanical medicines and serve as the basis for developing a common method to evaluate the safety of and collect clinical data on botanical medicines, given the information in Table 12.2. To facilitate this process, there should be an agreement on a standard data collection format, nomenclature, and reporting requirements that will help with improved traceability and postmarketing surveillance, ultimately leading to increased public trust and greater patient safety (Guillopé et al., 2011).

Achieving an agreed set of international standards will require strong political will and continued collaboration among governments, regulatory agencies, academic institutions, and industry stakeholders. Through their collaborative efforts to establish a single regulatory framework integrating scientific integrity with ethical accountability and cultural considerations, the herbal medicines sector can ultimately eliminate fragmented national systems and establish a unified global regulatory model that enhances innovation, facilitates easier access to herbal medicines, and protects public health (Wong et al., 2025) (Table 12.2).

12.5 Integrative and personalized botanicals

12.5.1 The role of botanicals in integrative and holistic healing

Healthcare professionals often utilize botanical medicine to enhance patient-centered care rather than as a stand-alone treatment option. Botanical medicine may provide symptom relief, support quality-of-life improvement, or be used together with conventional treatments (Ben-Arye et al., 2012). Many oncology patients utilize biologically based complementary therapies to relieve symptoms (e.g., fatigue) or support their overall health and well-being. Healthcare professionals have a responsibility to provide evidence-based therapeutic recommendations and to monitor for adverse effects, such as herb-drug interactions. The use of aromatherapy, using essential oils to treat anxiety, sleep disturbances, and mood, is also becoming a common component of supportive and palliative care and is often combined with both psychosocial and pharmacological interventions (Lim et al., 2011).

Botanical formulas not only relieve symptoms but also provide multifaceted support for patients' overall health. This includes, among other things, physical, mental, and social well-being. Many traditional systems of healing focus on restoring balance and strength rather than on eradicating specific symptoms of a disease as the primary goal of therapy. As such, botanical medicine aligns well with current models of whole-person, person-centered care. Botanicals are chosen for their effectiveness based on a patient's constitution, the stage of illness, and the patient's own expectations and preferences; therefore providing a means for individualization and flexibility of care delivery methods that will be more likely to keep patients compliant and happy if used in combination with standard forms of treatment (Fasinu et al., 2012).

Table 12.2 Regulatory evolution and global harmonisation of botanical medicines.

Regulatory dimension	Current challenges	Proposed regulatory actions	Expected impact
International standards and cooperation	Fragmented national regulations; duplicative testing; inconsistent quality and safety criteria	Harmonized quality standards (shared analytical methods, reference materials); standardized safety frameworks; coordinated pharmacovigilance	Faster regulatory review; reduced redundancy; safer global exchange of botanicals
Capacity building and equity	Limited analytical, clinical, and regulatory capacity in biodiversity-rich regions	Technical training, shared labs, regulatory mentoring, and institutional partnerships	Equitable implementation of standards; accelerated responsible innovation
Taxonomy and authentication	Misidentification; adulteration; inconsistent nomenclature	Voucher authentication; molecular identification with submissions	Improved product integrity and safety
Adaptive regulatory pathways	No clear pathways for multiconstituent products; one-size-fits-all approvals	Tiered, risk-proportionate frameworks; differentiated pathways for traditional vs. novel claims	Proportionate oversight; innovation without compromising safety
Evidence generation and surveillance	Limited feasibility of RCTs; weak postmarket monitoring	Use of RWE, pragmatic/registry studies (STROBE-aligned); active digital pharmacovigilance	Policy-relevant evidence; enhanced patient safety
Data interoperability and transparency	Disparate databases; poor traceability	Common databases, harmonized terminology, and reporting formats	Regulatory convergence; improved trust and traceability
Unified global framework	Disjointed national systems; limited cross-border collaboration	Coordinated global framework integrating tradition and science	Transparent pathways; improved access; public health protection

From a clinical standpoint, to properly incorporate botanicals into standard healthcare delivery systems, there must be an established, clear communication channel between the patient and the HCP. This is necessary because patients have varying levels of comfort sharing their use of herbal remedies with their nurse or doctor, which can increase the risk of missing an adverse reaction between a botanical and a drug or of failing to detect these interactions altogether. Implementing a screening question about botanical use as part of the patient's history and providing evidence-based guidance on how to make a shared decision about treatment will help eliminate these complications. In addition, it is equally important to provide education to HCPs regarding identifying common botanicals, their safety, and the best source(s) of information (Rousseau, 2020).

12.5.2 Personalized herbal therapy and consumer trends

Greater demand for personalized and natural therapeutic choices creates increasing challenges for personalization strategies that weigh patient preferences against clinical indication, safety, risk of polypharmacy, and evidence. Effective personalization should balance therapeutic benefit with risk minimization. The process of clinical phenotyping can help identify the most appropriate herbal agents for a given patient, taking into consideration their associated comorbidities, medications, and where applicable, genetic/metabolic risk factors. Clinicians can use digital tools and clinical decision support systems to identify interactions and create a physician-tailored regimen. Patient education should also be included in this process, along with shared decision-making (Christensen et al., 2009).

Consumer trends indicate that persons with chronic or persistent symptom burdens are more likely to use complementary and alternative medicine. This underscores the need for proactive clinician engagement and greater research attention to patient-centered outcomes in botanical therapy. Integrative botanical medicine also aligns with a movement in clinical care that emphasizes prevention and health support. Herbs with antioxidant, antiinflammatory, or adaptogenic effects are the focus of research on how stress can be managed, the immune system modulated, and the progression of chronic diseases prevented. If these uses are supported by solid scientific evidence and the patient's condition is closely monitored, they may lead to reduced symptoms, improved capabilities, and support in the treatment of chronic conditions. Hence, botanicals are instruments that the physician can use to make the patient's treatment more comprehensive and effective. The progress of precision health technologies is also contributing to personalized herbal therapy through more accurate patient stratification and outcome monitoring (Bishop and Lewith, 2010).

Pharmacogenomic screening, metabolomic profiling, and digital symptom-tracking platforms, among others, can provide deep insight into how an individual responds to botanical interventions. Such methods are effective tools for adaptive dosing and for identifying patients who can benefit most from the selected herbal preparations and those at risk of adverse effects. Additionally, integrating such tools with electronic health records and evidence-based guidelines enhances clinical oversight. It facilitates personalization of botanical therapies more safely and effectively (Kellogg et al., 2019). As a consumer, my access to, and use of, herbal products has changed dramatically with the availability of these products through the internet and other direct-to-consumer marketing methods, which gives me increased freedom of access; however, it also increases my risk of receiving false information about the product, the potential quality of the product being inconsistent, and the potential for using these products in an uncontrolled manner. Therefore to ensure that consumers do not receive products that do not meet the safety and therapeutic effectiveness of their needs, the levels of regulatory control must be increased, the amount of information can be provided with product labeling is enhanced, and that the level of health literacy is developed, so that consumers will not make ill-informed decisions about their use of products that have been generated by demand. Additionally, clinicians must provide consumers with the knowledge they need to make informed decisions and assist them in clarifying their preferences with evidence-based care (Ekor, 2014).

12.5.3 Future outlook for integrative and personalised botanical medicine

The future direction of integrative and personalized botanical medicine lies in merging traditional knowledge with biomedical research and digital innovation. The increasing shift of healthcare systems to patient-centered, value-based models of care will provide exceptional opportunities for botanicals to support interventions, prevention, and improvements in quality of life. High-quality clinical research will play an integral role; however, there is also a need for improved mechanistic understanding of multicomponent botanical formulations and their appropriate indications, thereby lending more confidence to clinicians' use of botanicals (Erratum, 2012).

The new technologies being created today are expected to revolutionize the delivery of personalized botanical care. AI, systems biology, and RWE platforms

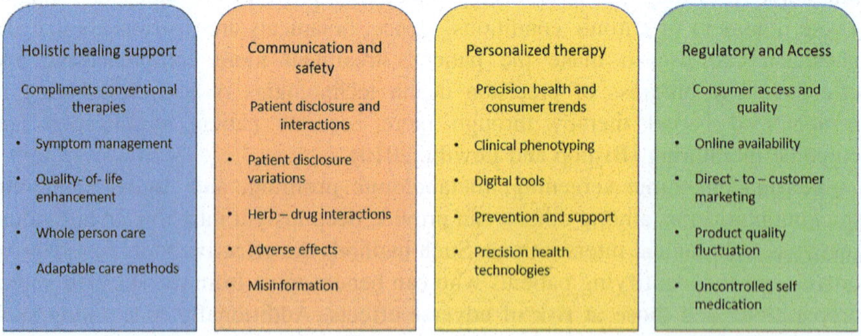

FIGURE 12.4

Personalized and integrative botanical treatments. Integrations in botanical treatments.

will provide better methods for developing personalized treatment plans that are phenotypically, genetically, and environmentally aligned. In addition, they will provide mechanisms for continuous assessment of outcomes and feedback loops, further enhancing the safety, efficacy, and patient compliance with botanical therapies (Li et al., 2014). Fig. 12.4 illustrates the growing importance of integrative and personalized approaches in botanical medicine.

12.6 Improving quality, safety, and transparency

12.6.1 Enhancing quality control and product integrity

To ensure the prompt delivery of high-quality, consistent herbal products, quality systems should be implemented to integrate all aspects of the product's manufacturing process, from sourcing raw materials to distributing the finished product (DeArmond and DiGoregorio, 2013). The primary objectives of establishing and maintaining high-quality standards include: ensuring that unqualified (adulterated) products are not introduced in the market; verifying the identity of the source of the unprocessed material used in the manufacture of the product; and ensuring that the product is accurately measured in terms of quality, potency, and effectiveness. To ensure the product meets the required standards, the following basic components should be implemented by the product manufacturer: GACP, GMP), molecular genetic identity of the plant material, multimarker chromatographic profile, and shelf-life stability testing (Marques and Pereira, 2019).

Moreover, new AI-based visual inspection and deep learning/neural-network monitoring systems should additionally support advanced manufacturing quality initiatives like Six Sigma and DMAIC, which can help optimize manufacturing processes and promote the reduction of defects (i.e., quality) during the manufacturing process through effective use of analytics. To promote traceability in the context of regulatory and clinical assessments, there is a need for batch-level certificates of analysis for complex formulations, with transparent labeling that identifies both active and inactive ingredients. In addition to their importance in meeting regulatory requirements (Lee et al., 2015). A robust quality assurance program should demonstrate a strong commitment to continuous monitoring, accurate documentation, and mechanisms for improvement to ensure product integrity. All requirements for identifying variations and quality failures will be met by testing every batch to verify audited supplier qualifications and managing all deviations using an appropriate system, thus preventing the possibility of manufactured products being delivered to consumers. The integration of a digital quality management system and a physical supply chain can provide traceability by linking data on raw materials, laboratory test results for manufactured goods, and manufacturing records. Establishing visibility at a minimum level of regulatory compliance will enhance consumer confidence in the safety, authenticity, and quality of herbal products (Li et al., 2019).

12.6.2 Consumer education and misinformation reduction

As consumers continue to seek legitimate, secure botanicals, they are frequently left without sufficient guidance to protect them from the dangers of misinformation, mistakenly prescribing themselves a botanical, or having a botanical interact with their prescription medications. By improving education and transparency, we could significantly reduce the risks associated with the safe and appropriate use of botanicals. Clinicians can receive training modules on commonly used botanicals, their interaction profiles, and the levels of evidence supporting their use in high-demand areas such as oncology and mental health, thereby improving clinical counseling and, consequently, patient safety. Public education efforts supporting standardized labels for products that provide information about where the product came from, how much to take, other drugs that may have interacted with the product, and how to report adverse events can further improve consumer awareness of the products and reduce the risk to the marketplace (Adebayo et al., 2018).

Digital communication channels, such as the internet and social media, will strongly influence consumer opinions and behaviors regarding herbal medicine. The significant increase in misinformation and promotional materials can offset the advantages these channels offer. Educational technology, developed based on scientific principles, for example, web-based information, decision-support technology, and smartphone applications developed in conjunction with HCPs, plays a very important role in countering misinformation by connecting consumers to credible sources of information. As a group, the ultimate goal that the various stakeholders, for example, the regulatory body, HCPs, and industry members, will be able to attain in terms of improving the quality and accuracy of the information available to consumers, increasing the level of health literacy, and ensuring that the consumer is provided with the necessary tools to be able to make informed decisions regarding the safe usage of herbal medicine and dietary supplements will be achieved by a collaborative effort to develop these solution strategies (Kantor et al., 2016).

12.6.3 Postmarketing surveillance, accountability, and continuous improvement

The monitoring of marketed products is a critical step in ensuring the safety and quality of botanical products remain high, even after they are made available to consumers. Because a large number of people use herbal products for treatment and because these products are often used in conjunction with other treatments, it is more important than ever to have a systematic and cohesive process in place to monitor their safety. An active pharmacovigilance system that collects adverse event reports, reports of drug-herb interactions, and reports of product quality issues will provide the information needed for regulatory authorities, manufacturers, and doctors to identify new risks and take measures to prevent them rapidly (Barnes, 2003).

Manufacturers who provide accountability for the complete lifecycle of an herbal product can help to build consumer trust and support a long-term commitment to quality improvement. Manufacturers are responsible for tracing supply chain components and investigating any public safety signals in a timely and thorough manner. In addition, any results of these investigations must be reported to regulatory authorities. HCPs must also document any adverse events that they suspect are associated with the use of herbal medicines and provide patients with detailed directions for using them safely. HCPs can utilize digital reporting and RWE systems to discover signals and communicate effectively with all members of the supply chain; thereby strengthening trust between consumers and manufacturers (Flower, 2013).

Ongoing use of surveillance data throughout the lifecycle of an herbal product can improve cultivation and manufacturing practices, enhance labeling accuracy, and support the development of clinical guidance. Information obtained through postmarketing monitoring should be utilized to strengthen both quality standards and educational initiatives, thereby improving the current regulatory framework for herbal medicines. The establishment of accountability and feedback loops throughout the entire production chain will enable responsible growth of the herbal industry and ensure that safety, efficacy, and public trust remain core values in the future development and integration of herbal medicine into contemporary healthcare practices (Li et al., 2010). A summary table of the integrated quality assurance strategies, consumer safety, transparency, and accountability throughout the lifecycle of an herbal product is provided in Table 12.3.

12.7 Economic and industry perspectives

12.7.1 Market growth, innovation, and global demand

The global market for herbal medicine is expanding, driven by growing consumer interest and policy support for integrative medicine approaches. This offers opportunities for innovation in formulation science and quality assurance, as well as in the identification of new therapeutic niches (Ekor, 2014).

Some of the most exciting areas of innovation are nanoformulations, biotech-derived active compounds, and engineered probiotics, all of which have the potential for greater efficacy, increased bioavailability, or targeted delivery with sufficient supporting evidence. The rapid advances in these fields show how quickly the botanical therapeutic field is changing and that there will always be a need to develop new products as evidence becomes available and regulatory agencies begin regulating them (Lee et al., 2016).

The global demand keeps rising, which is also changing the industry's structure and the focus of investment in the herbal medicine market. Big international pharmaceutical companies, biotech firms, and venture capital are getting involved,

Table 12.3 Framework for improving the quality, safety, and transparency of herbal medicines.

Focus area	Core challenges	Key measures implemented	Impact on herbal medicine systems
Quality control and product integrity	Adulteration, variable potency, inconsistent raw materials, weak traceability	GACP and GMP compliance; molecular botanical authentication; multimarker chromatographic profiling; validated stability testing	Ensures product identity, consistency, potency, and regulatory compliance
Advanced manufacturing and QA systems	Process variability, manufacturing defects, and documentation gaps	Six Sigma/DMAIC approaches; AI-based visual inspection; neural-network monitoring; digital quality management systems	Reduced defects, improved process control, enhanced batch-level traceability
Transparency and labeling	Poor disclosure of composition and quality attributes	Batch-specific certificates of analysis; transparent labeling of markers and excipients	Improved regulatory review, clinical confidence, and consumer trust
Consumer education and risk awareness	Misinformation, unsafe self-medication, herb–drug interactions	Clinician education modules; standardized consumer labeling; public awareness initiatives	Safer product use, informed decision-making, and reduced adverse outcomes
Digital communication and misinformation control	Spread of unverified claims through online platforms	Evidence-based digital tools, clinician-endorsed apps, verified information portals	Improved health literacy and mitigation of misleading claims
Postmarketing surveillance and accountability	Underreporting of adverse events, delayed risk detection	Active pharmacovigilance; digital reporting platforms; RWE databases	Early detection of safety signals and rapid corrective actions
Continuous improvement and feedback loops	Static standards, limited learning from real-world data	Integration of surveillance data into cultivation, manufacturing, labeling, and policy updates	Long-term system resilience, sustained quality, and public trust

and the pace of research and development has accelerated, along with higher product quality, clinical validation, and regulatory compliance. Traditional medicine practitioners, educational institutions, and industry stakeholders are partnering to create innovation pipelines that can translate ethnobotanical knowledge into commercially viable, scientifically validated products. The above-mentioned trends point to a slow shift away from the fragmented, small-scale market towards more organized, research-based industry structures that comply with global healthcare standards (Dessì et al., 2021).

12.7.2 Opportunities and challenges: industrial stakeholders

Establishing strict quality certification, sustainable credentials, and credible clinical evidence will allow organizations across the industry to differentiate through the use of botanicals. Establishing these commitments allows organizations to create an opportunity not only to establish a premium position in the marketplace, but also to create strategic alliances with healthcare systems (Zhu et al., 2020). Despite these rewarding opportunities, many barriers exist. Building quality control and clinical studies infrastructures, in addition to creating traceable supply chains, requires immense capital outlay from the outset of the company. Additionally, businesses will face challenges navigating a complex regulatory environment. They will be expected to manage the potential liability associated with poor-quality products or the presence or absence of credible therapeutic claims on their brand image (Fradera-Soler et al., 2022).

The botanical medicine industry faces operational and regulatory issues; however, the industry participants also have a daunting list of items to accomplish to build consumer trust, develop an educated and skilled workforce, and sustain innovation to meet consumer expectations over the long-term. As competition in this area heats up, firms' ability to hire skilled employees will be critical; employees with expertise in regulatory sciences, pharmacognosy, clinical research, and quality control will be especially in demand. In addition, clear communication with consumers and responsible marketing practices will be necessary to maintain the credibility of firms in this area, which often make exaggerated claims. Firms that, to have an adequately skilled workforce, use scientific rigor in their decision-making process for product development and to obtain ethical sources will be the firms most likely to experience sustainable growth and resilience in the global botanical medicine industry (Lum and Hirsch, 2006).

12.7.3 Investment trends, innovation ecosystems, and future industry outlook

Investment patterns in the botanical medicine sector are increasingly shaped by innovations grounded in evidence, a focus on sustainability, and preparedness for regulatory changes. Partnerships across public and private sectors, venture capital funding, and government-supported research initiatives are among the important contributors to continuously advancing clinical validation, biotechnological production methods, and the integration of digital health (PammolLi et al., 2011).

Innovation clusters formed around academic research centers, contract research organizations, manufacturers, and (HCPs) accelerate the development and market introduction of new products. Collaborative ecosystems result in improved flow of information and ideas, reduced risk from developing new products, and support a collaborative, cross-disciplinary approach to formulation science, quality, and clinical assessment. Collaborative, innovative ecosystems are likely to be globally recognized as the basis for botanical therapeutics, given their history of success. The sustainability and credibility of the botanical medicine industry as a whole will ultimately depend on the degree to which scientific integrity, ethical sourcing, and open governance are adopted as everyday business practices. Organizations that are committed to developing a robust research culture, operating in accordance with established regulatory requirements, and being good corporate citizens will face

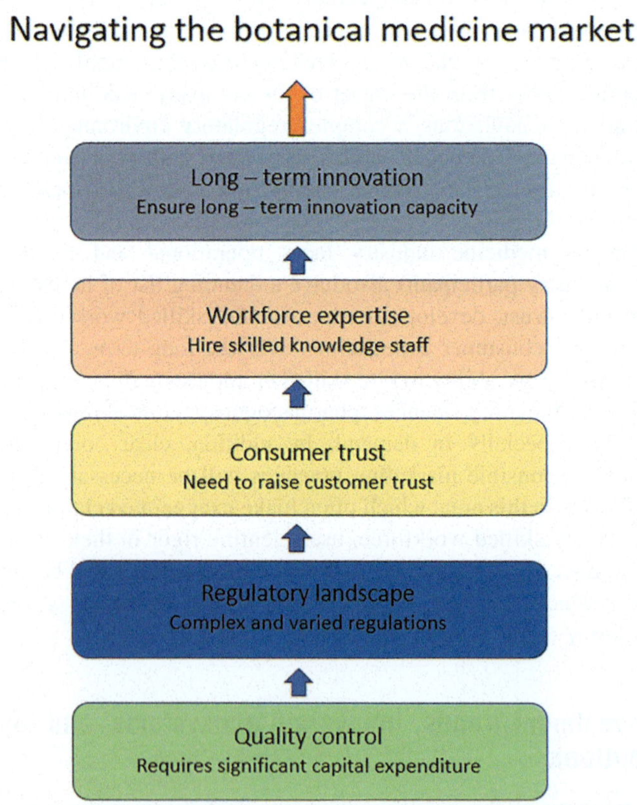

FIGURE 12.5

Navigating the botanical medicine market. Lab to market navigation.

fewer challenges in adjusting to consumer trends and policy changes over time. Once the botanical industry achieves a degree of stability and maturity, these principles will guide the direction of a vibrant and innovative industry that will ultimately be a meaningful contributor to healthcare systems worldwide (Newman and Cragg, 2020). Fig. 12.5 illustrates the key structural and strategic challenges involved in navigating the botanical medicine market.

12.8 Future research priorities in botanical sciences

12.8.1 Bridging gaps in clinical, pharmacological, and toxicity data

To move forward with botanical treatments, more focused studies should be conducted across several target areas. The priority would be to conduct high-quality clinical studies on priority indications; for example, to manage symptoms from cancer treatments or support mental health. The quality of clinical studies would help create guidelines for the use of botanical medicine and ensure patient safety. In addition, thorough studies examining how botanical medicines interact with other drugs are necessary to develop safety and long-term monitoring protocols properly. The development of validated biomarkers and patient-reported outcomes will not only provide additional information for efficacy assessments but also facilitate a more established introduction of botanical treatments into evidence-based practice (Zaza et al., 2020).

To progress within the continuum of pharmacology requires an affinity towards understanding the interactions between the biological mechanisms of action versus the pharmacokinetic characteristics of these complex medicinal botanicals. It is well known that a majority of botanical products produce beneficial effects on health through multiple targets and synergistic pathways, which traditional models for single compounds or single targets do not account for. Therefore implementing an integrative systems approach to pharmacology, utilizing network analysis with both sophisticated in vitro and in vivo models, can help shed light on these interactions and assist in the development of rational dosing approaches. By integrating mechanistic understanding with standardized evaluations of toxicological potential and prolonged observational trials, more reliable causal inferences can be made, which, in turn, will create a safer, more empirically-based approach to utilizing botanical therapies in "conventional" (mainstream) health care (Rothrock et al., 2011).

The future direction of research should also place strong emphasis on establishing extensive, comprehensive, well-documented botanical research repositories that can integrate ethnobotanical knowledge with phytochemical profiles, pharmacological information, clinical outcomes, and safety data. This extensive resource would facilitate metaanalyzes of botanical studies and comparative effectiveness research, and support the identification of research gaps across multiple therapeutic areas. The use of standardized data formats and open-access systems can further facilitate interinstitutional/cross-national collaboration, expedite the generation of evidence, and reduce redundancy in botanical research efforts. Another area of research that is extremely critical to the future of botanical research is examining the effects of botanicals on populations (at the macro level) and over time (long-term). Longitudinal cohort studies and pragmatic trials in clinical practice settings can provide important information on botanical efficacy, patient adherence to botanical therapies, botanical safety, and the economic impact of botanical therapies on healthcare costs. The importance of these types of studies is further underscored by the fact that, in most cases, patients with chronic diseases rely on long-term use of botanical therapies for their care and often use both botanical and conventional medicines concurrently (Been et al., 2013).

12.8.2 Expanding the multidisciplinary collaborations

To keep making progress in the field of plant medicine, we must come together as one. We must create partnerships among all stakeholders, including researchers in academia, clinicians, industry members, holders of Indigenous knowledge, and regulatory agencies. Together, we will share knowledge, expertise, data, and common research methods to use advanced analytical tools such as AI and systems biology; accelerate clinical translation; and maintain ethical and cultural integrity, as well as the scientific credibility of our research (He et al., 2021).

Collaborative models across multiple disciplines require shared infrastructure, governance structures, and processes to enable researchers from different fields to continue collaborating. These combinations of infrastructure and governance allow researchers to combine ethnobotanical information, lab results, clinical results, and insights into regulatory frameworks for the purpose of conducting joint research on a joint research platform, utilizing harmonized data and sharing agreements, and interoperating through interoperable databases with data generated by researchers using various means (e.g., laboratories, hospitals). It is also very important for research organizations constructing governance structures to take the necessary steps to ensure they can adequately address issues such as intellectual property rights, benefit-sharing obligations, data stewardship, etc. These issues are especially prevalent in research collaborations involving the use of traditional knowledge or other community-based resources (Laird et al., 2020). In addition, capacity-building efforts are necessary for the successful implementation of collaborative activities. Collaborative efforts should integrate pharmacognosy, clinical research, regulatory science, bioinformatics, and ethics into training programs; this will prepare future researchers and clinical practitioners to develop skills to work across various disciplines. Research results can be more easily transferred to practice through a better appreciation of the continuum of academic/research institutions, with exchanges, cross-sector fellowships, and other collaborative efforts across the public, academic/research, and private sectors (Dear et al., 2011).

Global collaboration is essential for tackling the many issues associated with the use of botanicals in research and therapeutic applications, including: (1) conservation of biological diversity; (2) the development of comparable rules and regulations across countries; and (3) the eventual elimination of barriers (e.g., economic, procedural, etc.) that presently limit individuals from accessing novel therapies when developing botanically-based medicines. Collaborative research between two or more countries provides the opportunity to establish regional/global databanks for the development of large-scale (regional and global) studies; utilize global study data to compare results; establish a universally accepted reference point for evidence in support of the development of botanically-based medicines; and create one or more internationally accepted sets of evidence standards for botanically-based medicines. In addition, if entities involved in developing botanically-based medicines (e.g., government agencies, nonprofit organizations, and for-profit companies) share their collaborative efforts with other entities that are committed to the common goal of creating an effective framework for the development of botanically-based medicines and appropriately manage these types of partnerships, there will be multiple opportunities for implementation of multidisciplinary collaborations that incorporate scientific rigor, ethical integrity, and global collaboration to advance the development of botanically-based medicine (Røttingen et al., 2013).

12.8.3 Building infrastructure, data integration, and research capacity

The establishment of robust basic scientific infrastructure and interconnected data systems has greatly enhanced ongoing improvements in plant science. For example, a dedicated plant science laboratory for research on its implementation, such as a new biobank for research in the clinical laboratory (analytical or clinical) environment. In this way, the creation and standardization of repositories to provide high-quality reference materials (plant species, plant extracts, and/or reference standards) can provide both an organizational and physical basis for high-quality (i.e., reproducible) plant science research. In addition, standardized repositories designed to store plant extracts and/or reference standards will promote consistency in the analysis of resulting plant extracts and/or reference standards for use in pharmacological and toxicological studies and/or clinical research (Fitzgerald et al., 2020).

The primary benefit will be the integration of heterogeneous data sources across compatible platforms, enabling researchers to conduct multidimensional analysis and aggregate evidence. The integration of ethnobotanical, phytochemical, pharmacological, clinical, and safety data will dramatically reduce the time required to generate and validate research hypotheses. Adoption of standardized data formats, metadata systems, and open science principles will create collaborative environments, increase the transparency of research findings, and enable better utilization of research investments across institutions and regions. Human capacity should be developed in parallel with physical and digital infrastructure. Sustained support for a variety of education/training programs in plant science, clinical research, bioinformatics, regulatory science, and ethical science will undoubtedly yield a sufficiently skilled workforce to propel the field forward. Large investments in infrastructure, data integration, and capacity building for research will empower the botanical sciences to more effectively address complex global health issues and deliver research innovations as safe and effective, evidence-based therapeutic interventions (Wilkinson et al., 2016).

12.9 Vision for a botanically empowered global health future

12.9.1 Long-term opportunities and transformative potential

With sufficient well-documented scientific evidence, ethically harvested, responsibly sourced products, and relevant regulatory frameworks (including certifications), botanical products can provide safe, culturally appropriate, and environmentally sustainable options in global health systems. Therefore botanical products have the potential to complement existing treatments for better symptom management. Additionally, new formulations combining botanicals with current therapies may be developed to improve adherence; further, botanicals may be used to support multiple lifestyle choices, including wellness (Thaker and Floyd, 2021).

In addition to supporting the primary healthcare system and aiding in the care of chronic illnesses in underresourced communities without being an insurmountable barrier to cultural practices or customary knowledge systems, well-executed botanical interventions would be instrumental in enabling the establishment of potent public health programs through affordable, locally available therapeutic modalities. Therefore if botanicals were integrated into national health structures and leveraged through appropriate training of health professionals, they could significantly contribute to developing robust and inclusive health systems that are both responsive to the needs and capabilities of the communities they serve (FirenzuoLi and Gori, 2007).

As a long-term growth opportunity, the most advantageous use of medicinal plants is to combine their therapeutic value with efforts to prevent and manage chronic disease, as well as to promote well-being by finding new ways to use them. The world is confronted with an increasing challenge from noncommunicable illnesses and the needs of an older population, and there is an opportunity for herbs that help with anxiety, inflammation, metabolism, and supporting the immune system to serve as important adjuncts or partners to health care providers. The role of botanical products in becoming scientifically validated therapies should be supported by additional development of personalized medicine, digital health initiatives, and environmentally sustainable production technologies to create long-lasting relationships between plants and humans, enhancing physical and mental health and happiness throughout one's life. Accomplishing this objective will require ongoing commitment to research across multiple disciplines, good governance, and global collaboration. Investment in education, facility construction, and consistency of standards will be the main drivers of changing ideas into products and services of value to be offered to people. By combining traditional knowledge and modern science while prioritizing public health, plant medicine will help create a more effective, humanistic healthcare system (Atanasov et al., 2021).

12.9.2 Recommendations and closing reflections

Several strategic actions will advance the responsible development of botanical drugs. Ethnobotanical evidence will dictate which species (and their clinical uses) take priority and will yield a strategic hierarchy for pursuing translational research. By investing in the development of analytical methods and systems for compliant manufacturing, educating physicians about botanical therapy, and establishing a comprehensive postmarket surveillance program, the safety and efficacy of botanical drugs can be improved. Furthermore, when combined with appropriate caution, ethical considerations, and scientific evaluation, the emerging technologies of artificial Intelligence and biotechnology will enhance both evidence development and supply chain systems. Finally, to continue building a solid foundation of ethical practice, long-term sustainability, and the integrity of the botanical industry itself, equitable benefit sharing and sound conservation commitments must continue to underpin botanical development (Santoro et al., 2018).

Promoting the responsible development of botanical therapeutics can occur through a set of strategically coordinated actions that align scientific rigor with the ethical and ecological aspects of the issue. The first call to action is to identify both plants and disease entities for which there is substantial ethnobotanical data, to facilitate translational research in an organized and strategic manner. There is also a need to establish advanced analytical methodologies, GMP, clinician training, and postmarketing surveillance methods to improve the safety, quality, and efficacy of botanical medicine. Newly emerging technologies have great potential to enhance the quality of evidence, improve the drug development process, and identify supply chain issues more quickly (examples include Artificial Intelligence [AI] and biotechnology). When used judiciously and within the framework of ethical principles, regulatory oversight, and scientifically sound evaluations, these technologies offer the potential to accelerate innovation while maintaining patient safety and recognizing the various cultures that contribute to the process. These new technologies should complement and/or enhance the traditional domains of pharmacological, clinical, and toxicological research rather than replace them completely (Wynberg and Laird, 2022).

12.10 Conculsion

Botanical therapeutic options play a unique and vital role in health care on a global scale, where their value is not only based on past use but also on the ever-growing body of scientific validation and clinical applications, as well as the preference of many patients for natural, culturally congruent, and sustainable treatment choices. To support these initiatives, coordinated efforts must be established across research, regulation, and industry, in collaboration with communities and partners. Evidence from this chapter suggests the future success of botanical medicine depends upon strengthening the entire evidence pipeline—from ethnobotanical documentation to phytochemical characterization and through to robust pharmacology, clinical trials, and postmarketing safety assessment. Advances in biotechnology, artificial intelligence, and quality system-based standardizations will aid in creating reliable botanical supply chains, expediting the development of new products, helping ensure that products are utilized in an ethically responsible manner, and fostering the creation of strong, high-quality, evidence-based data. The following programs will require robust regulatory structures to be implemented successfully. These regulatory frameworks support standardization to recommend consumer protection standards and to ensure that the clinical use of products is supported and validated. In addition, there must be an established pharmacovigilance system to monitor product safety as products are used. The regulations, standards, and systems used to administer these programs must work together to ensure safety and effective use of the products provided by the program. The foundational pillars of botanical knowledge are sustainability, ethical sourcing of funding, and equitable benefit sharing from the use of botanical materials. Integrative and personalized care models demonstrate how botanicals can enhance the effectiveness of conventional medical therapies and support symptom management and whole-person health and wellness. To develop safe, scientifically validated botanical therapy options for improved access to quality healthcare for individuals across the globe, a long-term strategic investment plan that includes collaboration among researchers and research institutions, and a partnership approach grounded in the stewardship of the earth's natural resources, will create resilient, sustainable local communities.

References

Adebayo OW, Salerno JP, Francillon V, Williams JR. A systematic review of components of community-based organisation engagement. Health and Social Care in the Community 2018;26(4):e474. Available from https://doi.org/10.1111/hsc.12533, http://onlinelibrary.wiley.com/journal/10.1111/(ISSN)1365-2524.

Asahina Y, Tanaka A, Uyama Y, Kuramochi K, Maruyama H. The roles of regulatory science research in drug development at the pharmaceuticals and medical devices agency of Japan. Therapeutic Innovation and Regulatory Science 2013;47(1):19–22. Available from https://doi.org/10.1177/2168479012469950.

Atanasov AG, Zotchev SB, Dirsch VM, Orhan IE, Banach M, Rollinger JM, et al. Natural products in drug discovery: advances and opportunities. Nature Reviews. Drug Discovery 2021;20(3):200–16. Available from https://doi.org/10.1038/s41573-020-00114-z, http://www.nature.com/nrd/index.html.

Barnes J. Pharmacovigilance of herbal medicines. Drug Safety [Internet] 2003;26(12):829–51. Available from https://doi.org/10.2165/00002018-200326120-00001.

Been JV, Nurmatov U, van Schayck CP, Sheikh A. The impact of smoke-free legislation on fetal, infant and child health: a systematic review and meta-analysis protocol. BMJ Open 2013;3(2):e002261. Available from https://doi.org/10.1136/bmjopen-2012-002261.

Ben-Arye E, Ali-Shtayeh MS, Nejmi M, Schiff E, Hassan E, Mutafoglu K, et al. Integrative oncology research in the Middle East: weaving traditional and complementary medicine in supportive care. Supportive Care in Cancer 2012;20(3):557–64. Available from https://doi.org/10.1007/s00520-011-1121-0.

Bhadra S, Pundir C, Mukherjee M, Kar A, Banerjee S, Charoensup R, et al. Reimagining ethnopharmacology with generative AI: towards inclusive, ethical, and data-driven traditional medicine. Pharmacological Research 2025;221:108002. Available from https://doi.org/10.1016/j.phrs.2025.108002.

Bishop FL, Lewith GT. Who uses CAM a narrative review of demographic characteristics and health factors associated with CAM use. Evidence-based Complementary and Alternative Medicine 2010;7(1):11–28. Available from https://doi.org/10.1093/ecam/nen023.

Burchard G. Infektionspatienten mit ansteckenden Krankheiten. Notfallmedizin up2date 2009;4(01):69–84. Available from https://doi.org/10.1055/s-0029-1185370.

Chen S-L, Yu H, Luo H-M, Wu Q, Li C-F, Steinmetz A. Conservation and sustainable use of medicinal plants: problems, progress, and prospects. Chinese Medicine 2016;11(1). Available from https://doi.org/10.1186/s13020-016-0108-7.

Christensen KB, Minet A, Svenstrup H, Grevsen K, Zhang H, Schrader E, et al. Identification of plant extracts with potential antidiabetic properties: effect on human peroxisome proliferator-activated receptor (PPAR), adipocyte differentiation and insulin-stimulated glucose uptake. Phytotherapy Research 2009;23(9):1316–25. Available from https://doi.org/10.1002/ptr.2782, http://www3.interscience.wiley.com/cgi-bin/fulltext/121661577/PDFSTART.

Company WAB, Collection HGGN and STC. Special wholesale prices on Burpee's sweet peas and other flower seeds. April 10, 1922 [Internet]. 1922. Available from: https://doi.org/10.5962/bhl.title.125657.

Dear RF, Barratt AL, Crossing S, Butow PN, Hanson S, Tattersall MHN. Consumer input into research: the Australian Cancer Trials website. Health Research Policy and Systems 2011;9. Available from https://doi.org/10.1186/1478-4505-9-30, http://www.health-policy-systems.com/content/9/1/30.

DeArmond PD, DiGoregorio AL. Rapid liquid chromatography-tandem mass spectrometry-based method for the analysis of alcohol ethoxylates and alkylphenol ethoxylates in environmental samples. Journal of Chromatography. A 2013;1305:154–63. Available from https://doi.org/10.1016/j.chroma.2013.07.017.

Dessì P, Rovira-Alsina L, Sánchez C, Dinesh GK, Tong W, Chatterjee P, et al. Microbial electrosynthesis: towards sustainable biorefineries for production of green chemicals from CO2 emissions. Biotechnology Advances 2021;46:107675. Available from https://doi.org/10.1016/j.biotechadv.2020.107675.

Ekor M. The growing use of herbal medicines: issues relating to adverse reactions and challenges in monitoring safety. Frontiers in Pharmacology 2014;4. Available from https://doi.org/10.3389/fphar.2013.00177.

Elhaddad, Hamam. AI-driven clinical decision support systems: an ongoing pursuit of potential. Cureus 2024;16(4). Available from https://doi.org/10.7759/cureus.57728.

von Elm E, Altman DG, Egger M, Pocock SJ, Gøtzsche PC, Vandenbroucke JP. The Strengthening the Reporting of Observational Studies in Epidemiology (STROBE) statement: guidelines for reporting observational studies. Lancet 2007;370(9596):1453–7. Available from https://doi.org/10.1016/S0140-6736(07)61602-X.

Englund JA, Boeckh M, Kuypers J, Nichols WG, Hackman RC, Morrow RA, et al. Brief communication: fatal human metapneumovirus infection in stem-cell transplant recipients. American College of Physicians, United States Annals of Internal Medicine 2006;144(5):344–9. Available from https://doi.org/10.7326/0003-4819-144-5-200603070-00010, http://annals.org/issues.aspx.

Erratum. British Journal of Clinical Pharmacology [Internet] 2012;74(1):224–5. Available from https://doi.org/10.1111/j.1365-2125.2012.04349.x.

Fabricant DS, Farnsworth NR. The value of plants used in traditional medicine for drug discovery. Environmental Health Perspectives 2001;109(suppl 1):69–75. Available from https://doi.org/10.1289/ehp.01109s169.

Fan C, Sun X, Wang X, Yu H. Therapeutic potential of the chemical composition of *Dendrobium nobile* Lindl. Frontiers in Pharmacology 2023;14. Available from https://doi.org/10.3389/fphar.2023.1163830.

Fasinu PS, Bouic PJ, Rosenkranz B. An overview of the evidence and mechanisms of herb-drug interactions. Frontiers in Pharmacology 2012;3. Available from https://doi.org/10.3389/fphar.2012.00069, http://www.frontiersin.org/Journal/DownloadFile.ashx?pdf=1&FileId=%201343&articleId=%2021203&Version=%201&ContentTypeId=21&FileName=%20fphar-03-00069.pdf.

Firenzuoli F, Gori L. Herbal medicine today: clinical and research issues. Evidence-Based Complementary and Alternative Medicine 2007;4(S1):37–40. Available from https://doi.org/10.1093/ecam/nem096.

Fitzgerald M, Heinrich M, Booker A. Medicinal plant analysis: a historical and regional discussion of emergent complex techniques. Frontiers in Pharmacology 2020;10. Available from https://doi.org/10.3389/fphar.2019.01480.

Flower RJ. Pharmacology 2.0. British Journal of Clinical Pharmacology [Internet] 2013;76(5):625–9. Available from https://doi.org/10.1111/bcp.12088.

Fradera-Soler M, Grace OM, Jørgensen B, Mravec J, Griffiths H. Elastic and collapsible: current understanding of cell walls in succulent plants. Journal of Experimental Botany 2022;73(8):2290–307. Available from https://doi.org/10.1093/jxb/erac054.

Fu PP, Xia Q, Lin G, Chou MW. Pyrrolizidine alkaloids - genotoxicity, metabolism enzymes, metabolic activation, and mechanisms. Drug Metabolism Reviews 2004;36(1):1–55. Available from https://doi.org/10.1081/DMR-120028426.

Gebru T, Morgenstern J, Vecchione B, Vaughan JW, Wallach H, Iii HD, et al. Datasheets for datasets. Communications of the ACM 2021;64(12):86–92. Available from https://doi.org/10.1145/3458723, http://dl.acm.org/citation.cfm?id=J79.

Glienke W, Esser R, Priesner C, Suerth JD, Schambach A, Wels WS, et al. Advantages and applications of CAR-expressing natural killer cells. Frontiers in Pharmacology 2015;6. Available from https://doi.org/10.3389/fphar.2015.00021, http://journal.frontiersin.org/article/10.3389/fphar.2015.00021/full.

Gröber U. Supportiver Einsatz mitotroper Mikronährstoffe: Mitochondriale Toxizität von Zytostatika. Zeitschrift für Orthomolekulare Medizin 2011;9(01):10–4. Available from https://doi.org/10.1055/s-0030-1270829.

Guillopé R, Escobar-Khondiker M, Guérineau V, Laprévote O, Höglinger GU, Champy P. Kaurenoic acid from pulp of *Annona cherimolia* in regard to annonaceae-induced parkinsonism. Phytotherapy Research 2011;25(12):1861–4. Available from https://doi.org/10.1002/ptr.3508.

Halder M, Jha S. Medicinal plants and bioactive phytochemical diversity: a fountainhead of potential drugs against human diseases. Sustainable Development and Biodiversity [Internet] 2023:39–93. Available from https://doi.org/10.1007/978-981-19-9936-9_2.

He X, Duan X, Liu J, Sha X, Gong Y, Lu W, et al. The antiinflammatory effects of Xuefu Zhuyu decoction on C3H/HeJ mice with alopecia areata. Phytomedicine: International Journal of Phytotherapy and Phytopharmacology 2021;81:153423. Available from https://doi.org/10.1016/j.phymed.2020.153423.

Heinrich M, Edwards S, Moerman DE, Leonti M. Ethnopharmacological field studies: a critical assessment of their conceptual basis and methods. Journal of Ethnopharmacology 2009;124(1):1–17. Available from https://doi.org/10.1016/j.jep.2009.03.043.

Hopkins AL. Network pharmacology: the next paradigm in drug discovery. Nature Chemical Biology 2008;4(11):682–90. Available from https://doi.org/10.1038/nchembio.118, http://www.nature.com/nchembio.

Howes MJR, Quave CL, Collemare J, Tatsis EC, Twilley D, Lulekal E, et al. Molecules from nature: reconciling biodiversity conservation and global healthcare imperatives for sustainable use of medicinal plants and fungi. Plants People Planet 2020;2(5):463–81. Available from https://doi.org/10.1002/ppp3.10138, nph.onlinelibrary.wiley.com/journal/25722611.

Jagger P, Luckert MMK, Duchelle AE, Lund JF, Sunderlin WD. Tenure and forest income: observations from a global study on forests and poverty. World Development 2014;64(1):S43. Available from https://doi.org/10.1016/j.worlddev.2014.03.004, http://www.journals.elsevier.com/world-development/.

Kantor ED, Rehm CD, Du M, White E, Giovannucci EL. Trends in dietary supplement use among US adults from 1999-2012. JAMA: The Journal of the American Medical Association 2016;316(14):1464. Available from https://doi.org/10.1001/jama.2016.14403.

Kellogg JJ, Paine MF, McCune JS, Oberlies NH, Cech NB. Selection and characterization of botanical natural products for research studies: a NaPDI center recommended approach. Natural Product Reports 2019;36(8):1196–221. Available from https://doi.org/10.1039/c8np00065d, http://pubs.rsc.org/en/journals/journal/np.

Kovalchuk I, Turdiev T, Mukhitdinova Z, Frolov S, Reed BM, Kairova G. New techniques for rapid cryopreservation of dormant vegetative buds. Acta Horticulturae 2014;1039:137–46. Available from https://doi.org/10.17660/ActaHortic.2014.1039.17, http://www.actahort.org/members/showpdf?session=12675.

Laird S, Wynberg R, Rourke M, Humphries F, Muller MR, Lawson C. Rethink the expansion of access and benefit sharing. Science (New York, N.Y.) 2020;367(6483):1200–2. Available from https://doi.org/10.1126/science.aba9609.

Lee J, Bagheri B, Kao HA. A cyber-physical systems architecture for Industry 4.0-based manufacturing systems. Manufacturing Letters 2015;3:18–23. Available from https://doi.org/10.1016/j.mfglet.2014.12.001, http://www.journals.elsevier.com/manufacturing-letters/.

Lee NYS, Khoo WKS, Adnan MA, Mahalingam TP, Fernandez AR, Jeevaratnam K. The pharmacological potential of *Phyllanthus niruri*. Journal of Pharmacy and Pharmacology 2016:953–69. Available from https://doi.org/10.1111/jphp.12565, http://www.interscience.wiley.com/jpages/0022-3573.

Li S, Han Q, Qiao C, Song J, Cheng CL, Xu H. Chemical markers for the quality control of herbal medicines: an overview. Chinese Medicine 2008;3. Available from https://doi.org/10.1186/1749-8546-3-7.

Li S, Zhang B. Traditional Chinese medicine network pharmacology: theory, methodology and application. Chinese Journal of Natural Medicines 2013;11(2):110–20. Available from https://doi.org/10.1016/s1875-5364(13)60037-0.

Li Y, Xie Y, He Y, Hou W, Liao M, Liu C. Quality markers of traditional chinese medicine: concept, progress, and perspective. Engineering 2019;5(5):888–94. Available from https://doi.org/10.1016/j.eng.2019.01.015.

Li YJ, Duan CL, Liu JX, Xu YG. Pro-angiogenic actions of Salvianolic acids on in vitro cultured endothelial progenitor cells and chick embryo chorioallantoic membrane model. Journal of Ethnopharmacology 2010;131(3):562–6. Available from https://doi.org/10.1016/j.jep.2010.07.040.

Li S, Fan TP, Jia W, Lu A, Zhang W. Network pharmacology in traditional Chinese medicine. Evidence-Based Complementary and Alternative Medicine: eCAM 2014;2014:138460. Available from https://doi.org/10.1155/2014/138460.Epub, 24707305.

Lim B, Ju H, Kim M, Kang C. Increased genetic susceptibility to intestinal-type gastric cancer is associated with increased activity of the RUNX 3 distal promoter. Cancer 2011;117(22):5161–71. Available from https://doi.org/10.1002/cncr.26161.

Liu S, Yang X, Zheng S, Chen C, Qi L, Xu X, et al. Research progress on the use of traditional Chinese medicine to treat diseases by regulating ferroptosis. Genes & Diseases 2025;12(3):101451. Available from https://doi.org/10.1016/j.gendis.2024.101451.

Liu X, Ding W, Jiang H. Engineering microbial cell factories for the production of plant natural products: from design principles to industrial-scale production. Microbial Cell Factories 2017;16(1). Available from https://doi.org/10.1186/s12934-017-0732-7, http://www.microbialcellfactories.com/home/.

Louie KB, Kosina SM, Hu Y, Otani H, de Raad M, Kuftin AN, et al. Mass spectrometry for natural product discovery. Elsevier BV; 2020. p. 263–306. Available from 10.1016/b978-0-12-409547-2.14834-6.

Lum MR, Hirsch AM. Molecular methods for the authentication of botanicals and detection of potential contaminants and adulterants. Acta Horticulturae 2006;720(720):59–72. Available from https://doi.org/10.17660/ActaHortic.2006.720.6.

Marques AV, Pereira H. A methodological approach for the simultaneous quantification of glycerol and fatty acids from cork suberin in a single GC run. Phytochemical Analysis 2019;30(6):687–99. Available from https://doi.org/10.1002/pca.2846.

Moher D, Naudet F, Cristea IA, Miedema F, Ioannidis JPA, Goodman SN. Assessing scientists for hiring, promotion, and tenure. PLoS Biology 2018;16(3):e2004089. Available from https://doi.org/10.1371/journal.pbio.2004089.

Morgera E., Tsioumani E., Buck M. Unravelling the Nagoya Protocol [Internet]. 2014. Available from https://doi.org/10.1163/9789004217188.

Newman DJ, Cragg GM. Natural products as sources of new drugs over the nearly four decades from 01/1981 to 09/2019. Journal of Natural Products 2020;83(3):770–803.

Available from https://doi.org/10.1021/acs.jnatprod.9b01285, http://pubs.acs.org/journal/jnprdf.

Pammolli F, Magazzini L, Riccaboni M. The productivity crisis in pharmaceutical R&D. Nature Reviews. Drug Discovery 2011;10(6):428–38. Available from https://doi.org/10.1038/nrd3405.

Pelkonen O, Xu Q, Fan TP. Why is research on herbal medicinal products important and how can we improve its quality? Journal of Traditional and Complementary Medicine 2014;4(1):1–7. Available from https://doi.org/10.4103/2225-4110.124323, www.jtcm.org/.

Phillips GC, Garda M. Plant tissue culture media and practices: an overview. In Vitro Cellular and Developmental Biology - Plant 2019;55(3):242–57. Available from https://doi.org/10.1007/s11627-019-09983-5, http://www.springerlink.com/content/1054-5476.

Porter-Bolland L, Ellis EA, Guariguata MR, Ruiz-Mallén I, Negrete-Yankelevich S, Reyes-García V. Community managed forests and forest protected areas: an assessment of their conservation effectiveness across the tropics. Forest Ecology and Management 2012;268:6–17. Available from https://doi.org/10.1016/j.foreco.2011.05.034.

Pringsheim T, Lam D, Ching H, Patten S. Metabolic and neurological complications of second-generation antipsychotic use in children. Drug Safety 2011;34(8):651–68. Available from https://doi.org/10.2165/11592020-000000000-00000.

Rahman SA, Jacobsen JB, Healey JR, Roshetko JM, Sunderland T. Finding alternatives to swidden agriculture: does agroforestry improve livelihood options and reduce pressure on existing forest? Agroforestry Systems 2017;91(1):185–99. Available from https://doi.org/10.1007/s10457-016-9912-4, www.wkap.nl/journalhome.htm/0167-4366.

Robinson N, Lorenc A, Ding W, Jia J, Bovey M, Wang XM. Exploring practice characteristics and research priorities of practitioners of traditional acupuncture in China and the EU - a survey. Journal of Ethnopharmacology 2012;140(3):604–13. Available from https://doi.org/10.1016/j.jep.2012.01.052, www.elsevier.com/locate/jethpharm.

Rodhouse TJ, Ormsbee PC, Irvine KM, Vierling LA, Szewczak JM, Vierling KT. Establishing conservation baselines with dynamic distribution models for bat populations facing imminent decline. Diversity and Distributions 2015;21(12):1401–13. Available from https://doi.org/10.1111/ddi.12372, http://onlinelibrary.wiley.com/journal/10.1111/(ISSN)1472-4642.

Rothrock NE, Kaiser KA, Cella D. Developing a valid patient-reported outcome measure. Clinical Pharmacology and Therapeutics 2011;90(5):737–42. Available from https://doi.org/10.1038/clpt.2011.195.

Røttingen J-A, Regmi S, Eide M, Young AJ, Viergever RF, Årdal C, et al. Mapping of available health research and development data: what's there, what's missing, and what role is there for a global observatory? The Lancet 2013;382(9900):1286–307. Available from https://doi.org/10.1016/s0140-6736(13)61046-6.

Rousseau PC. Recent literature. Journal of Palliative Medicine [Internet] 2020;23(11):1555–6. Available from https://doi.org/10.1089/jpm.2020.0564.

Santoro FR, Nascimento ALB, Soldati GT, Ferreira Júnior WS, Albuquerque UP. Evolutionary ethnobiology and cultural evolution: opportunities for research and dialog. Journal of Ethnobiology and Ethnomedicine 2018;14(1). Available from https://doi.org/10.1186/s13002-017-0199-y, https://ethnobiomed.biomedcentral.com/.

Saslis-Lagoudakis CH, Savolainen V, Williamson EM, Forest F, Wagstaff SJ, Baral SR, et al. Phylogenies reveal predictive power of traditional medicine in bioprospecting.

Proceedings of the National Academy of Sciences 2012;109(39):15835–40. Available from https://doi.org/10.1073/pnas.1202242109.

Setzer WN. The phytochemistry of cherokee aromatic medicinal plants. Medicines [Internet] 2018;5(4):121. Available from https://doi.org/10.3390/medicines5040121.

Sharma A, Adhikari D, Wagh VV. An ethnobotanical investigation of medicinal plants used among rural communities in Northeastern Madhya Pradesh, India: a quantitative study. Plant Biosystems - An International Journal Dealing with all Aspects of Plant Biology 2024;158(5):1160–81. Available from https://doi.org/10.1080/11263504.2024.2395845.

Sharma K., Ravi K., Wani A.W., Kumar S., Qureshi S. Introduction to Herbal Pharmacopoeia. In: Herbal Pharmacopoeia [Internet]. 2025. p. 1–18. Available from: https://doi.org/10.1201/9781003544418-1.

da Silva MM, Rocha TA, de Moura DF, Chagas CA, de Aguiar Júnior FCA, da Silva Santos NP, et al. Effect of acute exposure in swiss mice (*Mus musculus*) to a fibrinolytic protease produced by Mucor subtilissimus UCP 1262: an histomorphometric, genotoxic and cytological approach. Regulatory Toxicology and Pharmacology 2019;103:282–91. Available from https://doi.org/10.1016/j.yrtph.2019.02.009, http://www.elsevier.com/inca/publications/store/6/2/2/9/3/9/index.htt.

Thaker J, Floyd B. Shifting COVID-19 vaccine intentions in New Zealand: next steps in the vaccination campaign. The Lancet Regional Health - Western Pacific 2021;15:100278. Available from https://doi.org/10.1016/j.lanwpc.2021.100278.

Thamizhoviya G. Global integration of traditional and modern medicine: policy developments, regulatory frameworks, and clinical integration model. Future Integrative Medicine 2025. Available from https://doi.org/10.14218/fim.2025.00033.

Thorsen BJ, Strange N, Helles F. A comment on: V.C. Tassone et al., 2004: diverging incentives for afforestation from carbon sequestration: an economic analysis of the EU afforestation program in the south of Italy. Forest Policy and Economics 6, 567-578. Forest Policy and Economics 2006;9(2):107–8. Available from https://doi.org/10.1016/j.forpol.2005.06.005.

Tilburt JC, Kaptchuk TJ. Herbal medicine research and global health: an ethical analysis. Bulletin of the World Health Organization 2008;86(8):594–9. Available from https://doi.org/10.2471/BLT.07.042820, http://www.who.int/bulletin/volumes/86/8/07-042820.pdf.

Tsai FJ, Li TM, Cheng CF, Wu YC, Lai CH, Ho TJ, et al. Effects of Chinese herbal medicine on hyperlipidemia and the risk of cardiovascular disease in HIV-infected patients in Taiwan. Journal of Ethnopharmacology 2018;219:71–80. Available from https://doi.org/10.1016/j.jep.2018.03.006, www.elsevier.com/locate/jethpharm.

Wang R, Sun Y, Zong J, Wang Y, Cao X, Wang Y, et al. Remote sensing application in ecological restoration monitoring: a systematic review. Remote Sensing 2024;16(12):2204. Available from https://doi.org/10.3390/rs16122204.

Wilkinson MD, Dumontier M, Aalbersberg IJ, Appleton G, Axton M, Baak A, et al. Comment: the FAIR guiding principles for scientific data management and stewardship. Scientific Data 2016;3. Available from https://doi.org/10.1038/sdata.2016.18, www.nature.com/sdata/.

Wong YMA, Ahn S, Bana A, Dua PK, Eggers R, Kuruvilla S, et al. Policy implications of WHO's global traditional medicine strategy 2025–2034. Bulletin of the World Health Organization 2025;103(11):715–21. Available from https://doi.org/10.2471/BLT.25.293414, https://pmc.ncbi.nlm.nih.gov/articles/PMC12578523/pdf/BLT.25.293414.pdf.

Wu C, Lee SL, Taylor C, Li J, Chan YM, Agarwal R, et al. Scientific and regulatory approach to botanical drug development: a U.S. FDA perspective. Journal of Natural Products 2020;83(2):552–62. Available from https://doi.org/10.1021/acs.jnatprod.9b00949, http://pubs.acs.org/journal/jnprdf.

Wynberg R, Laird S, Access and benefit sharing and biodiversity conservation. In: Access and Benefit Sharing of Genetic Resources, Information and Traditional Knowledge (pp. 50-70) [Internet]. 2022. p. 50–70. Available from: https://doi.org/10.4324/9781003301998-6.

Yang Z, Chan KW, Abu Bakar MZ, Deng X. Unveiling drimenol: a phytochemical with multifaceted bioactivities. Plants 2024;13(17). Available from https://doi.org/10.3390/plants13172492, www.mdpi.com/journal/plants.

Zaza P, Matthieu R, Jean-Luc C, Charles K. Drug repurposing in Raynaud's phenomenon through adverse event signature matching in the World Health Organization pharmacovigilance database. British Journal of Clinical Pharmacology 2020;86(11):2217–22. Available from https://doi.org/10.1111/bcp.14322.

Zeng X, Jiang H, Yang G, Ou Y, Lu S, Jiang J, et al. Regulation and management of the biosecurity for synthetic biology. Synthetic and Systems Biotechnology 2022;7(2):784–90. Available from https://doi.org/10.1016/j.synbio.2022.03.005.

Zhang R, Li X, Zhang X, Qin H, Xiao W. Machine learning approaches for elucidating the biological effects of natural products. Natural Product Reports 2021;38(2):346–61. Available from https://doi.org/10.1039/d0np00043d.

Zhu W, Wang Z, Sun Y, Yang B, Wang Q, Kuang H. Traditional uses, phytochemistry and pharmacology of genus Syringa: a comprehensive review. Journal of Ethnopharmacology [Internet] 2020;266:113465. Available from https://doi.org/10.1016/j.jep.2020.113465.

Index

Note: Page numbers followed by "*f*" and "*t*" refer to figures and tables, respectively.

C

H